Biochemistry of the Elemental Halogens and Inorganic Halides

BIOCHEMISTRY OF THE ELEMENTS

Series Editor: Earl Frieden
Florida State University
Tallahasse, Florida

A Continuation Order Plan is available for this series. A continuation order will bring delivery of each new volume immediately upon publication. Volumes are billed only upon actual shipment. For further information please contact the publisher.

Biochemistry of the Elemental Halogens and Inorganic Halides

Kenneth L. Kirk

National Institutes of Health
Bethesda, Maryland

PLENUM PRESS • NEW YORK AND LONDON

Library of Congress Cataloging-in-Publication Data

Kirk, Kenneth L.
    Biochemistry of the elemental halogens and inorganic halides /
Kenneth L. Kirk.
        p.   cm. -- (Biochemistry of the elements ; v.9A)
    Includes bibliographical references and index.
    ISBN 0-306-43653-1
    1. Halogens--Physiological effect. 2. Halides--Physiological
effect. 3. Biochemistry.   I. Title.   II. Series.
QP533.K57 1991
574.19'214--dc20   90-25111
                                                            CIP

ISBN 0-306-43653-1

© 1991 Plenum Press, New York
A Division of Plenum Publishing Corporation
233 Spring Street, New York, N.Y. 10013

Printed in the United States of America

To my wife, Susan; and children, David and Rebecca

Preface

The elements in group 17 (VIIA) of the periodic table of elements—fluorine (F), chlorine (Cl), bromine (Br), and iodine (I)—were designated by Berzelius as "halogens" (Greek *hals*, sea salt; *gennao*, I beget) because of their propensity to form salts. In this first of the two volumes of *Biochemistry of the Halogens*, the biochemistry of the elemental halogens and inorganic halides is reviewed. Discovery, properties, and biochemistry of the elemental halogens are reviewed first (Chapter 1). This is followed by a review of the developments in the various areas of inorganic halide biochemistry (Chapters 2 through 5). The biochemistry of thyroid hormones is considered in Chapter 6, while biohalogenation, an important link between inorganic and organic halogen biochemistry, is reviewed in Chapter 7. Chapter 8 covers the biochemistry of products produced by human-inspired halogenation, in particular, polyhalogenated compounds that present environmental problems. In Chapter 9, the process is reversed and biodehalogenation is reviewed.

In each subject, the attempt has been made to find an appropriate balance between depth and breadth of treatment, since a thorough, in-depth review of this field would not be possible in a single volume. To provide readers not familiar with subjects with the necessary background to place subsequent discussions in perspective, brief historical developments of many of the topics are given. Following this, specific areas of research are discussed in a degree of detail such that those who themselves are active in related areas may find the material useful as a reference source. In much of the material, biochemical mechanisms are stressed—a reflection of my own interests.

The approach that I followed has presented certain problems—for example, decisions as to what topics should be treated in depth were necessary, and omissions of important areas become inevitable. Moreover, a single reviewer enters into a project of this scope armed with adequate knowledge only in his own field of expertise (in this case, synthesis and biochemistry of fluorinated organic compounds). This disadvantage, however, can be offset by the ability of a single author, after extensive review of the literature, to convey a broad perspective and, perhaps, by not being an expert, to communicate information in a manner appreciated by his fellow nonexperts. To compensate for the inability to review adequately all of the relevant work in a given area—both because of space constraints and possible inadvertent omissions—extensive literature citations to original articles and reviews are given.

A rapidly increasing volume of published scientific data has required that scientists become increasingly specialized in increasingly narrow fields—unfortunately often at the expense of breadth of knowledge. As noted above, preparation of this volume has necessitated the review of research from a wide range of scientific disciplines. This exercise in turn emphasized to me the increasing interdependence of various scientific specialties as rapid progress is made on a broad front in understanding biological functions. To the extent that this interdependence has been illustrated successfully in the two volumes of *Biochemistry of the Halogens*, scientists from several areas may find this work useful—for example, biochemists, organic chemists, medicinal chemists, pharmacologists, physiologists, and toxicologists.

As living organisms evolved in an environment where inorganic halide salts were abundant, many mechanisms were established to modulate potential deleterious effects of high halide concentrations on living organisms, and many of these mechanisms incorporate halides into biochemical processes essential to the existence of the organism. The roles of the individual halides vary greatly, a reflection of their relative abundance in the environment as well as their differing physicochemical properties. For example, chloride—a major constituent of seawater, the makeup of which had evolutionary influence on the composition of mammalian extracellular fluid—has by far the greatest number of natural biochemical functions. Thus, chloride transport and the regulation thereof are critical to such functions as cellular homeostasis, inhibitory neurotransmission, and a host of secretory processes (Chapter 3). Fluoride, in contrast, has a limited number of natural roles, although it may be an essential trace element because of its ability to effect development of mineralized tissues. However, as a mimic of hydroxide, fluoride can interact with a host of biological macromolecules, and a vast amount of biochemistry has been developed based on this relationship. A recent major development in this area is the recognition that fluoride, in the presence of aluminum salts, forms aluminum tetrafluoride, a species that appears to mimic many functions of the ubiquitous and important phosphate group (Chapter 2). Iodide also is a trace element but has assumed a critical role in mammalian development as a structural feature of the thyroid hormones. For this reason, the biochemistry of iodide deals in large part with thyroid hormone biosynthesis (Chapters 5 and 6). As the molecular mechanisms of thyroid hormone action have been elucidated, a possible link between thyroid hormone receptors and the toxicity of certain polyhalogenated aromatic hydrocarbons, including dioxin, has been found (Chapter 8). Bromide is abundant in nature, but, with one apparent exception, has no natural function in mammalian biochemistry (Chapter 4). However, bromide is the substrate for many haloperoxidase-catalyzed reactions and, as such, is important in the formation of a multitude of secondary metabolites, particularly those of marine origins (Chapter 7). Astatine (At), the fifth member of this group, is a rare radioactive element whose isotopes have fleeting existences. This element will be given only brief mention. The carbon–halogen bond is the target for enzymes that metabolize organic halogenated compounds, processing these to polar, excretable metabolites with loss of halide. However, these same biochemical reactions can also produce highly toxic, mutagenic intermediates from many halogenated compounds. These issues are discussed in Chapter 9.

Acknowledgments

Preparation of the two volumes that make up *Biochemistry of the Halogens* has been a rewarding, but labor-intensive, task. I wish to express my appreciation to friends and colleagues who have given me unwavering encouragement and stimulation during this effort. As writing became more intense, it became necessary for me to spend increasing amounts of time in the "pit" (my study) communicating with my PC rather than with my family, and so I am especially grateful to them for their love, encouragement, and remarkable patience. A message to my son David, who at one point early in the project asked why I was writing about "ignorant fluoride" (Chapter 2)—now we can build that tree house!

Contents

Biochemistry of the Elemental Halogens and Inorganic Halides

The Halogens: Discovery, Occurrence, and Biochemistry of the Free Elements

1.1 Introduction

The halogens—fluorine (F), chlorine (Cl), bromine (Br), iodine (I), and astatine (At)—make up group 17 (VIIA) of the elements. Because of their propensity to form salts [the standard potentials (E^0, volts) for the oxidation of halides ($2X^- \rightarrow X_2 + 2e^-$) are -3.06, -1.36, -1.07, and -0.54 for fluorine, chlorine, bromine, and iodine, respectively], Berzelius designated these elements as "halogens" (Greek *hals*, sea salt; *gennao*, I beget). The halogens make up a particularly well defined family, having the most regular gradation of physical properties of all the families of elements. Thus, an almost perfect doubling of atomic weights on going from one halogen to the next down the periodic table is accompanied by an increase in specific gravity, melting points, and boiling points, and a decrease in water solubility and chemical reactivity (Table 1-1) (Stokinger, 1981). The concentrations of the halogens found in seawater and in the lithosphere are given in Table 1-2.

Astatine (Greek *astatos*, unstable) has no stable natural or synthetic isotopes. Four isotopes belonging to natural radioactive decay chains each have half-lives of less than one minute, and the estimate has been made that at a given moment the amount of At in the Earth's crust does not exceed 50 mg, making it easily the rarest element (Banks, 1986).

In this chapter, the discovery of the elemental halogens will be described briefly, and a discussion of the properties of the elements and their industrial production and uses will be given. The discussion of the biochemistry of the elemental halogens given will be concerned largely with toxic effects of exposure to the free elements. An important aspect of the chemistry of the elemental halogens concerns their *in vivo* production by enzyme-catalyzed oxidation of halides. This oxidation is important in several biologically important processes such as biohalogenation and in leukocyte-mediated defense mechanisms. These topics will be discussed in

Table 1-1. Trends in Physical Properties
within the Halogen Family[a]

Halogen	Symbol	Atomic weight	Specific gravity	mp (°C)	bp (°C)	Water solubility
Fluorine	F	18.9984	1.69	−219.62	−188.14	soluble
Chlorine	Cl	35.453	3.214	−100.62	−34.6	310 kg/liter (10 °C)
Bromine	Br	79.904	3.119	−7.2	58.78	41.7 g/liter (0 °C)
Iodine	I	176.905	4.93	113.5	184.35	0.29 g/liter (20 °C)

[a] Stokinger, 1981.

later chapters in this volume. Two aspects of the use of halogen and halogen-derived species for disinfection will be considered because of the worldwide impact on public health involved. Thus, recent concerns associated with chlorination of public water supplies will be discussed, as will the use of iodine as a disinfectant.

1.2 Fluorine

1.2.1 Isolation of Fluorine

An entertaining and thorough history of the isolation of fluorine recently has been published in an issue of the *Journal of Fluorine Chemistry* (Vol. 33, No. 1–4) commemorating the centenary of the discovery of fluorine by Moissan (Banks, 1986; Flahaut and Viel, 1986). The systematic study of fluorine chemistry was initiated in 1764 when the German phlogistonist A. S. Marggraf treated fluorspar (calcium fluoride) in a glass retort with sulfuric acid to produce the reaction products HF and silica.

Table 1-2. Concentrations of Halogens in Seawater
and the Lithosphere[a]

Element	Seawater (ppm)	Lithosphere (ppm)
Fluorine	1.4	770
Chlorine	18,980	550
Bromine	65	1.6
Iodine	0.05	0.3

[a] Stokinger, 1981.

Subsequent research by several chemists followed, and in 1809 Gay-Lussac and Thenard succeeded in preparing a highly concentrated aqueous solution of HF. In 1813, Sir Humphry Davy concluded from his research with HF that there must exist a "peculiar substance" that combines strongly with metallic bodies and hydrogen. Because of this strong affinity and "high decomposing agencies," he concluded further that this substance would be hard to isolate and examine in pure form. Following the suggestion of Ampère, he gave to this substance the name *le fluor* (*fluorine*), a term descriptive of its presence in fluorspar and coined to harmonize with *le chlor* (chlorine). [*Fluor* was derived from the Latin *fluo* ("I flow") because of the fusibility of fluorspar and because of its use in flux.] There followed several unsuccessful attempts to isolate F_2, using reasonable strategies that included the electrolysis of HF, chemical and electrochemical oxidation of fluorides, and thermal or electrochemical decomposition of fluorides. During this period, the several chemists who followed the suggestion of Ampère and pursued the electrolysis of HF succeeded only in decomposing water (Banks, 1986, and references therein).

In 1884, the French chemist Henri Moissan addressed the fluorine problem. A major step in the isolation of fluorine had come in 1856 when Fremy succeeded in the preparation of pure anhydrous HF. (KHF_2, precipitated from an impure solution of HF, was heated in a platinum apparatus to produce KF and anhydrous HF.) From a careful study of the unsuccessful attempts of others, and from his own experiments with fluorides, Moissan concluded that electrolysis of fluorides at low temperature provided the only possible route to the isolation of F_2. After unsuccessful experiments with arsenic trifluoride, Moissan, on June 26, 1886, attempted again the electrolysis of pure HF using a platinum electrode. A gas escaped from the anode compartment which in every respect behaved as Moissan had predicted for the elusive F_2, including supporting the combustion of silicon. Subsequent research confirmed that this gas was elemental F_2 and that Moissan had finally succeeded in the isolation of this elusive and reactive element. Later work revealed that the key to the initial success was the presence in HF of traces of KF, required to conduct current (pure, anhydrous HF is a nonconductor). For this research, Moissan was awarded the Nobel Prize in chemistry in 1906 (Flahaut and Viel, 1986).

1.2.2 Properties

Fluorine is the most electronegative and most reactive of all the elements, capable of forming binary compounds with all other elements except helium, neon, and argon. These properties reflect the position of F in the periodic table. The atomic configuration of F $[(He)2s^22p^5]$ has an unshielded high nuclear charge that strongly attracts the surrounding electrons. The acquisition of an additional electron to complete the neon core $[(He)2s^22p^6]$, through formation of a covalent bond or through formation of F^-, is an extremely favorable process energetically. In addition, the $F-F$ bond in F_2 is very weak (158.8 kJ/mol), a result of the strong repulsion of lone pair electrons of the individual F atoms in this small molecule. In contrast, because the small size of the fluorine atom allows good overlap of atomic orbitals, fluorine forms strong bonds with other elements, either ionic or covalent. The low $F-F$ bond energy and the high energy of bonds formed make reactions of F_2 with other elements or with compounds extremely exothermic, and often explosive. In fact, an important area of fluorine chemistry involves the design of procedures to control the reactivity of F_2. Recent important progress in this area has contributed to the enormous increase in compounds available for biological studies (see Vol. 9B of this series). Unlike other halogens, fluorine cannot expand its octet, and thus the chemistry of fluorine is limited to an oxidation state of -1 (Banks, 1986; see Gould, 1955, for a concise treatment of the physical basis for the trends in properties of the halogens).

F_2 is a pale greenish yellow, highly toxic gas with a penetrating odor. Because of its high reactivity, F_2 does not occur free in nature. Fluoride salts, however, are widely distributed, predominantly as fluorspar and phosphate rock, and, of the elements, fluorine ranks 13th (chlorine is 20th) in abundance in the Earth's crust.

1.2.3 Industrial Production and Uses of Fluorine

A major impetus to the production of F_2 came in the early 1940s during research to produce an atomic bomb. This project required the separation of ^{235}U from natural uranium, a task that was accomplished by the diffusion of uranium hexafluoride on a very large scale. Uranium hexafluoride is almost as reactive as elemental fluorine, and the requirement for equipment compatible with this reactivity led to extensive research into the production of inert fluorocarbons, a process that required substantial amounts of F_2 (reviewed by Goldwhite, 1986; Meshri, 1986). Fluorine gas now is produced by the electrolysis of HF containing KF at a temperature of 100 °C (Ellis and May, 1986).

Several aspects of inorganic and organic fluorine chemistry have important industrial uses. The preparation of Freons in the 1930s was a major advance in the field of refrigeration. Another important use of fluorine gas is in the production of perfluorocarbons, stable polymers having a host of applications. As methods to control the reactivity of F_2 have been developed, F_2 has also become an increasingly important laboratory reagent for the synthesis of selectively fluorinated molecules, many having important biological and medical applications.

1.2.4 Biochemistry and Toxicology

As would be expected from its reactivity, F_2 has devastating effects on animal tissue. As part of the Manhattan Project, several toxicity studies were carried out to study the effects of inhalation of and skin exposure to F_2. Animals exposed to lethal concentrations (for example, 3 h at 200 ppm) in air died of respiratory damage with pulmonary edema. At 5 to 10 ppm, dogs exhibited irritation of the eyes and nasal and buccal mucosa, along with irrational and sometimes fatal seizures. A tolerated exposure was regarded as 1 ppm $(1.7 \, mg/m^3)$. Later studies with human volunteers showed that short-term exposures of up to $15 \, mg/m^3$ could be tolerated without discomfort. Exposure of the skin to low concentrations of F_2 causes serious cutaneous burns. Fluorine oxide (F_2O), once considered a good candidate as an oxidizer for missile propellant systems, is even more acutely toxic than is F_2, having a threshold limiting value (TLV) of 0.05 ppm (reviewed by Stokinger, 1981).

Because of the high reactivity of F_2 gas, research on its biochemical and pharmacological behavior relates primarily to toxicological issues, as discussed briefly above. In contrast, the research on the interaction of fluoride (F^-) with biological systems has produced a wealth of information and useful medical applications (Chapter 2).

1.3 Chlorine

1.3.1 Isolation of Chlorine

Chlorine was the first of the halogens to be isolated and identified. In 1774, the Swedish apothecary and phlogistonist Scheele oxidized concentrated hydrochloric acid (marine or muriatic acid; from the Latin *muria*, brine) with manganese dioxide and produced a greenish yellow gas. He called this gas dephlogisticated marine acid, a name changed later to oxymuriatic acid after the overthrow of the phlogiston theory by Lavoisier.

This gas originally was thought to contain oxygen, since Lavoisier had thought that all acids contained oxygen. In 1810, research by Sir Humphry Davy proved the elemental nature of this gas and disproved Lavoisier's assumption about the nature of acids. Davy named the new element *chlorine* from the Greek *chloros*, greenish yellow (reviewed by Banks, 1986).

1.3.2 Properties

Chlorine (Cl_2) is a toxic gas with an irritating, suffocating odor (the physical properties are given in Table 1-1). The reactivity of Cl_2 lies between that of F_2 and Br_2, and thus Cl_2 can react with bromide and iodide salts to produce the corresponding free halogens. Because of its high reactivity, Cl_2 is not found free in nature.

1.3.3 Industrial Production and Uses of Chlorine Gas

Chlorine gas is produced commercially by electrolysis of NaCl brine in asbestos diaphragm cells or mercury cathode cells. Of the halogens, Cl_2 has the widest industrial use and was ranked eighth among large-volume chemicals manufactured in the United States in 1979. Main uses include the production of organic compounds (70% of Cl_2 produced); use in bleaches for paper, pulp, and textiles and in sanitation and disinfection of water supplies (20%); and production of inorganic chemicals (10%) (Greenwood and Earnshaw, 1984).

1.3.4 Biochemistry and Toxicology of Elemental Chlorine

Chlorine gas is highly toxic, and its many industrial uses place thousands of workers at risk to potential exposure. In addition, many humans were exposed to the lethal effects of Cl_2 during its use as a poison gas during World War I. The odor threshold for Cl_2 is about 0.3 ppm, the threshold limit value is 1 ppm, sustained exposure at 5–8 ppm causes acute to mild distress, 15 ppm results in immediate upper respiratory tract irritation, 30 ppm causes serious damage to the respiratory tract, and 100 ppm may be fatal (Stokinger, 1981; Conradi Fernandez and Inclan Cuesta, 1983).

Years of widespread exposure to Cl_2 have produced much information on its acute effects on humans, but much of this information has been derived from accidental exposures where concentrations have not been determined. The acute effects of exposure to Cl_2, which acts as a primary irritant to the mucous membranes of the eyes, nose, and throat and the lining of the respiratory tract, have been reviewed by Stokinger (1981). Symptoms include pulmonary congestion and edema that develop several hours after exposure. Postexposure studies on casualties of a recent industrial accident that exposed 28 chemical workers in Bombay to 66 ppm of Cl_2 revealed several cytopathic changes in bronchiepithelial cells (Chandralekha et al., 1988). In a study designed to determine potential physiological damage due to long-term chronic exposure to Cl_2, Klonne et al. (1987) exposed rhesus monkeys to concentrations of up to 2.3 ppm for 6 h per day, 5 days per week. The results of this study showed that chronic exposure to 2.3 ppm of Cl_2 caused upper respiratory irritation in monkeys, whereas exposure to 0.5 and 1.0 ppm induced changes that are of questionable clinical significance.

The high reactivity of Cl_2 precludes selective interactions of exogenous sources of the element with biochemical macromolecules, and elemental Cl_2 has no direct medical uses. However, Cl_2 (or related species having electrophilic reactivity) generated in vivo by the oxidation of Cl^- is very important in certain biological functions. These include biological chlorination reactions, to be discussed in Chapter 7, and roles in host defense mechanisms through activation of neutrophils, a topic to be discussed in Chapter 3.

1.3.5 Water Chlorination

Both elemental Cl_2 and hypochlorous acid (HOCl), the product formed by the rapid hydrolysis of Cl_2 in water, are potent germicidal agents. In an important exploitation of this activity, continuous application of Cl_2 to the water supply of Jersey City in 1908 initiated a public health measure designed to decrease the incidence of waterborne infections. Water chlorination of community water supplies is now common practice, and currently approximately 170,000,000 people in the United States drink chlorinated water. The public health significance of this practice was made evident early by the precipitous drop in typhoid fever deaths during the first third of the century (Aken and Hoff, 1985, and references therein).

The biochemistry related to chlorinated water is more precisely the biochemistry of products of the hydrolysis of Cl_2 in water (HOCl or ClO^-). At pH 8, ClO^- is the predominant species; at pH 6, the potent microbicidal agent HOCl is the major species, representing more than 90%

of the total chlorine in solution, while Cl_2 is present only in small amounts (Fukayama *et al.*, 1986, and references therein).

1.3.5.1 Water Chlorination and Trihalomethane Production

Recently, concern has arisen over the extensive use of Cl_2 as a disinfectant of community water supplies. Recognition that the public health benefits resulting from water chlorination might be accompanied by hazards to health came in 1974 with reports of the presence of chloroform and other trihalomethanes in chlorinated drinking water (Bellar *et al.*, 1974; Rook, 1974). The discovery that chloroform induced hepatocellular carcinomas in mice and renal tumors in rats led the United States Environmental Protection Agency to limit the levels of halomethanes in drinking water in 1979 (Meier, 1988, and references therein). Mechanisms of toxicity of simple halogenated compounds are discussed in detail in Chapter 9.

Much research has been conducted to determine the source of trihalomethanes and of other halogenated compounds in chlorinated water. Being highly reactive species, Cl_2 and hypochlorite will react with organic compounds that may be present in water. The resulting uptake of halogenating species, known as the *chlorine demand* of the sample being treated, will decrease the bactericidal efficacy of the added Cl_2. Natural aquatic humic material—mixtures of polymeric phenols of high molecular weight formed by decomposition of organic material—has been identified as an important source of trihalomethanes (de Leer *et al.*, 1985, and references therein). Since nitrogenous compounds make up as much as one-third of the total organic content of natural waters, the contribution of proteins and amino acids to trihalomethane production has also been investigated (Scully *et al.*, 1985, and references therein). In a recent study to determine the contribution of halogenation of proteins to the production of trihalomethanes in natural and model water systems, the level of formation was found to be quite low (Scully *et al.*, 1985).

Related chlorine-containing volatile compounds that have been identified following the chlorination of humic acid include chlorinated acetones, propenals, and dichloroacetonitrile (Meier, 1988, and references therein). Carcinogenic and mutagenic activity detected for certain of these by-products indicated that proper evaluation of risks associated with water chlorination would require studies beyond trihalomethanes (reviewed by Bull and Robinson, 1985).

1.3.5.2 Nonvolatile Mutagenic By-Products of Water Chlorination

The discovery of trihalomethanes in chlorinated water prompted closer scrutiny of treated water supplies. After concentration, mutagenic activity (Ames test) was found in drinking water samples taken from a variety of locations, and evidence indicated that chlorination was responsible for the majority of this activity (Loper, 1980; Meier, 1988; Kronberg and Christman, 1989; and references therein). Cheh *et al.* (1980) provided early evidence that chlorination of raw water leads to formation of nonvolatile mutagens, while Bull *et al.* (1982) were the first to demonstrate that mutagenic activity was produced by chlorination of humic substances. Production of mutagenicity subsequently was shown to be dependent on the concentration of humic acid and chlorine and to be pH dependent and appeared to be associated primarily with nonvolatile compounds (Meier *et al.*, 1983).

Recent research by several groups has focused on fractionation and identification of mutagens formed by chlorination of humic acid. Included in the products identified are the potent mutagens 3-chloro-4-(dichloromethyl)-5-hydroxy-2(5*H*)-furanone (MX) (**1**) and the isomeric *E*-2-chloro-3-(dichloromethyl)-4-oxobutenoic acid (EMX) (**2**) (Fig. 1-1). MX, one of the most potent mutagens ever tested in the Ames assay, accounted for about 40% of the mutagenic activity of chlorinated humic acid solution while EMX contributed about 10% of this activity (Meier, 1988, and references therein). The association of MX and EMX with mutagenic activity found in drinking water supplies was strengthened by the detection of MX and EMX in samples of drinking water collected from different localities in Finland (Kronberg and Christman, 1989) and the United States (Meier *et al.*, 1987).

Many other compounds identified in chlorinated drinking water have been examined for possible mutagenic or carcinogenic activity, including chlorinated polycyclic aromatic hydrocarbons, chlorinated phenols, and organic *N*-chloroamines (Meier, 1988; Bull and McCabe, 1985; and references therein). Meier (1988) has summarized the enormity of the problems associated with identification of potentially harmful compounds in water supplies. For example, nearly 1100 organic compounds, most at concentrations of <1 μg/liter, were identified in a survey of drinking water from five U.S. cities, and nearly twice as many were detected and not iden-

Figure 1-1. Structures of MX (**1**) and EMX (**2**), potent nonvolatile mutagens found in chlorinated drinking water.

tified. Furthermore, in genotoxicity studies, almost 30% of those compounds identified showed activity in one or more test systems. Difficulties in determining accurately the extremely low concentrations of most of these compounds and lack of quantitative data on their mutagenicity are among the complications that make difficult assessment of their contribution to mutagenic activity of drinking water.

1.3.5.3 Risk Assessment and Summary

There is no argument that the use of Cl_2 to disinfect drinking water and sewage has done much to ensure the potability of water supplies provided to a huge proportion of the human population. However, this treatment also is the source of mutagenic and carcinogenic compounds in water supplies, creating concern over possible risks of long-term exposure to these compounds (Davis and Roberts, 1985). A vast amount of research has been done on possible relationships between the use of chlorine and other disinfectants for water purification, mutagenic activity found in drinking water supplies, and potential increased incidences of cancer and other diseases. Indeed, epidemiological associations have been found between drinking water and cancer of the colon, bladder, and rectum (Zierler et al., 1988, and references therein). This association has been considered weak (Crump and Guess, 1982) although the estimates of the risk have been suggested to be low (Zierler et al., 1988). Evaluation of alternative disinfectants, for example, monochloramine, chlorine dioxide, and ozone, has revealed that these agents also may have potentially harmful direct effects or produce potentially harmful by-products (Bull and McCabe, 1985).

In summary, a public health measure of unquestioned value has introduced unexpected risks of uncertain magnitude. An inescapable fact is the necessity of having some procedure for the disinfection of water supplies, particularly in view of the increasing world population that increasingly depends on surface water sources. The discussion presented above represents only a small fraction of the research that has been done in this area, and continuing research efforts are providing an increasing amount of information that will help in making future decisions regarding any changes that may be necessary in existing procedures or regarding institution of new procedures. This clearly is a complex and important issue and any attempt to make risk/benefit evaluations in this volume would be inappropriate. Many thoughtful discussions are available and are recommended (for example, Bull and McCabe, 1985; Davis and Roberts, 1985; Crump and Guess, 1982).

1.4 Bromine

1.4.1 Isolation of Bromine

Bromine was isolated by the French chemist A.-J. Ballard in 1826. After isolation of sodium chloride and sodium sulfate from waters of the Montpellier salt marshes by crystallization, Ballard treated the mother liquors (rich in $MgBr_2$) with chlorine water. After noting the formation of a deep yellow color, Ballard extracted the solution with KOH and ether. Reaction of the resulting material (KBr) with sulfuric acid and manganese dioxide produced the element as a red liquid. The similarity to chlorine and iodine was apparent, and its elementary nature was soon established. The name *bromine* was given to this element by the French Academy because of its foul odor (Greek *bromos*, stink). Because Br_2 is the only nonmetal that is liquid at room temperature, the suggestion has been made that this element would have been more deserving of the name "fluorine," had this not already been given to element number 9. Several years earlier, von Liebig had in hand a sample of Br_2 but had misidentified it as iodine monochloride (Greenwood and Earnshaw, 1984).

1.4.2 Industrial Production and Uses of Bromine

Br_2 is prepared commercially by the oxidation of bromide with Cl_2. World production of Br_2 is about one-hundredth that of Cl_2. The production of 1,2-dibromoethane as a scavenger for lead in gasoline had dominated the industrial uses of Br_2. However, a decrease in the use of leaded gasoline and concern over the toxicity of 1,2-dibromoethane (Chapter 9) have greatly curtailed the production of this compound. Use of bromine-containing compounds as pesticides and as fire retardants constitutes another important industrial application of Br_2 (Chapter 8) (Greenwood and Earnshaw, 1984).

1.4.3 Biochemistry and Toxicology

Elemental Br_2 is a highly corrosive liquid. The vapor is extremely irritating to the eyes, skin, and mucous membranes. A limit of 0.1 to 0.15 ppm has been established as the maximal allowable concentration for prolonged exposure to the vapor. Brief exposure to 40 to 60 ppm poses a serious hazard, and 1000 ppm is rapidly fatal. Inhalation of small amounts of Br_2 leads to coughing, nosebleed, a feeling of oppression, vertigo, and headache. Delayed effects can include abdominal pain, diarrhea, and asthma-like respiratory difficulties, sometimes accompanied by measles-like eruptions on the skin. Necropsy of animals following a 3-h exposure to 300 ppm of Br_2 revealed edema of the lungs, pseudomembranous deposits on the trachea and bronchi, and hemorrhages of the gastric mucosa. Brief topical exposure to Br_2 leads to formation of vesicles and pustules, and longer exposure causes deep slow-healing ulcers (reviewed by Alexandrov, 1983; Stokinger, 1981).

Solutions of Br_2 in water have comparable hydrolytic behavior to that of solutions of Cl_2. The HOBr that is formed has potent antimicrobial activity, reacting readily with such biologically important species as sulf-hydryl groups, thioethers, aromatic compounds, amines, and amino acids. The reactions of HOBr often are faster than reactions of HOCl, and HOBr is a faster bleaching agent than is HOCl (Weiss *et al.*, 1986, and references therein). Another bromine-derived bleaching agent, potassium bromate ($KBrO_3$), has been used widely as a food additive for the maturation of flour in the bread-making process; it has also been employed in permanent-wave formulations. Bromate salts are toxic when ingested orally, with death attributed to renal failure (Lichtenberg *et al.*, 1989), and carcinogenic (Kurokawa *et al.*, 1983). However, $KBrO_3$ is added to bread at levels of <50 ppm and is essentially all converted to KBr in the normal baking process.

The *in vivo* biochemistry of elemental Br_2 concerns the enzymatic oxidation of Br^- to brominating species (Br_2, HOBr, or enzyme-bound hypobromite). These species are involved in biobromination reactions, discussed in Chapter 7, and in host defense mechanisms mediated by eosinophils, covered in Chapter 4.

1.5 Iodine

1.5.1 Isolation of Iodine

Iodine was first isolated from kelp—seaweeds that have been burned at low temperature—in 1811 by B. Courtois, a French manufacturer of saltpeter. An aqueous extract of kelp was used to convert calcium nitrate to the potassium salts. Courtois concentrated this aqueous extract in order to identify an unknown substance that was contributing to erosion of the copper vats used in this process. After removal of precipitated potassium sulfate, sodium sulfate, sodium chloride, and sodium carbonate, he heated the remaining liquid with sulfuric acid in a retort. "A vapor of superb violet color" condensed in the beak of the retort to form brilliant crystals. In 1814, Gay-Lussac established the elementary nature of this substance, which he named *iodine* (Greek *ioeides*, violet). Many of these results were confirmed by Sir Humphry Davy at about the same time (Greenwood and Earnshaw, 1984).

1.5.2 Industrial Production and Uses of Iodine

I_2 is isolated from brine by Cl_2 oxidation followed by air blowout and purification by sublimation. Alternatively, precipitation as the silver salt is followed by treatment with scrap iron to produce metallic silver and ferrous iodide in solution. This solution is treated with Cl_2 to release I_2. Iodine present as iodate in Chilean saltpeter is isolated by a process which involves reduction of a portion of the iodate to I^- with bisulfite and use of the I^- so produced in a redox reaction with iodate which produces I_2 and water (Greenwood and Earnshaw, 1984).

There is no predominant commercial outlet for I_2. About 50% of I_2 produced is used in organic synthesis, and 15% each for the production of purified I_2, KI, and other inorganic chemicals having a variety of applications. World production in 1979 was 10,900 tons (Greenwood and Earnshaw, 1984).

1.5.3 Biochemistry and Toxicology

Iodine is an essential trace element, required for the biosynthesis of thyroid hormones, and, accordingly, the biochemistry and pharmacology of this element have been studied extensively. The biological processing of iodine is initiated by the uptake and oxidation of I^-, and this subject will be covered in Chapter 5. The biochemistry and pharmacology of thyroid hormones is the topic of Chapter 6, while the incorporation of iodine into organic molecules will be discussed in Chapter 7. Toxicology and biochemistry of elemental I_2 will be considered here.

1.5.3.1 Toxicity of Elemental Iodine

Stokinger (1981) has reviewed the toxic properties of I_2 vapor, which has been cited as being more irritating and corrosive than Cl_2 or Br_2. Lacrimation and a burning sensation of the eyes result from low levels. Excessive exposures can cause pulmonary edema comparable to that seen in Cl_2 gassing. Human response studies have shown that exposures of 0.57 ppm are tolerated for 5 min without eye irritation, while 1.63 ppm caused irritation after 2 min. Topical toxicity, characterized by acute dermatitis, can accompany application of strong iodine solutions to the skin. Oral toxicity is quite low, with an oral dose between 2 and 3 g/kg body weight considered lethal. Cases of industrial I_2 poisoning are rare.

1.5.3.2 Iodine as a Disinfectant

In many respects, I_2 is an ideal topical disinfectant. Iodine is bactericidal, sporicidal, fungicidal, protozoacidal, cysticidal, and virucidal yet has relatively low toxicity when applied to intact skin. I_2, as tincture of iodine, was first used as a disinfectant for topical control of infections by a French surgeon in 1839 and also was used to treat battle wounds in the U.S. Civil War. I_2 preparations still remain disinfectants of choice in many situations.

For antiseptic use, solutions of I_2 are prepared in such solvents as water, alcohol, and glycerol, or combinations of these, containing I^- to aid solubility. For example, tincture of iodine contains 2% I_2 and 2.4% NaI in 1:1 aqueous ethanol. The I^{3-} that is formed in these solutions serves as a reservoir for I_2 and HOI. While the solution chemistry of I_2 is complex—I_2 dissolved in pure water produces seven different iodo species (I_2, I^-, H_2OI^+, HOI, OI^-, IO_3^-, and I_3^-)—HOI, H_2OI^+ (present in small amounts), and I_2 are the microbicidal agents. Reactions of these iodinating species with critical biological molecules produce the microbicidal action. Free amino groups of basic amino acids (lysine, histidine, arginine) and the bases of nucleotides, sulfhydryl groups, tyrosine residues, and carbon–carbon double bonds of unsaturated fatty acids are assumed to be important targets. The therapeutic efficacy of I_2 relative to the other halogens is increased because I_2 reacts more slowly with proteins than does Cl_2 or Br_2 but still is sufficiently reactive to kill living microbes rapidly (reviewed by Harvey, 1985; Gottardi, 1983).

In addition to solutions of I_2 and I^- (and I_3^-), there are several preparations that consist of a complex of I_2 and carriers of high molecular weights. A notable example is the iodophor polyvinylpyrrolidone–iodine (povidone–iodine), a water-soluble complex that serves as a reservoir for the slow release of I_2 (Fig. 1-2) (Schenck et al., 1979). Povidone–iodine was introduced as an antiseptic in 1956 and now is widely used in several clinical applications. The level of free I_2 that is present in dynamic equilibrium with the complex depends on the concentration of the complex, having a value of 1 ppm in a 10% solution and 10 ppm in a 1% solution and reaching a maximum of 24 ppm in a 0.7% solution. This limitation of the concentration of free I_2 has led to the term "tamed iodine" for povidone–iodine. As I_2 is released, it undergoes the reactions typical of the element, and more I_2 is released to maintain equilibrium until all the available I_2 is exhausted. By virtue of an affinity of the polymer for cell

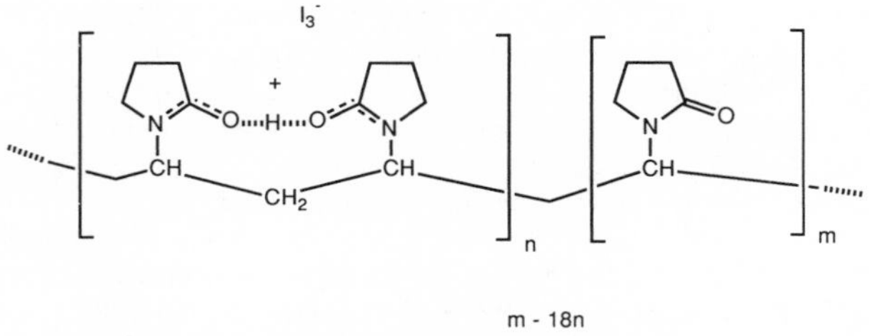

Figure 1-2. Proposed structure of povidone–iodine. After Schenck *et al.* (1979); adapted with permission of the copyright owner, the American Pharmaceutical Association.

membranes, povidone–iodine can deliver free I_2 directly to bacterial cell surfaces, an effect that apparently is important in its antibacterial activity. Thus, despite a much lower concentration of free I_2, povidone–iodine has higher bactericidal activity than does Lugol's solution—an aqueous solution containing 5% I_2 and 10% KI (reviewed by Zamora, 1986).

Disinfection of the skin, both prophylactic and therapeutic, remains the most important medical application of iodine. Because of their lower tendency to produce irritation and other unwanted reactions, iodophors, most notably povidone–iodine, to a large extent have replaced the previously used tinctures and aqueous solutions. In addition, low-molecular-weight povidone–iodine is considered to be a suitable reagent for intravenous use (Zamora, 1986; Gottardi, 1983).

1.6 Summary

The initial chapter in this volume has served to introduce the halogens—their discovery, properties, production, and uses. High reactivity precludes selective interactions of exogenous sources of the free elements with biological moieties, and toxicity studies dominate their biochemical literature. The biochemistry of free halogens produced *in vivo* will be discussed in later chapters. In contrast to the elemental halogens, halide ions are the subjects of an enormous volume of biochemical literature encompassing essentially all aspects of living processes. In the following four chapters, the attempt is made to summarize important issues and recent developments related to the biochemistry of the individual halides.

References

Aken, E. W., and Hoff, J. C., 1985. Microbiological risks associated with changes in drinking water disinfection practices, in *Water Chlorination. Chemistry, Environmental Impact and*

Health Effects, Vol. 5 (R. L. Jolley, R. J. Bull, W. P. Davis, S. Katz, M. H. Roberts, and V. A. Jacobs, eds.), Lewis Publishers, Chelsea, Michigan, pp. 99–110.

Alexandrov, D. D., 1983. Bromine and compounds, in *Encyclopaedia of Occupational Health and Safety*, Vol. 1, 3rd ed. (L. Parmeggiani, ed.), International Labor Organization, Geneva, pp. 326–329.

Banks, R. E., 1986. Isolation of fluorine by Moissan: Setting the scene, *J. Fluorine Chem.* 33:3–26.

Bellar, T. A., Lichtenberg, J. J., and Kroner, R. C., 1974. The occurrence of organohalides in chlorinated drinking water, *J. Am. Water Works Assoc.* 66:703–706.

Bull, R. J., and McCabe, L. J., 1985. Risk assessment issues in evaluating the health effects of alternate means of drinking water disinfection, in *Water Chlorination. Chemistry, Environmental Impact and Health Effects*, Vol. 5 (R. L. Jolley, R. J. Bull, W. P. Davis, S. Katz, M. H. Roberts, Jr., and V. A. Jacobs, eds.), Lewis Publishers, Chelsea, Michigan, pp. 111–130.

Bull, R. J., and Robinson, M., 1985. Carcinogenic activity of haloacetonitrile and haloacetone derivatives in the mouse skin and lung, in *Water Chlorination. Chemistry, Environmental Impact and Health Effects*, Vol. 5 (R. L. Jolley, R. J. Bull, W. P. Davis, S. Katz, M. H. Roberts, Jr., and V. A. Jacobs, eds.), Lewis Publishers, Chelsea, Michigan, pp. 221–227.

Bull, R. J., Robinson, M., Meier, J. R., and Stober, J., 1982. Use of biological assay systems to assess the relative carcinogenic hazards of disinfection by-products, *Environ. Health Perspect.* 46:215–227.

Chandralekha, P. S., Khade, M. V., and Srinivasan, M., 1988. Respiratory cytopathology in chlorine gas toxicity: A study of 28 subjects, *Diagnostic Cytopathol.* 4:28–32.

Cheh, A. M., Skochdopole, J., Koski, P., and Cole, L., 1980. Nonvolatile mutagens in drinking water: Production by chlorination and destruction by sulfite, *Science* 207:90–92.

Conradi Fernandez, L. C., and Inclan Cuesta, M. I., 1983. Chlorine and inorganic compounds, in *Encyclopaedia of Occupational Health and Safety*, Vol. 1, 3rd ed. (L. Parmeggiani, ed.), International Labor Office, Geneva, pp. 454–459.

Crump, K. S., and Guess, H. A., 1982. Drinking water and cancer, *Annu. Rev. Public Health* 3:339–357.

Davis, W. P., and Roberts, M. H., Jr., 1985. Water chlorination: Crossroad of uncertainties and decisions, in *Water Chlorination. Chemistry, Environmental Impact and Health Effects*, Vol. 5 (R. L. Jolley, R. J. Bull, W. P. Davis, S. Katz, M. H. Roberts, Jr., and V. A. Jacobs, eds.), Lewis Publishers, Chelsea, Michigan, pp. 3–4.

de Leer, E. W. B., Damste, J. S. S., and de Galan, L., 1985. Formation of aryl-chlorinated aromatic acids and precursors for chloroform in chlorination of humic acid, in *Water Chlorination. Chemistry, Environmental Impact and Health Effects*, Vol. 5 (R. L. Jolley, R. J. Bull, W. P. Davis, S. Katz, M. H. Roberts, Jr., and V. A. Jacobs, eds.), Lewis Publishers, Chelsea, Michigan, pp. 843–857.

Ellis, J. F., and May, G. F., 1986. Modern fluorine generation, *J. Fluorine Chem.* 33:133–146.

Flahaut, J., and Viel, C., 1986. The life and scientific work of Henri Moissan, *J. Fluorine Chem.* 33:27–42.

Fukayama, M. Y., Tan, H., Wheeler, W. B., and Wei, C.-I., 1986. Reactions of aqueous chlorine and chlorine dioxide with model food compounds, *Environ. Health Perspect.* 69:267–274.

Goldwhite, H., 1986. The Manhattan project, *J. Fluorine Chem.* 33:109–131.

Gottardi, W., 1983. Iodine and iodine compounds, in *Disinfection, Sterilization, and Preservation* 3rd ed. (S. S. Block, ed.), Lea and Febiger, Philadelphia, pp. 183–196.

Gould, E. S., 1955. *Inorganic Reaction and Structure*, Holt, Rinehart, and Winston, New York, pp. 205–227.

Greenwood, N. N., and Earnshaw, A., 1984. *Chemistry of the Elements*, Pergamon Press, Oxford, pp. 921–1041.

Harvey, S. C., 1985. Antiseptics and disinfectants; fungicides; ectoparasiticides, in *The Pharmacological Basis of Therapeutics*, 7th ed. (A. G. Gilman, L. S. Goodman, T. W. Rall, and F. Murad, eds.), Macmillan, New York, pp. 959–979.

Klonne, D. R., Ulrich, C. E., Riley, M. G., Hamm, T. E., Jr., Morgan, K. T., and Barrow, C. S., 1987. One-year inhalation study of chlorine in rhesus monkeys (*Macaca mulatta*), *Fundam. Appl. Toxicol.* 9:557–572.

Kronberg, L., and Christman, R. F., 1989. Chemistry of mutagenic by-products of water chlorination, *Sci. Total Environ.* 81/82:219–230.

Kurokawa, Y., Hayashi, Y., Maekawa, A., Takahashi, M., Kokubo, T., and Odashima, S., 1983. Carcinogenicity of potassium bromate administered orally to F344 rats, *J. Natl. Cancer Inst.* 71:965–971.

Lichtenberg, R., Zeller, W. P., Gatson, R., and Hurley, R. M., 1989. Bromate poisoning, *J. Pediat.* 114:891–894.

Loper, J. C., 1980. Mutagenic effects of organic compounds in drinking water, *Mutat. Res.* 76:241–268.

Meier, J. R., 1988. Genotoxic activity of organic chemicals in drinking water, *Mutat. Res.* 196:211–245.

Meier, J. R., Lingg, R. D., and Bull, R. J., 1983. Formation of mutagens following chlorination of humic acid. A model for mutagen formation during drinking water treatment, *Mutat. Res.* 118:25–41.

Meier, J. R., Knohl, R. B., Coleman, W. E., Ringhand, H. P., Munch, J. W., Kaylor, W. H., Streicher, R. P., and Kopfler, F. C., 1987. Studies on the potent bacterial mutagen, 3-chloro-4-(dichloromethyl)-5-hydroxy-2($5H$)-furanone: Aqueous stability, XAD recovery and analytical determination in drinking water and in chlorinated humic acid solutions, *Mutat. Res.* 189:363–373.

Meshri, D. T., 1986. The modern inorganic fluorochemical industry, *J. Fluorine Chem.* 33:195–226.

Rook, J. J., 1974. Formation of haloforms during chlorination of natural waters, *Water Treat. Exam.* 23:234–243.

Schenck, H.-U., Simak, P., and Haedicke, E., 1979. The structure of polyvinylpyrrolidone–iodine (povidone–iodine), *J. Pharm. Sci.* 68:1505–1509.

Scully, F. E., Jr., Kravitz, R., Howell, G. D., Speed, M. A., and Arber, R. P., 1985. Contribution of proteins to the formation of trihalomethanes on chlorination of natural waters, in *Water Chlorination. Chemistry, Environmental Impact and Health Effects*, Vol. 5 (R. L. Jolley, R. J. Bull, W. P. Davis, S. Katz, M. H. Roberts, Jr., and V. A. Jacobs, eds.), Lewis Publishers, Chelsea, Michigan, pp. 807–820.

Stokinger, H. E., 1981. The halogens and the nonmetals boron and silicon, in *Patty's Industrial Hygiene and Toxicology*, Vol. 2B, 3rd ed. (G. D. Clayton and F. E. Clayton, eds.), Wiley, New York, pp. 2937–2965.

Weiss, S. J., Test, S. T., Eckmann, C. M., Roos, D., Regiani, S., 1986. Brominating oxidants generated by human eosinophils, *Science* 234:200–203.

Zamora, J. L., 1986. Chemical and microbiological characteristics and toxicity of povidone–iodine solutions, *Am. J. Surg.* 151:400–406.

Zierler, S., Feingold, L., Danley, R. A., and Craun, G., 1988. Bladder cancer in Massachusetts related to chlorinated and chloraminated drinking water: A case-control study, *Arch. Environ. Health* 43:195–200.

Biochemistry of
Inorganic Fluoride

2

2.1 Introduction

The ionic radii of fluoride (F^-) (1.33 Å) and the hydroxide ion (1.29 Å) are comparable, and each has a primary hydration number of 5. In contrast, the ionic radii of chloride (Cl^-) (1.81 Å), bromide (Br^-) (1.97 Å), and iodide (I^-) (2.23 Å) are significantly higher, and these anions have lower hydration numbers (2, 2, and 1, respectively). As a result of its similarity to the hydroxide ion, F^- readily substitutes for hydroxide in many biochemical transformations, often with profound consequences. In addition, as a consequence of these same physicochemical properties, the biological behavior of F^- differs dramatically from that of the other halogens. For example, unlike Cl^-, F^- (as HF) readily enters cells by passive transport, F^- is the only halogen incorporated into the crystal lattice of mineralized tissue, and F^- does not compete with I^- in the thyroid gland.

This chapter will consider the biochemistry of F^- in soft tissue, including a discussion of the effects of F^- on enzyme activity and cellular function, and the biochemistry of F^- in mineralized tissue. There is much material from which to choose in preparing a survey of the biochemistry of inorganic F^-. Several reviews are available and will be cited extensively (Banks, 1986; Hodge and Smith, 1981; Wiseman, 1970; Banks and Goldwhite, 1966; Smith, 1966).

2.2 Occurrence and Distribution of Inorganic Fluoride

Fluorine is widely distributed in nature, making up 0.065% of the Earth's crust, primarily as fluorspar (CaF_2), cryolite (Na_3AlF_6), and fluorapatite [$3Ca_3(PO_4)_2 Ca(F, Cl_2)$]. Fluorine ranks 13th (chlorine is 20th) in abundance among elements in the Earth's crust. Chlorine is much more abundant in seawater and, with respect to total accessible halogen, is about 2.7 times more abundant than fluorine (Banks, 1986). As noted in Chapter 1, because of its high oxidation potential, fluorine is not found naturally in its uncombined, elemental form. While a vast number of synthetic organofluorine compounds are now available, relatively few are found in nature–fluoroacetic acid is a notable example.

2.3 Biological Uptake, Distribution, and Metabolism of Inorganic Fluoride

The occurrence of F^- in biological material was noted over 200 years ago with the discovery by Berzelius of F^- in teeth. F^- is a normal, albeit minor, component of food (fish, tea, and sea salt have especially high F^- content), with dietary F^- amounting to 0.3–0.5 mg per day (Smith, 1966). Fluoridation of water probably accounts for an additional 1 mg per day (Section 2.15). At levels of 1–20 mg per day, approximately one-half of ingested F^- is excreted quickly, and the remainder is absorbed by the gastrointestinal tract. F^- blood levels peak within 30–60 min after ingestion. While there is no apparent accumulation in soft tissue, 35–50 % of absorbed F^- is taken up by skeletal tissue. Thus, F^- can be considered a normal constituent of bone, its concentration being dependent on age and amount of F^- ingested (Section 2.15) (Smith, 1966).

In soft tissues, normal extracellular concentrations are thought to be in the low micromolar range (Guy et al., 1976). In mammals, intracellular concentrations are expected to be even lower. This results from the passive

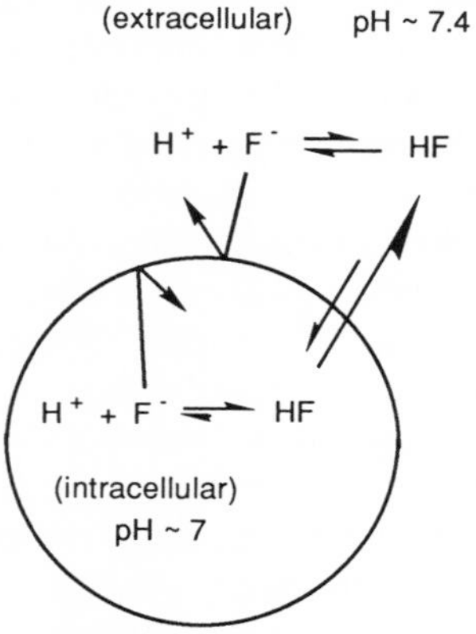

Figure 2-1. The effect of pH on the passive transport of HF into the cell.

mechanism established for the transport of F^- across cellular membranes. Whereas cells are impermeable to ionic F^-, the undissociated, nonpolar HF molecule is transported by passive diffusion. In mammalian cells, the intracellular pH (~ 7) is lower than the extracellular pH (7.4), leading to a relative higher concentration of extracellular F^-, as shown in Fig. 2-1. Plasma ionic F^- shows a linear relationship to F^- content in drinking water, and concentrations of 4.3 μM were found in individuals whose drinking water had the relatively high-concentration of 5.6 ppm (Guy *et al.*, 1976). Thus, under most circumstances, low micromolar concentrations are present in mammalian soft tissue, although concentrations several times higher probably result from chronic F^- poisoning (Section 2.15.4). Even so, it must be noted that many *in vitro* manifestations of biological actions of F^- are apparent only at millimolar concentrations.

2.4 Overview of Effects of Fluoride on Enzyme Activity

The concentrations of F^- required to inhibit enzyme activity are often sufficiently high to cast doubt on the physiological significance of this inhibition. Nonetheless, an enormous number of enzyme systems have been investigated, and it is instructive to review briefly generalizations that can be made from these studies. Fortunately, several extensive reviews are available which have tabulated and characterized much of the earlier work (Wiseman, 1970; Hodge and Smith, 1965; Borei, 1945).

In one approach to the organization of this subject, Wiseman (1970) grouped enzymes into four categories, based on their responses to F^- inhibition. The first group includes enzymes requiring metal ions for activation, the inhibition of which by F^- requires phosphate (P_i). Enolase, a notable member of this group, has been selected for more thorough discussion in this chapter (Section 2.5). Other enzymes listed in this limited group include succinic dehydrogenase, phosphoglucomutase, and lecithinase. A large number of enzymes belong to the second of Wiseman's categories—enzymes inhibited by F^- which require divalent metal ions, but which do not require P_i. Included in this list are a number of phosphatases that often require magnesium as an activating metal. Low to moderate millimolar concentrations of F^- are usually required for inhibition. Seven kinases are listed in this group, all requiring millimolar F^- concentrations for inhibition, as well as several other enzymes. Another group consists of enzymes that are inhibited by F^- but have no metal requirement, and a fourth group (catalase, peroxidases, and myeloperoxidases) requires trivalent ions for activation.

Enolase and inorganic pyrophosphatase have been chosen for review

here because these enzymes are relatively sensitive to F^-, because of potential physiological significance of this inhibition, and because results of
detailed mechanistic studies are probably relevant to interaction of F^-
with other systems. In particular, these two cases illustrate the effects of
substitution of F^- for hydroxide in the coordination sphere of metals
involved with enzyme catalysis. Cholinesterase, also reviewed, is quite sensitive to F^-, a surprising fact in view of the absence of metal requirement.
Finally, the recent advances in the understanding of the role of regulatory
proteins in a host of physiological processes, and parallel advances in the
understanding of the mechanism by which F^- influences these proteins,
will be reviewed.

2.5 Inhibition of 2-Phospho-D-Glycerate Hydrolyase (Enolase)

F^- is a potent inhibitor of enolase, a glycolytic enzyme that catalyzes
the elimination of water from 2-phosphoglyceric acid (2-PG) to form
phosphoenolpyruvic acid (Fig. 2-2). This inhibition has had historial
significance, in that inhibition of enolase by F^- was used in many classical
biochemical studies that revealed the details of the metabolic pathways of
carbohydrates.

2.5.1 Properties of Enolase

Although enolases from several sources have been isolated and characterized, the ease of purification and isolation from yeast have made enzyme
from this source the object of many studies. The properties of yeast enolase
have been reviewed thoroughly by Brewer (1981). A feature of the enzyme
that has been of particular interest to biochemists is the absolute requirement for a divalent cation for activity. The yeast enzyme is composed of
two identical subunits.

2.5.2 Fluoride Inhibition of Enolase: Dependence on Inorganic Phosphate

Enolase was isolated from yeast and crystallized by Warburg and
Christian (1942), who found that F^- alone gave no inhibition of enolase

Figure 2-2. Reversible addition of water to phosphoenolpyruvate catalyzed by enolase.

but that, in the presence of inorganic P_i, inhibition was strong. Based on their data, they reasoned that a magnesium fluorophosphate intermediate ($MgFPO_4$) must have been formed *in situ* and that this, rather than F^-, was the true inhibitory species. This mechanism was widely accepted until an obvious, and critical, experiment was performed. Peters *et al.* (1964) found that mixtures of F^- and P_i, both at $5\,mM$, strongly inhibited enolase, whereas fluorophosphate, at $5\,mM$ and $10\,mM$, exhibited no inhibitory activity. Thus, the requirement for P_i again required an explanation.

The first detailed kinetic study was carried out by Cimasoni (1972), a study that confirmed that F^- alone exhibited very weak (uncompetitive) inhibition. However, inhibition was strong and competitive in the presence of P_i. Cimasoni suggested that, in the absence of P_i, F^- binds to the substrate–enzyme complex but, in the presence of P_i, F^- binds directly to the enzyme and thus displays competitive inhibition. In another kinetic study, Wang and Himoe (1973) examined the effects of varying concentrations of Mg^{2+}, Mn^{2+}, F^-, and P_i on the rate of enolase-catalyzed dehydration. With Mg^{2+} as cation, F^- alone was noncompetitive at high concentrations and P_i was competitive with respect to 2-PG, whereas F^- and P_i together showed greatly enhanced competitive inhibition.

2.5.3 Cation Requirements for Enolase

Wang and Himoe (1973) also demonstrated that the degree of F^- inhibition was affected by the identity of the activating metal cation. The Mg^{2+}-activated enzyme was inhibited 40 times more strongly than the Mn^{2+}-activated enzyme, whereas the Zn^{2+}-activated enzyme showed no inhibition by F^-. This represents a direct relationship between the effectiveness of inhibition of specific metal-activated enolase by F^- and the relative order of effectiveness of the metals in activating the enzyme. Three types of metal binding sites have been described for enolase (Brewer, 1981). These metal binding sites include the "conformational" site, which binds several divalent cations very tightly, a catalytic site, which exists only in the presence of substrate, product, or, presumably, substrate analogues, and an inhibitory binding site, which becomes evident only at higher metal concentrations. Certain cations, most effectively Mg^{2+} and also Mn^{2+} and Zn^{2+}, bind to the conformational site and produce an active enzyme (Brewer *et al.*, 1983). A distorted, "precatalytic" configuration of the substrate induced by the conformational requirements of the bound metal is thought to produce activity in the enzyme–metal complex (Brewer and Ellis, 1983). Other cations, for example, Ca^{2+}, bind but lead to an inactive enzyme.

2.5.4 Interaction of Fluoride with Enzyme-Bound Metal

According to the mechanism of enolase-catalyzed dehydration of 2-PG, the C-3 hydroxyl group of the substrate must assume a negative charge as the reaction proceeds. The enzyme active site presumably is designed to stabilize this charge (Wang and Himoe, 1973). Using nuclear relaxation experiments, Nowak *et al.* (1973) showed that Mn^{2+} bound to enolase is coordinated directly to a water molecule that becomes the 3-OH group of 2-PG in the reverse reaction (Fig. 2-3). This coordinated water molecule is stabilized by the phosphate group of the substrate. Wang and Himoe (1973) suggested that F^-, acting as a hydroxide analogue, could take the place of enzyme-bound water. Stabilization of this fluoride–metal complex through hydrogen bonding with a P_i also complexed with the enzyme would explain the cooperative inhibitory effects of P_i and F^-.

2.5.5 Synergistic Binding of Fluoride and Phosphate to the Enzyme–Metal Complex

Formation of ligand complexes of enolase have been studied by ultraviolet spectroscopy and ^{19}F- and ^{31}P-NMR spectroscopy and with a fluoride-specific electrode (Maurer and Nowak, 1981; Nowak and Maurer, 1981; Bunick and Kashket, 1982). Dissociation constants obtained for the quaternary complex E–Mn–F–P_i reflect the strong cooperative effects of

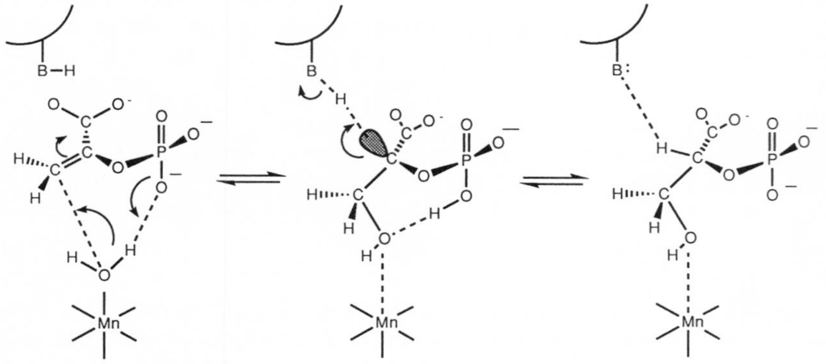

Figure 2-3. Proposed catalytic mechanism of enolase. Mn^{2+} bound to enolase is coordinated directly to a water molecule that becomes the 3-OH group of 2-PG in the hydration of phosphoenolpyruvate. Adapted, with permission, from T. Nowak, A. S. Mildvan, and G. L. Kenyon, Nuclear relaxation and kinetic studies of the role of Mn^{2+} in the mechanism of enolase, *Biochemistry* 12:1690–1701. Copyright 1973 American Chemical Society.

F^- and P_i in binding (K_d for $F^- = <10\,mM$ and K_d for $P_i = 0.5\,\mu M$), which had been manifested previously in the inhibitory behavior of these anions. The cooperative effect of the ligands is quite large, with dissociation constants decreasing by more than two orders of magnitude as the tightly bound inactive quaternary complex is formed. The conclusion was reached that P_i, F^-, and Mn^{2+} form a tightly bound quaternary complex with enolase wherein F^- may replace the hydroxide ion necessary for initiation of the enolase-catalyzed reaction. According to this proposal, a conformational change accompanying quaternary complex formation renders the coordinated metal inaccessible to surrounding solvent, greatly increasing the stability of the complex. Whereas F^- is bound directly to the metal, the P_i binding site is separated from the Mn^{2+} by 8 Å (Fig. 2-4) (Maurer and Nowak, 1981; Nowak and Maurer, 1981).

Bunick and Kashket (1982) used a fluoride-specific electrode to measure binding of F^- to the Mg^{2+}–enolase complex, both in the presence

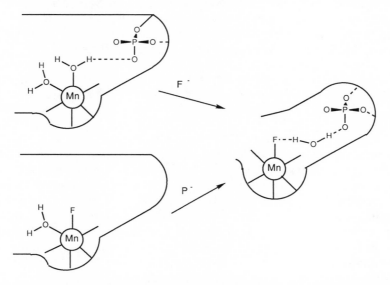

Figure 2-4. Structures of the enolase–Mn ternary and quaternary complexes. In the E–Mn–P_i complex, P_i is bound in a second–sphere complex to the Mn^{2+} at the catalytic site, the site occupied by the phosphoryl group of the substrate. In the ternary E–Mn–F complex, the F^- ligand interacts in the first coordination sphere of the Mn^{2+}, displacing one of the two water molecules. In the quaternary E–Mn–F–P_i complex, a conformational change occurs that facilitates the binding of Mn^{2+}, F^-, and P^i. Bound Mn^{2+} no longer has free access to the solvent, and the P_i moves further from the Mn^{2+}. Adapted, with permission, from T. Nowak and P. J. Maurer, Fluoride inhibition of enolase 2. Structural and kinetic properties of the ligand complexes determined by nuclear relaxation rate studies, *Biochemistry* 20:6901–6911. Copyright 1981 American Chemical Society.

and absence of P_i. In the presence of P_i, sequential formation of ternary and quaternary complexes was observed, with strong synergistic binding of F^- and P_i again evident.

Brewer and Ellis (1983) used ^{31}P-NMR spectroscopy to measure binding of substrate to enolase under a number of conditions and found that P_i and F^- together lead to additional metal binding sites on the enzyme. Brewer *et al.* (1983) also found fundamental differences in behavior toward F^- between a metal (Ca^{2+}) that binds to the conformational site but produces an inactive enzyme and a metal (Mg^{2+}) that activates the enzyme. First, the enzyme–Ca^{2+} complex is not susceptible to synergistic binding of F^- and P_i, possibly because of improper orientation of bound F^-. Also, in the presence of P_i, proton release accompanying binding of Ca^{2+} to enolase is unaffected by the presence of F^-. In the parallel experiment, a comparable release seen with binding of Mg^{2+} is abolished by the addition of F^- and, in fact, is replaced by an uptake of protons into the complex. This indicates that a group involved with Mg^{2+}, but not Ca^{2+}, binding is being protonated in the presence of F^- and P_i. Brewer *et al.* note that a synergism produced by hydrogen bonding as proposed by Nowak and Maurer (1981) requires an uptake of protons.

2.5.6 Substrate-Dependent Inhibition of Yeast Enolase by Fluoride in the Absence of P_i

Studies on the effect of F^- on enolase have been dominated by attempts to understand the synergistic inhibitory activities of F^- and P_i. The assumption that this is a physiologically important phenomenon is based on the drastically lower F^- concentrations required for inhibition in the presence of P_i. Spencer and Brewer (1982) have suggested that inhibition in the absence of P_i, noted in previous studies, may be an even more important process, particularly where F^- poisoning is involved. Since *in vivo* P_i levels are expected to be in the micromolar range, millimolar concentrations of F^-, which result in toxicity, could lead to inhibition of enolase independent of the synergistic effects of P_i. A corollary of this is that the low *in vivo* concentrations of P_i may be an important unexpected safety factor in water fluoridation (Section 2.15.6.6).

2.6 Inhibition of Inorganic Pyrophosphatase

Inorganic pyrophosphatase, a dimeric enzyme comprised of two identical subunits, catalyzes the reversible hydrolysis of pyrophosphate (PP_i) to orthophosphate (P_i). Yeast inorganic pyrophosphatase was purified by Bailey and Webb (1944) and shown to have a strict requirement for Mg^{2+} for activity and to be strongly inhibited by F^- ($K_i^{50} = 2 \times 10^{-5}\, M$). Extensive studies at Moscow State University have revealed many mechanistic details concerning this inhibition (Baikov et al., 1984; Smirnova and Baikov, 1983; and references therein). The enzyme is ubiquitous, but enzymes from different sources can vary significantly in their properties, including sensitivity to F^- inhibition. At pH 7.5, Bakers' 6.7 μM F^- is required for 50% inhibition of yeast pyrophosphatase, making it one of the most sensitive enzymes to F^- inhibition known [50 μM F^- is required for 50% inhibition of enolase (Wang and Himoe, 1973)]. In contrast, certain bacterial pyrophosphatases are insensitive to F^-.

2.6.1 Bakers' Yeast Pyrophosphatase: Mg^{2+} and Substrate-Dependent Inhibition by Fluoride

Several aspects of the inhibition of Bakers' yeast pyrophosphate are noteworthy. Inhibition is seen only in the presence of the activating metal (Mg^{2+}) and substrate (or product), and the establishment of this inhibition requires a period of time on the order of tens of minutes. A syncatalytic mechanism was suggested, since inactivation by F^- requires substrate- and metal-induced sensitization of the enzyme to the inhibitor. The rate of inactivation is proportional to the amount of the enzyme–$MgPP_i$ complex present, and substantial stabilization of the enzyme–metal–substrate complex accompanies F^- inactivation ($K_{dis} = \sim 1\, h^{-1}$). Analysis of fluoride-inactivated pyrophosphatase, following removal of excess F^-, Mg^{2+}, and PP_i by gel filtration, revealed the presence of two Mg^{2+}, one PP_i, and one F^- per active site. However, a covalent attachment of the substrate to the enzyme apparently is not involved (Bakuleva et al., 1981).

2.6.2 Metal Binding Sites and Fluoride Inhibition

Kinetic studies have shown that there are at least three metals per active center participating in pyrophosphatase catalysis (Baykov [Baikov] and Avaeva, 1974; Springs et al., 1981). Based on rates of formation and dissociation of enzyme–substrate and enzyme–substrate–fluoride complexes, a two-step mechanism of inactivation by F^- was proposed. In this scheme, the slow step in the inhibition process involves migration of F^- initially bound to Mg^{2+} associated with a low-affinity metal binding site to Mg^{2+} bound to a high-affinity site or to Mg^{2+} bound to a new site associated with the substrate in the inhibited complex. While the initial step in F^- binding causes a drop in activity of the enzyme, the full inhibition is presumed to result from displacement of OH^- by F^- from a critical site in the coordination sphere of the Mg^{2+}; this OH^- is presumably required for nucleophilic attack on the bound pyrophosphate (Fig. 2-5) (Smirnova and Baikov, 1983).

Other mechanistic studies have demonstrated that the active substrate in the pyrophosphatase-catalyzed reaction is the bidentate Mg^{2+} complex of the pyrophosphate tetraanion $MgPP_i$. Enzyme active site-mediated proton transfer converts a phosphoryl group into a good leaving group and $Mg(P_i)_2$ is produced (Haromy et al., 1982) by hydrolytic cleavage prior to, or synchronous with, proton transfer (Lin et al., 1986; Knight et al., 1983, 1981) (Fig. 2-6).

2.6.3 Biological Significance of Inhibition of Inorganic Pyrophosphatase by Fluoride

Baikov et al. (1984) isolated inorganic pyrophosphatase from mammalian mitochondrial membrane and found that this enzyme displayed a sensitivity to F^- comparable to that of the yeast enzyme and that it was inhibited by the same syncatalytic mechanism. At least two physiological consequences of pyrophosphatase inhibition could be important. First, the pyrophosphate anion is a product of many enzymatic reactions, particularly those involved in biosynthetic pathways. Pyrophosphatase-catalyzed hydrolysis of high-energy bonds provides a mechanism to shift the equilibrium of pyrophosphate-releasing reactions to completion. Pyrophosphatase inhibition thus could indirectly affect many reactions requiring this energy-producing step. In addition, accumulation of PP_i in

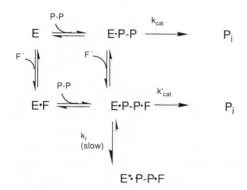

Figure 2-5. A two-step mechanism proposed for the inhibition of pyrophosphatase by F^- (Smirnova and Baikov, 1983).

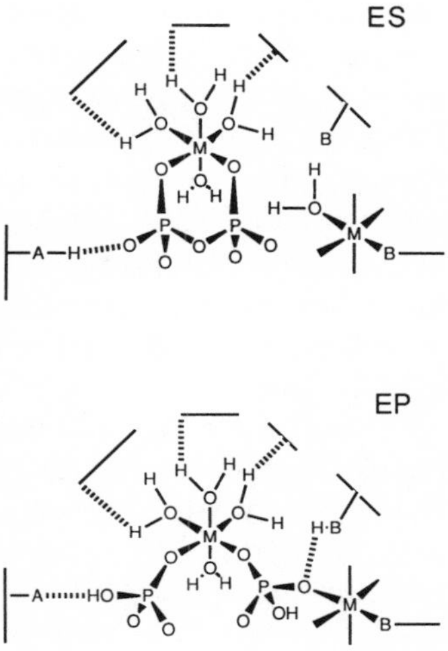

Figure 2-6. Proposed mechanism of action of pyro-phosphatase involving a bidentate Mg^{2+} complex of the pyrophosphate tetraanion, $MgPP_i$ (ES). Enzyme active site-mediated proton transfer converts a phosphoryl group into a good leaving group, and $Mg(P_i)_2$ (EP) is produced by hydrolytic cleavage prior to, or synchronous with, proton transfer. Adapted, with permission, from W. B. Knight, S. W. Fitts, and D. Dunaway-Mariano, Investigation of the catalytic mechanism of yeast inorganic pyrophosphate, *Biochemistry* 20:4079–4086. Copyright 1981 American Chemical Society.

the cytoplasm and mitochondria could have a direct toxic effect, since this moiety can function as an inhibitor of certain enzymes, including aminoacyl-tRNA synthases (Schelling and Cohen, 1987).

2.7 Inhibition of Acetylcholinesterases and Butyrylcholinesterase

Acetylcholine stimulates postsynaptic cholinergic receptors of neuromuscular junctions. The immediate response to receptor stimulation is the opening of cation gates to permit a rapid flow of both sodium and potassium, leading to transient depolarization of the postsynaptic membrane. In the proper functioning of cholinergic neurotransmission, the rapid deactivation of acetylcholine required to restore membranes to their resting potential following depolarization is accomplished by the serine hydrolytic enzyme, acetylcholinesterase. This enzyme is quite sensitive to F^- inhibition. Although the inhibition of cholinesterases by F^- has been studied for many years, details of the mechanism are still unclear (for a recent review, see Wilson, 1985). The fact of inhibition is itself unusual, since the enzymes are quite sensitive to F^- [concentrations leading to 50% inhibition (pI_{50}) were 1.6–2.9 mM at pH 8 for acetylcholinesterase and 0.4 mM for butyrylcholinesterase], yet they do not have a metal as part of the structure nor is a metal required for activity (Cimasoni, 1966).

The mechanism of hydrolysis by cholinesterases involves the formation of an enzyme–substrate complex, the formation of an acyl enzyme upon turnover of the substrate, and deacylation of the enzyme to regenerate active acetylcholinesterase (Fig. 2-7). Deacylation of the acyl enzyme intermediate normally is a very rapid process. However, in reactions with carbamates, an intermediate carbamoyl enzyme is formed that is much more stable to hydrolysis, permitting ready study of the rate of hydrolysis. Although F^- markedly decreased the rate of carbamylation, no discernible effect was seen on decarbamylation, suggesting that inhibition occurs in the first stage of the process. However, because of the much slower rate of decarbamylation, these data do not necessarily rule out an effect of deacetylation during hydrolysis of the natural substrate (Greenspan and Wilson, 1970).

Cholinesterase is strongly inhibited by arsenite, despite the absence of free sulfhydryl groups, the structural feature normally associated with arsenite sensitivity, making arsenite inhibition as unexpected as F^- inhibition. F^- and arsenite appear to be mutually exclusive in their interaction with the enzyme, suggesting that a common binding site is involved (Page and Wilson, 1983). A model was proposed wherein F^- forms hydrogen bonds with a serine (or threonine or tyrosine) hydroxyl group and one or more "ammonium ions" also derived from the protein. Formation of an arsenite ester with the same hydroxyl group would explain the competition between F^- and arsenite. Subsequent stopped-flow kinetic measurements showed that reaction of F^- with the enzyme was actually quite slow for a small symmetrical anion ($5 \times 10^3\ M^{-1}\ s^{-1}$), supporting the idea that a significant energy of activation is involved in breaking and making hydrogen bonds during F^- binding (Froede and Wilson, 1985).

Hydrolysis by butyrylcholinesterase decreases in the order butyrylcholine > acetylcholine > benzoylcholine (butyrylcholine is a very poor substrate for acetylcholinesterase). The much greater sensitivity of butyrylcholinesterase, as compared to acetylcholinesterase, to F^- is caused not by stronger binding of F^- to the free enzyme, but mainly by a high degree of cooperativity in binding of F^- with several substrates. This

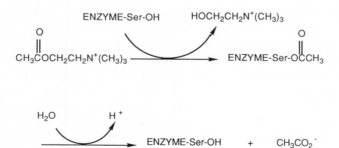

Figure 2-7. Catalytic mechanism of acetylcholinesterase.

cooperativity in binding was shown to be dependent on the nature of the acyl group in the substrate (butyryl, propionyl, and benzoyl were highly cooperative, acetyl only slightly), apparently ruling out electrostatic attraction between F^- and the ammonium group of the substrate as the cause. The physiological role of butyrylcholinesterase is still uncertain, even though the enzyme is quite abundant. Thus, the *in vivo* significance of this F^- sensitivity is also unclear (Page *et al.*, 1985).

The inhibition of the cholinesterases demonstrates that metal ions are not necessary for strong interactions of F^- with macromolecules. The strong inhibition of carboxyesterase by F^- represents another example of this phenomenon (Haugen and Suttie, 1974). The mechanism of cholinesterase inhibition has not been defined clearly. Future research should reveal important information regarding the effectiveness of F^- hydrogen bonding interactions in altering conformations and in stabilizing F^- binding in the absence of metals. Consideration of a potential role of aluminum, a likely contaminant of solutions used, would now seem proper in all biochemical studies of F^-, in view of the finding of Sternweis and Gilman (1982) (Section 2.9), although the consistent kinetic results reported would seem to allay such concerns in the studies discussed here. An example of aluminum-induced enhancement of F^- inactivation of enzyme activity is given in the next section.

2.8 Inhibition of ($Na^+ + K^+$)-Dependent ATPase (ATP Phosphohydrolase) by Fluoride—Influence of Aluminum

To maintain cells at their resting potential, Na^+ and K^+ are transported against concentration gradients across cell membranes by the action of the ($Na^+ + K^+$)-dependent ATPase. F^- inhibits this Na^+/K^+ pump in a Mg^{2+}-dependent first-order process that is facilitated by the presence of K^+ but diminished by the presence of Na^+. Advantage was taken of this inhibition in defining functionally distinct classes of K^+ sites on the enzyme (Robinson, 1975).

To explain the observation of Sternweis and Gilman (1982) that aluminum is required for F^- to activate adenylate cyclase (Section 2.9), Bigay *et al.* (1985) have proposed that AlF_4^- is formed and mimics the properties of PO_4^{3-}. Robinson *et al.* (1986) have investigated the effects of aluminum on the inhibition of ($Na^+ + K^+$)-dependent ATPase by F^- and found that low concentrations of $AlCl_3$ significantly increase the rate of inactivation. Similarities in the behavior of F^- (in the presence of aluminum), orthovanadate, and orthophosphate support the idea that the inhibitory species is AlF_4^-.

2.9 Stimulation of ATP Pyrophosphate-Lyase (Cyclizing) (Adenylate Cyclase)

Cyclic adenosine 3′,5′-monophosphate (cAMP) is a ubiquitous regulatory molecule, controlling a host of metabolic processes initiated by such species as adrenergic agents (norepinephrine, epinephrine), glucagon, thyrotropin-releasing hormone (TRH), and adenosine. In their early work leading to the isolation of cAMP produced by hormone (epinephrine or glucagon)-induced enzymatic synthesis in tissue particulate preparations, Sutherland and Rall (1958) found that yields could be increased by inclusion of 10 mM NaF in the incubation mixtures. At the time, they suggested that this enhanced yield resulted from the inhibition by F^- of the release of P_i from ATP. Subsequent research has shown that the process by which cAMP is produced through hormonal activation of membrane-bound receptors is quite complex. In these studies, F^- has played an important role because of its ability to stimulate adenylate cyclase (the enzyme responsible for the synthesis of cAMP from MgATP) through a process independent of receptor activation.

2.9.1 Mechanism of Hormonal Activation of Adenylate Cyclase

Despite early recognition of the significance of the role of cAMP in modulation and amplification of hormonal response, details of the mechanism of the adenylate cyclase system are still being elucidated. Factors that complicated early work include the fact that the cyclase activity is associated with an enzyme complex that is quite hydrophobic and labile, the involvement of a multiple-component system, and the extremely low concentration of the enzyme complex (Gilman, 1984a, b).

The overall process by which activation of receptors coupled to adenylate cyclase produces a cellular response was made much clearer by the recognition of the role of regulatory proteins. Several reviews are available (Gilman, 1987, 1984b; Lefkowitz *et al.*, 1984; Stiles *et al.*, 1984; Rodbell, 1980). The adenylate cyclase system consists of a receptor protein (R), a catalytic protein (C), and guanine nucleotide-binding regulatory proteins (G-proteins). The actions of the G-proteins are mediated through the binding of GTP. The G-protein required for stimulation of adenylate cyclase (G_s) is a heterotrimer, the subunits of which are designated α, β, and γ. The essential inhibitory G-protein (G_i) is also a heterotrimer (α, β, and γ subunits) (Bokoch *et al.*, 1983; Hildebrandt *et al.*, 1984; and references therein). Interaction of an agonist with the receptor produces an active form of the receptor protein. The activated receptor is coupled to and activates the G_s-protein, an event that is associated with dissociation of the α_s subunit and an exchange of an α_s-bound GDP with GTP. Binding of GTP produces the active α_s protein, which activates the catalytic subunit (C*), which, in turn, catalyzes the synthesis of cAMP from MgATP. Deactivation of the α_s-protein is carried out by hydrolysis of GTP by GTPase activity resident in the α subunit, forming the inactive GDP form. Reassociation of α_s with the $\beta\gamma$-protein forms the inactive G_s-protein, thus completing the cycle. Cellular responses to cAMP are mediated through cAMP activation of protein kinases. Because a single C* can produce many cAMP molecules and a single cAMP molecule can activate many kinase molecules before it is deactivated by phosphodiesterase-catalyzed hydrolysis, the cyclase system provides a mechanism for amplification of the hormonal message. Kinase-catalyzed phosphorylation of cellular enzymes results in the proper modulation of their activities to produce the ultimate response to hormonal activation of the receptor. Agonist binding to "inhibitory" receptors inhibits adenylate cyclase activity through a sequence involving the dissociation of α_i from G_i. Inhibition of cyclase activity is thought to occur through direct action of α_i on the catalytic protein, and also through $\beta\gamma$-protein-induced inhibition of dissociation of G_s. A diagram of the cyclase system is given in Fig. 2-8.

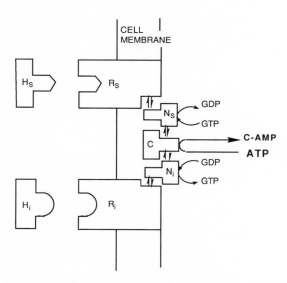

Figure 2-8. Schematic model of hormone-sensitive adenylate cyclase system. C, Catalytic unit; H, hormone; i, inhibitory; N, guanine nucleotide regulatory component; R, receptor; s, stimulatory. After Stiles *et al.* (1984), with permission.

2.9.2 Effects of Fluoride on Adenylate Cyclase Activation and Deactivation

2.9.2.1 Site of Fluoride Action

An important tool for characterization of the components of the adenylate cyclase system was a variant of an S49 lymphoma that lacked receptor-mediated stimulation of adenylate cyclase activity because of the absence of the G_s regulatory proteins. In reconstitution experiments using this system, Howlett *et al.* (1979) found that addition of purified regulatory protein, pretreated with F^-, to membrane-bound catalytic subunit resulted in cyclase activity, even though the F^- concentration at the time of addition was well below that required for activation. These and other data demonstrated that F^- does not act directly on the catalytic subunit, but rather exerts its action indirectly through the regulatory proteins.

2.9.2.2 Fluoride Response Is Enhanced by the Presence of Al^{3+}

In attempting to find the source of inconsistent results in experiments on F^- stimulation of purified regulatory proteins, Sternweis and Gilman (1982) suspected the presence of an unknown factor in variable concentrations in the assay mixtures. The identification of this factor as aluminum chloride, present in micromolar concentrations as a contaminant in samples of ATP or derived from glassware, was timely, in that interpretation of subsequent experiments involving F^- almost certainly would otherwise have been enormously complicated. Sternweis and Gilman suggested that the active species may be the tetrafluoroaluminate anion (AlF_4^-). Most studies with F^- and cyclase now combine F^- with aluminum chloride for consistent and enhanced activity. However, the implications of the observation that low concentrations of Al^{3+} can fundamentally alter the biological properties of F^- clearly are far-reaching, in view of the enormous number of studies that have been done with no appreciation of this possible complication. Furthermore, the low concentrations (5–10 μM) of Al^{3+} that are effective are well within the range of physiological concentrations.

2.9.2.3 Fluoride Stimulation and the Dissociative
Model for Activation of Adenylate Cyclase

The demonstration that addition of the β subunit greatly accelerated the deactivation of G_s previously activated with F^- (in the presence of Mg^{2+} and Al^{3+}), or guanine nucleotides in the absence of activating ligands, was taken as evidence that the β subunit serves as an inhibitory factor in the activation of G_s by F^- and guanine nucleotides (Northup *et al.*, 1983a). The finding that the α subunit was capable, by itself, of activating adenylate cyclase was further evidence for a dissociative mechanism of activation. Acceleration of the rate of deactivation of α subunit-induced activation by the β subunit was consistent with this model (Fig. 2-9) (Northup *et al.*, 1983b). The ability of F^- to stimulate dissociation of the subunits of G_s (and of G_i) (Bokoch *et al.*, 1984; Katada *et al.*, 1984a; Northup *et al.*, 1983b) thus represents a mechanism for fluoride activation of the catalytic subunit.

The functional role of the inhibitory protein, G_i, is also consistent with the dissociative model. The β subunit of G_i is indistinguishable from the β subunit derived from the stimulatory (G_s) protein, and thus is capable of deactivating the $α_s$ subunit. Fluoride-induced dissociation of the inhibitory protein, by producing the β subunit, should be reflected in an inhibitory activity of F^-. This was verified by Katada *et al.* (1984b), who demonstrated that, whereas F^- stimulates the basal activity of cyclase, it markedly inhibits the enhanced activity produced by prostaglandin E_1 stimulation.

2.9.2.4 Evidence for Fluoride Stimulation through a Nondissociated Complex

Much of the mechanistic work discussed above was done with detergent-solubilized preparations. Recent radiation inactivation studies

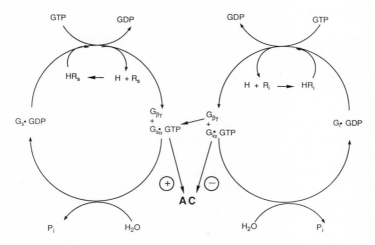

Figure 2-9. Dissociative model for adenylate cyclase activation. Adapted from Gilman (1987); reproduced, with permission, from *Annual Review of Biochemistry*, Vol. 56. Copyright 1987 Annual Reviews, Inc.

have allowed analyses of dissociative processes in the more physiologically relevant intact membrane. Based on this work, a steady-state model for cyclase activation, also involving subunit dissociation, has been proposed (Skorecki *et al.*, 1986). Target analysis also was used to study the mechanism of nonhormonal (including F^-) activation of adenylate cyclase. In contrast to results obtained from studies done previously with detergent-solubilized regulatory protein, analyses of target size suggested that an undissociated complex formed between F^- and G_s interacts with the catalytic subunit (C) to form a fluoride–G_s–C complex (G_sFC). According to this model, dissociation of this complex produces the active catalytic subunit (C^*), F^-, and G_s protein (Verkman *et al.*, 1986).

2.9.2.5 Effect of Fluoride plus Al^{3+} on G-Protein GTPase

The possibility that F^- in the presence of Al^{3+} could inhibit intrinsic G_s GTPase activity, with resultant stimulation of cyclase activity, was investigated by Brandt and Ross (1986). Such a mechanism of activation would parallel the action of cholera toxin, which, by irreversible inhibition of the GTPase activity of the α_s subunit, produces a permanently activated state of this protein, and of adenylate cyclase. In contrast to results obtained with transducin (Section 2.10.2), no inhibition of G_s GTPase activity by F^- was detected in this study.

2.10 Stimulation of Photoreceptor Phosphodiesterase I by Fluoride

2.10.1 G-Proteins and the Visual Process

Many additional energy transduction systems, similar in many ways to the adenylate cyclase system, have been identified. Thus, the involvement of cyclic nucleotides in the visual process is indicated by the fact that illumination of rod outer segments (ROSs) produces a 5–20-fold stimulation of cGMP phosphodiesterase (Miki *et al.*, 1973). In fact, the photoexcitation of one molecule of rhodopsin results in the hydrolysis of more than 10^5 molecules of cGMP. Rhodopsin, activated by light, catalyzes the exchange of GTP for GDP bound to a protein present in the rod outer segment (Fung and Stryer, 1980). This protein, given the name "transducin" because of its role in energy transduction, is a heterotrimer, with subunits designated $T\alpha$, $T\beta$, and $T\gamma$, having molecular masses of 39,000 Da, 36,000 Da, and 10,000 Da, respectively (Kuhn, 1980; Fung *et al.*, 1981). During activation of transducin by GTP, Tα becomes dissociated from T$\beta\gamma$ and activates cGMP phosphodiesterase. Amplification of the signal (photon) is provided by the ability of one activated receptor to catalyze many GTP–GDP exchanges, with the resulting dissociation of many Tα subunits, before it is deactivated and by the fact that a single activated phosphodiesterase can rapidly catalyze the hydrolysis of many cGMP molecules before deactivation. Ultimately, the information is translated into the rapid hyperpolarization of the plasma membrane of the ROS.

2.10.2 Effect of GTP and of Fluoride on Subunit Dissociation and on Transducin GTPase Activity

The similarities in structure and function of guanine nucleotide-binding proteins involved in the visual process and in the adenylate cyclase system were extended by results from studies of the effect of F^- on subunit properties in a reconstituted transducin system. Thus, the presence of 2.5 mM NaF and 20 μM AlCl$_3$ caused a 45 % reduction in binding of Tα (in the presence of T$\beta\gamma$) to rhodopsin. Since association of Tα with T$\beta\gamma$ had been shown to be a requirement for efficient binding of Tα to rhodopsin (Fung, 1983), this provided evidence that F^- (in the presence of Al^{3+}), similarly to GTP, promotes subunit dissociation. In addition, F^- alone, but more effectively in the presence of 20 μM AlCl$_3$, inhibited transducin GTPase activity in a dose- and time-dependent fashion (compare Section 2.9). The proposal was made that the inhibition of transducin-catalyzed hydrolysis of GTP in the presence of F^- results from enhanced dissociation of Tα from T$\beta\gamma$. Resulting inhibition of GTP–GDP exchange would account for the observed decrease in GTP hydrolysis (Kanaho *et al.*, 1985). More recently, in experiments in which the concentration of free Al^{3+} was controlled precisely, Al^{3+} by itself was found also to have an inhibitory effect on transducin GTPase activity, apparently replacing Mg^{2+} at the exchangeable nucleotide site on the protein. This replacement causes inhibition of GDP–GTP nucleotide exchange, and this inhibition in turn causes the lower GTPase activity. However, the overall effect of Al^{3+} alone is to inhibit activated rhodopsin signal transduction, an effect in opposition to the effect of Al^{3+} plus F^- (Miller *et al.*, 1989).

A similarity between the transducin–phosphodiesterase system and the G-protein–adenylate cyclase system is apparent. This similarity becomes more striking with the finding that the T$\beta\gamma$ subunit of transducin has inhibitory properties toward the G$_s$-protein of the adenylate cyclase system. Thus, activation of G$_s$ by F^- (10 mM) was inhibited 52 % by the presence of T$\beta\gamma$ (Boeckaert *et al.*, 1985).

2.10.3 Mechanism of Fluoride-Induced Activation of cGMP Phosphodiesterase

2.10.3.1 Fluoride Activation Is Mediated through Transducin

Stein *et al.* (1985) compared the release of transducin from ROS membranes catalyzed by GTP and by F^-. While illumination was required for GTP-stimulated release, fluoride-induced release was more efficient in the absence of light. The resulting activation of phosphodiesterase by F^- in the absence of rhodopsin again parallels the F^- activation of adenylate cyclase in the absence of hormone and receptor. Tryptic digestion of transducin released either by F^- or by GTP provided evidence that similar conformational changes occurred during each process, supporting the proposal that the activation of phosphodiesterase by F^- is mediated through transducin.

2.10.3.2 AlF_4^- May Mimic the γ-Phosphate Group of GTP

A mechanism for the activation of phosphodiesterase by F^- based on the obligatory role of aluminum has been proposed. Dose–activation curves of purified $T\alpha GDP$ with NaF and $AlCl_3$, run at various concentrations of $T\alpha$, revealed that full activation of $T\alpha$ was invariably achieved when the concentration of AlF_4^- was equal to the concentration of $T\alpha$. Furthermore, it was shown that AlF_4^- had no effect on $T\alpha$ when the nucleotide binding site was empty, or when it was occupied by the nonhydrolyzable analogue $GDP\beta S$. The effect of F^- on guanine nucleotide-binding proteins (G_s, G_i, or T) can be envisioned as imparting structural and functional properties of the GTP-bound protein to the GDP-bound one. Noting structural similarities between AlF_4^- and PO_4^{3-}, Bigay *et al.* (1985) proposed that AlF_4^- mimics the γ-phosphate group of GTP. In this model, binding of AlF_4^- in the GDP-containing nucleotide-binding proteins produces a quasi-GTP structure, in which AlF_4^- is more tightly bound than the phosphate group released by GTPase hydrolysis of GTP (Bigay *et al.*, 1985). Further support for this proposal came from the demonstration that activation of purified $T\alpha$–GDP by AlF_4^- (30 μM $AlCl_3$ in the presence of 10 mM NaF) does not require the presence of the $\beta\gamma$ subunit, that GDP or a GDP analogue unsubstituted on its β phosphate (e.g., $GDP\alpha S$) is required for AlF_4^- activation, and that this activation is reversed by dilution of AlF_4^- and partially reversed by addition of $T\beta\gamma$. Similar results were obtained with BeF_3^- (30 μM $BeCl_2$ in the presence of 10 mM NaF). According to this proposal, the β-phosphate group of GDP displaces one F^- from AlF_4^- and becomes coordinated to aluminum in the nucleotide binding site. Interactions of F^- with protein or with Mg^{2+} also present in this site would stabilize the structure (Fig. 2-10) (Bigay *et al.*, 1987).

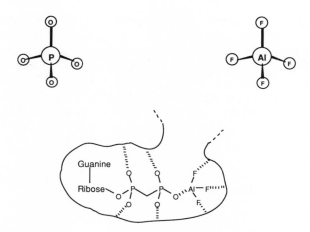

Figure 2-10. AlF_4^- as a phosphate mimic. See text for details. After Bigay *et al.* (1985), with permission.

This model for F^- activation of guanine nucleotide-binding proteins readily explains certain characteristics of the activation, including the requirement for the presence of GDP or a suitable analogue of GDP. Bigay *et al.* (1987) characterized AlF_4^- complemented with GDP as a new type of nonhydrolyzable, yet reversibly bound, GTP analogue. They further suggested that this model may have implications that reach far wider than the growing family of G-proteins. Indeed, a requirement for aluminum has been demonstrated in inhibition of hepatic glucose-6-phosphatase by F^- (Lange *et al.*, 1986), who suggested that AlF_4^- functioned as a high-affinity analogue of $H_2PO_4^-$. Robinson *et al.* (1986) invoked the model of Bigay *et al.* to explain a requirement for aluminum and beryllium for inhibition of $(Na^+ + K^+)$-dependent ATPase by F^- (Section 2.). Bigay *et al.* (1987) suggested that effects of F^- on many enzymes may result from a mimicking of the ubiquitous phosphate group by AlF_4^-.

2.11 Stimulation of Polyphosphoinositide Phosphodiesterase by Fluoride

There has accumulated much evidence that the action of many calcium mobilizing agents is related to the stimulation of phosphoinositide turnover (reviewed by Berridge and Irvine, 1984). The search for the link between membrane-bound phosphatidylinositol 4,5-bisphosphate (PI-4,5-P_2) and intracellular calcium stores has revealed that inositol trisphosphate (IP$_3$), a product of polyphosphoinositide phosphodiesterase (PPI-pde)-catalyzed hydrolysis of PI-4,5-P$_2$, serves as an intracellular second messenger in the process by which stored calcium is mobilized. A second product of PI-4,5-P$_2$ hydrolysis, diacylglycerol (DAG), activates protein kinase C. Recent reviews of these interactions, summarized in Fig. 2-11, are provided by Litosch and Fain (1986) and by Berridge (1987). Control over a variety of important cellular processes are mediated by IP$_3$ and DAG, including secretion, metabolism, and cell proliferation; accordingly, research in this area is extremely active (Berridge, 1987).

2.11.1 GTP Analogues and Fluoride Stimulate PPI-pde Activity

Examples of agonists that can stimulate PI-4,5-P$_2$ hydrolysis include muscarinic cholinergic agonists, vasopressin, α_1-adrenergic agonists, and the chemotactic peptide fMet-Leu-Phe. Agonist occupation of calcium-mobilizing receptors is translated into activation of the catalytic unit of PPI-pde. The demonstration that human neutrophil membrane PPI-pde could be activated by the addition of nonhydrolyzable analogues of GTP indicated an intermediary role for GTP-binding proteins in this transduction process (Cockcroft and Gomperts, 1985). A subsequent study revealed that F$^-$ (with added AlCl$_3$) and the hydrolysis-resistant analogue of GTP, guanosine 5'-[γ-thio]triphosphate (GTPγS), have very similar effects on activation of hepatocyte membrane PPI-pde. The GDP analogue, guanosine 5'-[β-thio]diphosphate inhibited stimulation by both F$^-$ and GTPγS. PPI-pde activation by these reagents parallels the similar stimulation of adenylate cyclase and cGMP phosphodiesterase (Cockcroft and Taylor, 1987).

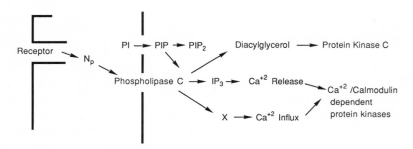

Figure 2-11. Phosphatidylinositol 4,5-bisphosphate turnover. Adapted, with permission, from I. Litosch and J. N. Fain, Regulation of phosphoinositide breakdown by guanine nucleotides, *Life Sci.* 39:187–194. Copyright 1986 Pergamon Press plc.

2.11.2 Cellular Effects of Fluoride and PPI-pde Stimulation

Most of the research on the stimulation of adenylate cyclase with F^- has been done with broken cell preparations, and the millimolar concentrations of F^- required for stimulation have placed the physiological significance of intracellular interaction of F^- with regulatory proteins in doubt. On the other hand, tissue adenylate cyclase activity as well as increased cAMP levels has been detected in animals given F^- in drinking water (Susheela and Singh, 1982; Kleiner and Allmann, 1982). The added sensitivity of regulatory proteins to F^- caused by the presence of low levels of $AlCl_3$ may lead to more ready detection of responses of whole cells to F^-. For example, in one study, several parameters which reflect hepatocyte PPI-pde activation were affected by NaF in a concentration-dependent manner. Thus, there was observed activation in phosphorylase, inactivation of glycogen synthase, efflux of calcium, increase in myo-inositol-1,4,5-P_3 levels, decrease in PI-4,5-P_2 levels, and increase in DAG levels. The presence of micromolar concentrations of $AlCl_3$ once again potentiated the effects of F^-. These results also further implicate guanine nucleotide-binding proteins (termed N_p in these studies) in this PPI-pde system and again show the parallels with the adenylate cyclase and photoreceptor cGMP phosphodiesterase systems (Blackmore et al., 1985).

As noted above, the PPI-pde system is involved in the control of many critical cellular functions. The recognition of the stimulatory action of F^- on PPI-pde has led to the suggestion this may be the biochemical basis for many cellular responses to F^- (Cockcroft and Taylor, 1987). For example, the fluoride-induced stimulation of platelet aggregation and release of ATP from platelets was associated with increased formation of inositol phosphates and release of stored calcium, indicative of PPI-pde activation (Kienast et al., 1987). The ability of F^- to stimulate the "respiratory burst" in human neutrophils also is associated with a rise in intracellular calcium (Strnad and Wong, 1985) as is release of histamine from fluoride-activated mast cells (Kuza and Kazimierczak, 1982).

2.12 Additional Regulatory Proteins That Interact with Fluoride

Much current research suggests that there are many additional biological systems in which F^-, or AlF_4^-, may elicit a significant response. In fact, activation by F^- is now considered strong evidence that mediation of signal transduction occurs by a G-protein (Humphreys and Macdonald, 1988). Since the identification of the essential role of guanine nucleotide-binding protein in the regulation of adenylate cyclase, it has become increasingly apparent that similar regulatory proteins are required for a host of biological processes (reviewed by Lochrie and Simon, 1988; Gilman, 1987). For example, a regulatory protein, G_o, isolated from bovine brain interacts with muscarinic, α_2-adrenergic, and GABA receptors and has been shown to have structural similarities to the cyclase G_s protein. Interaction of the stable GTP analog GTPγS with G_o in the presence of Mg^{2+} causes an increase in tryptophan fluorescence. A nearly identical spectral change was observed as a result of treatment of G_o with F^- in the presence of Al^{3+} and Mg^{2+}. This suggests that F^-, under these conditions, indeed mimics the action of the GTP analogue and supports the suggestion of Bigay et $al.$ (1985, 1987) that AlF_4^- functions as an isostere of the γ-phosphate group of GTP, producing a long-lasting activated state of the regulatory protein (Higashijima et $al.$, 1987). In contrast, tubulin—which requires GTP and an associated divalent cation for assembly into microtubules—was not activated by F^- or AlF_4^-. However, F^- alone inhibited intrinsic GTPase activity and stimulated microtubule formation, possibly as a result of the large sphere of hydration associated with F^-—an ordering of water molecules that could decrease the concentration of water available for GTPase-catalyzed hydrolysis (Humphreys and Macdonald, 1988).

Several other GTP-binding proteins have been identified. Regulatory proteins involved in protein synthesis will be discussed in Section 2.13.3. Of particular current interest is the identification of a GTP-binding regulatory protein (ras p21 protein) that may play a role in transducing signals from cell surface-acting growth factors to the rest of the cell and that possesses activated oncogenic mutants (Newbold, 1984; McGrath et $al.$, 1984).

2.13 Effects of Fluoride on Cellular Function

The effects of F^- on cellular function have been studied in many systems. In the majority these studies, however, F^- has been used as a metabolic inhibitor to probe the relationship between specific energy-producing pathways and cellular function, and the biochemical interactions of F^- thus have not been the primary concern. The review by Matthews (1970) describes the extensive use of F^- in classic studies of intermediary metabolism. Unfortunately, in many cases, the relatively high concentrations required to produce a given cellular response place physiological significance of these actions in doubt (Messer, 1984). Nonetheless, three examples are discussed below which illustrate the emerging recognition of the ability of F^- to interact with regulatory proteins in intact cells.

2.13.1 Effect of Fluoride on Platelet Function

The PPI-pde and adenylate cyclase systems are both involved in the response of platelets to outside hormonal signals. The binding of certain agents to platelet receptors leads to responses that include shape change, release of dense granules, and aggregation, while the binding of other agents, such as prostacyclin or prostaglandin E_1, produces a state of unresponsiveness to stimulatory signals. This "negative response" in platelets results primarily from increased cAMP levels.

Exposure to NaF also activates intact platelets, as manifested by dense granule release, but the mechanism of this response has not been obvious. Recent studies now have produced evidence that F^- can serve both to stimulate and to inhibit platelet function (e.g., Kienast et al., 1987; Poll et al., 1986). The nature of the response to F^- is concentration dependent. Stimulation of function (aggregation, ATP release) at concentrations of 30–40 mM is associated with increased formation of inositol phosphates, a rise in cytosolic free Ca^{2+}, and increased kinase activity, reflecting increased PPI-pde activity. Concentrations greater than 40 mM produced a transient rise in cAMP level that led to reduction in Ca^{2+} mobilization, aggregation, and ATP release. Inositol phosphate levels and kinase activity were not lowered by increasing F^- concentration. This concentration-dependent biphasic response to F^- was attributed to simultaneous modulation of PPI-pde and adenylate cyclase activity (Kienast et al., 1987).

2.13.2 Effect of Fluoride on Neutrophil Function

Neutrophils, the predominant circulating leukocytes in humans, provide a major defense mechanism against infections. Microorganisms that contact these, or other phagocytic cells, are ingested into phagocytic vacuoles. As part of the killing mechanism, lysosomes fuse with these vacuoles to form phagosomes and then discharge lytic and digestive enzymes. At the same time, an NADPH oxidase system is activated, resulting in a rapid increase in oxygen consumption known as the "respiratory burst." Superoxide radicals (HOO ·) are formed and rapidly converted to hydrogen peroxide in the phagosomes. Hydrogen peroxide is bactericidal and, furthermore, is converted enzymatically to very potent bactericidal hypohalous acids (see Chapters 3, 4, and 7), providing an effective additional killing mechanism. More complete discussions of this process can be found in papers by McPhail et al. (1985), Gabler and Hunter (1987), and Prince and Gunson (1987).

F$^-$ has been shown to effect leukocyte function by at least two mechanisms. All of the processes involved depend on glucose metabolism, primarily through the monophosphate shunt. In activated polymorpho-nuclear leukocytes, 0.1–1.0 mM F$^-$ markedly inhibited superoxide genera-tion and glucose metabolism, whereas at higher concentrations, >10 mM, increased superoxide production was observed, peaking at 20 mM (Gabler and Leong, 1979). Inhibition of human neutrophil phagocytosis also was observed in human neutrophils treated with 2.5–10 mM F$^-$ (Gabler and Hunter, 1987). In the responses at lower F$^-$ concentration, F$^-$ acts as a metabolic inhibitor. At concentrations above 10 mM, as noted by Gabler and Leong, F$^-$ is also an effective activator of the respiratory burst in a process independent of phagocytosis or degranulation (Curnutte et al., 1979). The similarities in the responses to F$^-$ and to receptor-dependent agonists, such as chemotactic peptides, suggest again activation of a guanine nucleotide-binding regulatory protein. The accompanying increase in intracellular Ca^{2+} implicates a persistent activation of the G-protein responsible for PPI-pde stimulation as the mechanism for F$^-$ activation of the respiratory burst (Strnad and Wong, 1985).

2.13.3 Effect of Fluoride on Protein Synthesis

Among the several actions F$^-$ has on cellular function, inhibition of protein synthesis, studied in a variety of cells and cell-free systems, is notable for its close correlation with cytotoxicity (Holland, 1979a, b). Reticulocytes have been used extensively since they produce predominantly one molecular species, hemoglobin. Even though the concentration required to cause inhibition is relatively high, in the 1–10 mM range, F$^-$ has been used extensively in the study of the complex sequence of enzyme-catalyzed events involved (Holland, 1979b). The inhibitory properties of F$^-$ have been particularly revealing in the study of the initiation phase of protein synthesis.

2.13.3.1 Biosynthesis of Proteins—A Brief Review

In the nucleus of eucaryotic cells, DNA-directed RNA polymerases produce RNA molecules, termed primary transcripts. These are modified in the nucleus and enter the cytosol as mature rRNA, tRNA, and mRNA molecules. mRNA molecules, the base sequences of which are coded for the biosynthesis of specific polypeptides, associate with polyribosomes containing the synthetic machinery for assembly of polypeptides. It is to the polyribosomes that tRNA molecules deliver their specific amino acids. In the initiation phase of protein synthesis, Met-tRNA, mRNA, and ribosomal subunits associate to form translationally active ribosomes. The dissociation of the 80S monomeric ribosome to 40S and 60S subunits is a prerequisite for this initiation process.

Recently, a series of protein factors, termed eucaryotic initiation factors (eIF), and required for this and other steps in initiation, have been identified (Pain, 1986). Subsequent to dissociation of the 80S monomer, in a process dependent on GTP and eIF-2, a 43S preinitiation complex is formed from the 40S ribosomal subunit, Met-tRNA, eIF-2, and GTP (Fig. 2-12). Through mediation of at least three initiation factors (eIF-4A, eIF-4B, and eIF-4F) and ATP, this complex binds to messenger RNA at the 5' end. The complex then moves along the mRNA until it recognizes the base sequence AUG (the code for Met-tRNA). The next step, which requires another initiation factor (eIF-5), involves the release of eIF-2 and eIF-3 from the 40S subunit and attachment of the 60S ribosomal subunit to form the active 80S initiation complex as part of the polyribosomal structure. Regulation of protein synthesis is thought to involve steps in formation of the initiation complex as part of the polyribosomal structure.

In the elongation and termination phases of protein synthesis, the anticodon of the appropriate acylated tRNA molecule is recognized by the three-base code present at the vacant A (amino acid) site on the polyribosome-bound mRNA. Binding of the amino acid-charged tRNA to the mRNA is followed by peptide bond formation catalyzed by peptidyl transferase. GTP-dependent translocation of the complex along the mRNA chain (5' to 3' direction) exposes a new empty A site. Another acylated tRNA binds to the newly vacated P site, and the process continues until the termination sequence is recognized by a release factor. Activation of peptidyl transferase at this stage results in hydrolytic cleavage of the completed peptide from mRNA.

2.13.3.2 Fluoride Inhibits Protein Synthesis by Blocking Initiation

Marks et al. (1965) found that inhibition of protein synthesis in reticulocytes produced by 10 mM F$^-$ was accompanied by a dissociation of polyribosomes. The accompanying decrease in both [^{14}C]-glucose metabolism and ATP concentration led them to postulate that polyribosome dissociation and inhibition of protein synthetis were events secondary to the inhibition of formation of high-energy compounds such as ATP and GTP. Several lines of evidence subsequently emerged, however, that showed that F$^-$ does have a direct inhibitory action on protein syn-

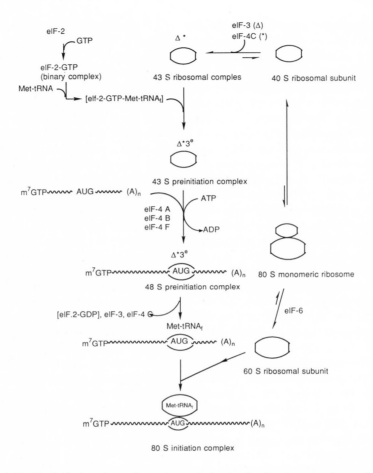

Figure 2-12. The initiation phase of protein synthesis, showing involvement of several initiation factors (see text). Adapted, with permission, from Pain (1986).

thesis. From this work, it became clear that F^- had little effect on completion of nascent peptide chains bound to ribosomes but that initiation of new chains was inhibited. The initial research papers in this area have been analyzed in considerable detail in the review by Matthews (1970).

Events subsequent to the fluoride-induced dissociation of polyribosomes were studied in several systems. In reticulocytes incubated with NaF, a decrease in ribosomal subunits was shown to precede inhibition of protein synthesis (Colombo et al., 1968). Removal of F^- led to restoration of protein synthesis accompanied by the reestablishment of normal levels of subunits, indicating that dissociation of the 80S ribosome to 40S and 60S ribosomal subunits, necessary for initiation of protein synthesis, is F^- sensitive.

Two apparent specific actions of F^- on ribosomes—an inhibition of dissociation of the 80S ribosome and a blocking of the binding of the 60S subunit to a preinitiation complex—are consistent with the results of several investigations. For example, in mammalian cells, the 40S subunit fraction contains a major 40S subunit and a minor 43S subunit, having the initiator $tRNA^{Met}$ molecule bound to it (Baglioni, 1972). In HeLa cells incubated with 15 mM NaF, only the 43S subunit is found (Baglioni et al., 1969). It is this subunit, normally present in limiting amounts, that participates in peptide chain initiation. Since both utilization of the 43S subunit and dissociation of the 80S to the 40S subunit are blocked by F^-, only the 43S subunit can accumulate (see Fig. 2-12).

The structure and function of initiation complexes that accumulate during inhibition of protein synthesis by F^- have been studied further by Godchaux and Atwood (1976).

2.13.3.3 Biochemical Mechanisms and Fluoride Inhibition

The determination of the site of action of F^- in inhibition of protein synthesis provides no direct evidence for the biochemical mechanism of this inhibition, although the involvement of regulatory protein factors required for the initiation process seems likely. As part of a study of the relationship between protein synthesis and initiation factor phosphorylation, Duncan and Hershey (1987) found that exposure of HeLa cells to 10 mM NaF for 30 min inhibited protein synthesis by greater than 80%. Analysis of the effect on initiation factors (eIF-2α, eIF-3, eIF-4A, and eIF-4B) showed that inhibition was accompanied by up to 30% phosphorylation of eIF-2α and eIF-4B. Since this degree of phosphorylation of eIF-2α is known to inhibit the activity of this initiation factor, the proposal was made that such phosphorylation is important in the inhibitory action of F^- on protein synthesis. The increased proportion of phosphorylated factors, in turn, would result from the action of F^- as a phosphoprotein phosphatase inhibitor.

Finally, the important role of a guanine nucleotide exchange factor (GEF) required in the eIF-2 activation cycle (Pain, 1986) is reminiscent of the role of other G-proteins discussed above. Interaction of F^- (or AlF_4^-) with the GTP-binding site of this protein would seem worthy of investigation.

2.14 *In Vivo* Toxicity of Fluoride

Several interrelated factors are important in considering mechanisms of F^- toxicity. As stressed by Messer (1984), a knowledge of intracellular F^- concentrations, and mechanisms by which these concentrations are controlled, is of paramount importance in any extrapolation of *in vitro* cellular response to *in vivo* toxicity. Thus, levels of 0.1–10 mM F^- normally are required to inhibit cell growth *in vitro* (Matthews, 1970; Hongslo *et al.*, 1980). The increase of blood and tissue F^- concentrations to millimolar levels following ingestion of large quantities of F^- show that the acute illness and death resulting from acute F^- toxicity may reflect inhibition of several cellular functions (Hodge and Smith, 1981). Included in the symptoms of acute toxicity are hypocalcemia, possibly stemming from inhibition of production of lactate required for extrusion of Ca^{2+} from bone—although other explanations have been advanced (Messer, 1984)—and hyperkalemia, which has been suggested as the mechanism of sudden death associated with acute F^- toxicity (McIvor *et al.*, 1985). McIvor *et al.* suggest that elevated serum potassium levels reflect diminished Na^+/K^+-ATPase activity brought about by diminished ATP levels as a consequence of glycolysis inhibition. The direct action of F^- by irreversible inhibition of this enzyme, as discussed in Section 2.8, presumably could also be a causative factor in hyperkalemia.

2.15 Effects of Fluoride on Mineralized Tissue

2.15.1 Introduction

The widespread natural occurrence of F^- together with the exceptionally high affinity of F^- for mineralized tissue has produced a plethora of medical issues. Effects of F^- ingestion can be either beneficial or harmful, with the consequences dependent on several factors. Thus, the recognition that mottled tooth enamel associated with abnormally high F^- intake is accompanied by a low incidence of dental caries led to the use of F^- in caries prevention—first through water fluoridation and then through topical application of F^-. Excessive bone growth is also associated with chronic F^- poisoning. Recognition of this has led to investigation of F^- as a potential drug to reverse the bone degeneration characteristic of osteoporosis. Extensive research encompassing several disciplines has resulted in a clearer understanding of the mechanisms by which F^- exerts its influence on mineralization, though several issues are yet to be resolved.

The effects of fluoride on mineralized tissue will be reviewed in the following sections.

2.15.2 Biochemistry of Mineralized Tissue—An Overview

2.15.2.1 Structure of Bone

Bone mineral is composed primarily of hydroxyapatite having the molecular formula $Ca_{10}(PO_4)_6 (OH)_2$. Mineralization occurs through precipitation of this salt from the surrounding supersaturated solution of calcium and phosphate ions. Bone is a self-repairing tissue, and, in the process of bone remodeling, hydroxyapatite is dissolved from the inner surface of compact bone. In a normally functioning state, bone formation and bone resorption are coupled, resulting in the proper degree of calcification and bone strength, avoiding both excessive deposition of bone and excessive dimineralization. The estimate has been made that in the skeleton of a human adult, 3.5% of total bone is being remodeled at any given time (for a broader treatment of this subject, see, e.g., Mundy, 1989; Revell, 1986; Vaughan, 1981).

2.15.2.2 Bone Cells

Three types of cells are found within bone. The osteoblasts cover the bone surface and are responsible for bone formation. An organic matrix, the osteoid, is deposited by the osteoblasts, and this matrix becomes mineralized by the precipitation of hydroxyapatite. As bone crystallization proceeds, the osteoblasts are incorporated into bone matrix as osteocytes, a second class of bone cells. The function of the osteocytes is a subject of current debate, particularly with regard to a possible role in bone resorption. Osteoclasts are the third type of bone cell, and these cells, found on the inner surface of compact bone, are responsible for bone resorption (Peck and Woods, 1988; Barnes, 1987; Revell, 1986).

2.15.2.3 Skeletal Regulation of Calcium Levels

Calcium has many critical physiological functions in addition to its structural role in mineralized tissue—for example, serving as a neuromodulator—and the serum level of this cation must be controlled precisely. This is accomplished through the action of the parathyroid hormone and by calcitonin. Absorption of dietary calcium through the gut is mediated by vitamin D [more precisely, 1,25-dihydroxyvitamin D_3 (1,25-

OH$_2$-D$_3$)] (see Chapter 3 in Vol. 9B of this series). In the presence of this vitamin, the rate of absorption is dependent almost entirely on the calcium ion concentration in the extracellular fluid, with small changes in concentration resulting in large changes in absorption rate. This sensitivity of the rate of calcium absorption to concentration is mediated by the action of the parathyroid hormone, whose presence in the kidney stimulates 1-hydroxylation of 25-hydroxyvitamin D$_3$ (25-OH-D$_3$) to produce active 1,25-(OH)$_2$-D$_3$. Parathyroid hormone secretion, in turn, is controlled by calcium ion concentration, increasing with decreasing extracellular calcium. Parathyroid hormone also serves to decrease the rate of calcium excretion. Calcitonin, the presence of which inhibits osteoclast-mediated bone resorption, also controls calcium levels, through a more long-term process.

2.15.3 Uptake of Fluoride by Bone

2.15.3.1 Factors Influencing Fluoride Uptake

Ingestion and absorption of fluoride were discussed in Section 2.3, and the high affinity of dietary F$^-$ for bone was noted. Accumulation in mineralized tissue is most rapid during bone growth, but the amount contained in bone continues to increase as long as the rate of F$^-$ intake is greater than the rate of loss by bone resorption.

2.15.3.2 Mechanism of Incorporation of Fluoride into Mineralized Tissue

F$^-$ enters apatite crystals by a hetero-ionic exchange process. The similarity of the F$^-$ ion to hydroxide ion (ionic radius, hydration number) permits F$^-$ to enter the hydration shell of hydroxyapatite. At this stage, F$^-$ and the other ions of the hydration shell are in dynamic equilibrium with the surrounding tissue fluid and with the apatite crystal. F$^-$ in the hydration shell subsequently exchanges with a monovalent group at the surface of the apatite crystal and becomes more firmly bound. While some of the F$^-$ may migrate further into the crystal by recrystalization, this is a slow process; F$^-$ ion content of bone accordingly is highest at the bone surface. The overall process, formation of fluorohydroxyapatite, is described by the reaction

$$Ca_{10}(PO_4)_6(OH)_2 + nF^- \rightarrow Ca_{10}(PO_4)_6 F_n(OH)_{2-n} + nOH^-$$

Overall displacement of hydroxide by F$^-$ is normally quite low, on the order of 1000–5000 ppm.

There is considerable evidence that F$^-$ incorporation into bone occurs preferentially during crystallization of apatite during the mineralization process. Thus, freshly ingested F^- is found to accumulate in areas of rapid bone growth (Bang, 1978). The exchange of F^- for hydroxide in fully grown apatite crystals is of minor importance in the incorporation of F^- (Eanes, 1982).

2.15.3.3 Effect of Fluoride Incorporation on Bone Structure

Physicochemical changes in bone resulting from F^- substitution include increased crystallinity and lower solubility (reviewed by Eanes, 1982; Eanes and Reddi, 1979). A diagram of the fluorohydroxyapatite lattice structure is given in Fig. 2-13 (Eanes and Reddi, 1979). As can be

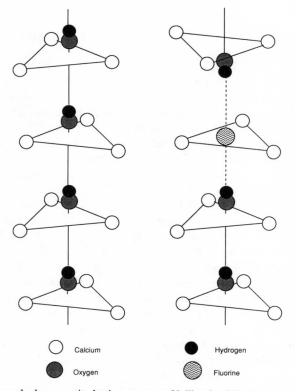

Figure 2-13. Fluorohydroxyapatite lattice structure. Unlike the OH^- ions that are displaced from the plane of the nearest triangle, the F^- ion fits snugly in the center of the triangle. This closer coordination of F^- by neighboring calcium ions largely accounts for the greater stability of fluorapatite compared to hydroxyapatite. Adapted, with permission, from Eanes and Reddi (1979).

seen, F^- is situated in the center of a triangular array of calcium ions while the hydroxide ion is somewhat off-center. This more stable geometric configuration of F^- is thought to be a major contributing factor to the ease with which F^- is taken up into bone. In addition, a decrease in solubility of fluorohydroxyapatite relative to hydroxyapatite (Moreno *et al.*, 1977) may be related to this more stable crystal structure, although such factors as different hydration energies of F^- and hydroxide may play a part (Eanes, 1982).

Incorporation of F^- into apatite leads to increased crystallinity of calcified tissue. X-ray diffraction studies have shown that crystal size increases and crystal strain decreases as F^- content increases. Improved crystallinity appears to result primarily from the former factor (Eanes, 1982).

2.15.4 Fluorosis

2.15.4.1 Historical Background

Dietary and other environmental exposure to F^- is unavoidable and can be beneficial. That excessive exposure must be avoided is apparent from the acute toxicity of F^- when ingested in large amounts [6–9 mg/kg body weight is given as one estimate of the lethal dose (Dreisbuch, 1980)]. Bone F^- content can be elevated up to 2500 ppm without detectable changes in bone structure, from 2500 to 5000 ppm bone mottling can occur, and at levels greater than 5000 ppm, clinical manifestation of bone fluorosis occurs (Riggs, 1983).

Several reviews are available detailing early developments as medical problems associated with overexposure to F^- became recognized [Weatherell, 1966 (pp. 155–157); Hodge and Smith, 1965b]. Early awareness of skeletal effects associated with chronic F^- toxicity came in the early 1930s from a study of Danish workers exposed to cryolite dust (Na_6AlF_3). Increased density of some bone material was observed, along with loss of structural detail. Similar disorders observed in India were caused by consumption of water having high F^- content ("hydric fluorosis") (Hodge and Smith, 1965b). Bone formed under the influence of F^- stimulation can have varying degrees of abnormalities. These include osteosclerosis (increased density), osteoporosis (decreased density), osteomalacia (softening), and architectural deformities (Riggs, 1983).

2.15.4.2 Effect of Chronic Fluoride Toxicity on Bone Structure

The primary lesion of fluorosis appears to be excessive and abnormal deposition of bone. This is often accompanied by increased resorption of bone. Since serum levels of calcium must remain optimal, the suggestion has been made that bone resorption accompanying fluorosis becomes necessary to meet the increased demand for calcium caused by fluoride-induced excessive bone growth. A drop in serum calcium levels would trigger parathyroid hormone production, which, in turn, would stimulate bone resorption. However, experiments done with parathyroidectomized rats showed that neither F^--induced bone formation nor bone resorption can be attributed to the action of parathyroid hormone (Baylink et al., 1983).

While excessive F^- stimulates bone growth, it inhibits mineralization of the newly formed bone. This results from an increased lag time between deposition of the osteoid matrix and onset of mineralization (Baylink et al., 1970). This decreased mineralization has serious consequences with respect to treatment of osteoporotics with F^- (see below). Since F^- also stimulates matrix production (a result of an F^--induced increase in osteoblast number and activity; see below), the net effect on the amount of mineralized bone is considerably diminished (reviewed by Baylink et al., 1983). F^- stimulates bone growth in both cortical and trabecular bone, but the effect is greater in the latter. The fact that turnover of trabecular bone is more rapid than that of cortical bone appears to be only a partial explanation for this phenomenon (Baylink et al., 1983).

2.15.4.3 Cellular Mechanisms of Fluoride-Induced Stimulation of Bone Growth

The mechanism by which F^- stimulates bone growth has been the focus of much research. Recently, Farley et al. (1983) demonstrated that $10 \mu M$ NaF induced an increase in cell proliferation of cultured osteoblasts from embryonic chick bone and that this stimulated growth was unique to this cell type. An increase in alkaline phosphatase (ALP) activity of 262 % over control in the culture medium and 435 % over control in the cells was also effected by exposure to $10 \mu M$ NaF. Exposure of an organ culture system to F^- also demonstrated enhanced growth and mineralization of embryonic chick bone. These results demonstrate that osteoblasts are particularly sensitive to F^- (most other biochemical effects of F^- require millimolar concentrations). Recent data from several groups support the idea that osteogenesis by osteoblasts is stimulated by F^- (Dandona et al., 1989).

The increased differentiation and proliferation of osteoblasts together with stimulated ALP activity, occurring at concentrations consistent with in vivo activity, support the view that an important factor in the stimulation of bone growth by NaF involves direct action on the bone-forming cells. Increased ALP activity has long been associated with increased bone growth. Thus, the observation was reported in 1923 that ALP caused precipitation of calcium phosphate when incubated with soluble calcium salts and organic phosphate esters (Robinson, 1923). Some evidence supports Robinson's initial hypothesis that ALP-catalyzed hydrolysis of phosphate causes increased inorganic phosphate concentration, which increase could promote ossification. However, other physiological roles of ALP seem likely to be important in mineralization (Wuthier and Register, 1984). Thus, there is evidence that crystal-growth inhibitors can be destroyed by ALP (Neuman et al., 1951). The suggestion has also been made that ALP may serve as calcium-binding protein in the calcification process (Vittur and deBernard, 1973). Evidence that ALP can act as a tyrosine-specific phosphoprotein phosphatase (Swarup et al., 1981; Burch et al., 1985) suggests that ALP may play a regulatory role in cell division during mineralization. However, evidence against such a role comes from recent cell-culture growth experiments (Whyte and Vrabel, 1987). ALP-mediated inorganic phosphate transmembrane transport has also been postulated.

2.15.4.4 Noncellular Mechanisms of Fluoride-Induced Stimulation of Bone Growth

Physical and mechanical factors also appear to be operative in the stimulation of bone formation by F^- (that is, a net gain of formation relative to resorption). Examples include lower solubility of fluoride-containing hydroxyapatite as discussed above (Section 2.15.3.3). An enhanced piezoelectric effect in fluorohydroxyapatite has been suggested as an additional factor. Bone crystals, by acting as transducers of mechanical energy to electrical energy, generate a piezoelectric current in bone (Bassett, 1968). Stimulation of osteoblasts by this current has been proposed as a mechanism by which mechanical stress can influence bone growth. Since fluorohydroxyapatite has an altered crystal structure, response to mechanical stress should also be altered. A resulting increased current could result in increased osteoblast stimulation (Riggs, 1983).

2.15.5 Fluoride in the Treatment of Osteoporosis

2.15.5.1 Osteoporosis

The normal functioning of bone tissue requires a balance of bone formation and bone resorption. If demineralization outpaces bone formation, a net loss of bone occurs. Osteoporosis describes a clinical state in which the absolute amount of bone present has been reduced to the point that even minor stress can result in fracture. Several reviews describe clinical aspects and approaches to treatment of osteoporosis (see, e.g., Riggs and Melton, 1988; Riggs, 1983). Management of the disease by agents that act through an inhibition of bone resorption (e.g., calcitonin, calcium, steroids) has met with problems, since an initial net gain in bone formation is often followed by a new steady state in which bone resorption again is greater than bone formation.

2.15.5.2 Fluoride and Osteoporosis

Recognition that increased (if abnormal) bone growth is one of the manifestations of chronic F^- poisoning prompted investigation of the use of F^- in the treatment of osteoporosis (Rich and Ensinck, 1961). The several abnormalities of fluorotic bone, together with the fact that fluorosis is often accompanied by bone resorption, emphasize the difficulties inherent in this approach. Indeed, early clinical trials gave mixed, often disappointing results (Baylink and Berstein, 1967).

Subsequent studies demonstrated the importance of controlling dosage and the duration of F^- treatment as well as the use of supplementary calcium. The therapeutic dose range for F^- was found to be in the relatively narrow range of 50–80 mg of NaF per day (Jowsey et al., 1972). The recognition of a lag time of several months before calcium retention ensues was found to be important in assessing early results of treatment (Rich et al., 1964). In F^- therapy, a balance must be found between the beneficial effects of stimulation of osteoid production and the harmful inhibition of mineralization. Fortunately, it has been found that the mineralization defect can often be offset by treatment with calcium supplement (Jowsey and Kelley, 1968). Calcium also serves to inhibit increased resorption during stimulated growth.

2.15.5.3 Effectiveness of Fluoride Treatment

Unfortunately, even under optimally defined therapeutic regimes, a significant percentage (25–50%) of patients do not respond satisfactorily to F^- treatment. While the causes of this individual variability remain to be determined, the suggestion has been made that the nonresponders represent an important subgroup of osteoporotics having impaired osteoblastic activity (Riggs et al., 1982).

Side effects complicating the use of NaF treatment include gastric irritation (Riggs et al., 1982). Monofluorophosphate (MFP), $F(PO)(OH)_2$, is a nonionic form of inorganic F^- that has been suggested as a substitute for NaF for the treatment of osteopororis (Ericsson, 1983). MFP is relatively nontoxic. Furthermore, because it is hydrolyzed in the intestine rather than in the stomach, it should presumably produce less gastric irritation (Ericsson, 1983). Recently, MFP has been shown to be a substrate for alkaline phosphatase, including embryonic chick intestinal alkaline phosphatase, but not for acid phosphatase. In the same study, MFP was shown to mimic the osteogenic actions of F^- in chick embryonic skeletal tissues, a result dependent on skeletal alkaline phosphatase hydrolysis of MFP (Farley et al., 1987).

The use of F^- in the treatment of osteoporosis appears to be a promising approach for the management of this disease. Extensive research on the mechanistic aspects of the osteogenic activity of F^- continues, and progress appears to be rapid. Recent reviews provide evaluations of the efficacy of this therapy (Eriksen et al., 1988; Riggs, 1984). New techniques for the analysis of bone tissue (Reinbold et al., 1986), the development of an animal model for F^--stimulated bone growth (Lundy et al., 1986), and the possibility of using alternate modes of delivery, such as MFP, are examples of continuing advances in this field. In a recent development, a treatment procedure using intermittent slow release of F^- has been shown to maintain serum F^- concentrations at therapeutic levels with minimal side effects, and response to the treatment has been good (Pak et al., 1989).

2.15.6 Fluoride in Dental Tissue

Many aspects of the discussion in the previous section also pertain to dental tissue. However, the subject of F^- in dental health raises additional issues that deserve separate consideration. Included in the extensive research in this area are classic epidemiolgical studies, investigations on the nature of the chemical reactions of F^- with enamel, cellular effects relating to mineralization, structural alterations of dental tissue caused by F^- incorporation, and biochemical studies of the effects of F^- on bacteria present on dental tissue. A particularly detailed historical account of the observations and research that led to the use of F^- in dental caries prevention is given by Hodge and Smith (1965). This history will be summarized in the following sections.

2.15.6.1 *"Denti di Chiaie"*

J. M. Eager, a surgeon for the U.S. Public Health Service stationed in Naples, Italy, submitted a report in 1901 describing a "dental peculiarity" among local inhabitants. The condition was known locally as "denti di Chiaie" after Professor Chiaie of Naples, who first described the defect. The condition was particularly severe in the residents of Pozzuoli, a town in a highly volcanic area near Naples. Eager noted that the residents of this village could be recognized by their "distinguishing characteristic of black teeth (denti neri), apparently strong and serviceable, but devoid of enamel and hideously dark" (Eager, 1901). Environmental factors, including consumption of water from volcanic springs and inhalation of noxious fumes, were implicated by Eager. While the prevalent theory in Italy invoked a selectively harmful effect of the volcanic gases on enamel formation in early childhood, Eager suspected drinking water.

2.15.6.2 Fluoride and Mottled Enamel

Several years elapsed before further information became available concerning this abnormality in enamel. In 1916, the term "mottled enamel" was first used by G. V. Black in the next scientific report on this condition. Subsequent animal experiments in the next decade, designed to study biological effects of NaF in the diet, led to the observation that rats in these experiments developed abnormalities in their teeth. Further implication of F^- came in 1931 from spectrographic analysis that demonstrated high F^- content in water from high mottling areas. Another study showing a relationship between sources of drinking water and mottled enamel in various districts strenghtened the conclusion that defects in enamel were associated with water supplies (reviewed by Hodge and Smith, 1965).

2.15.6.3 Fluoride and Inhibition of Dental Caries—Animal Experiments

In 1933, a direct connection between F^- in drinking water and mottled enamel was made by comparing the teeth of white rats given drinking water from a mottled enamel area with those of rats given water containing NaF. The observation was made that teeth of rats in this experiment, while showing signs of dental fluorosis, appeared to have a low incidence of caries. In a subsequent experiment (1938), a 90% inhibition of dental caries in rats was realized in rats fed 250 ppm of F^- in their diet plus 4.2 ppm of F^- in drinking water. The cariostatic action of F^- was attributed to an antimetabolic effect on the bacteria found on the tooth surface (Hodge and Smith, 1965).

2.15.6.4 Epidemiological Studies of Dental Fluorosis

A series of epidemiological studies of "chronic endemic dental fluorosis" were carried out by H. T. Dean and E. Evolve of the U.S. Public

Health Service in the late 1930s. Clinical examinations were limited to children who were born in and had always resided in the cities surveyed. Analyses of water samples permitted the computation of an arithmetic mean annual F^- content in communal water supplies. This F^- concentration was then correlated with the severity of dental fluorosis. These studies demonstrated that the severity of dental fluorosis could be correlated directly with the degree of ingestion of toxic amounts of F^- from community water supplies.

2.15.6.5 Fluoride and Inhibition of Dental Caries—Epidemiology

In addition to examinations of the severity of mottling, a study of other defects was carried out in certain of the communities. Analyses of the data indicated that residence in a community having a relatively toxic amount of F^- in the drinking water gave a degree of protection against dental caries. Further evidence came from a direct comparison of neighboring Illinois cities. The cities of Galesburg and Monmouth, with 1.8 and 1.7 ppm of F^- in their water supplies, had one-half to one-third the incidence of childhood dental caries (permanent teeth) of the cities of Macomb and Quincy, with 0.2 ppm of F^-.

These and many other data made it increasingly clear that the presence of F^- in drinking water had profound effects on teeth. At high levels, dental fluorosis occurs, with its characteristic mottling and other disfiguring features. At lower levels, the presence of F^- imparts protection against dental caries. Of particular significance was the evidence that maximum protection against caries could be achieved at concentrations in the range of 1 ppm, while dental fluorosis was not evident at these low concentrations (Hodge and Smith, 1965).

2.15.6.6 Water Fluoridation

By the mid-1940s, numerous epidemiological studies had demonstrated that consumption of fluoride-containing drinking water during the formative stages of teeth led to a 60–65% reduction in dental caries. A major event was the permission given by Grand Rapids, Michigan, to have F^- artificially added to its water supply as part of the continued study of this phenomenon. Muskegon, Michigan, was chosen as a control city, having low F^- (0.1 ppm) in its water supply, while Aurora, Illinois, having a natural F^- content of 1.2 ppm, was included for comparison of the effects of natural versus artificially added F^-. Long-term evaluation of these experiments confirmed that F^- added to community water supplies simulates the cariostatic action of F^- contained as a natural constituent.

The success of these and similar experiments conducted in other cities has led to widespread implementation and acceptance of fluoridation of community water supplies as a public health policy (Hodge and Smith, 1965). Topically applied F^- rinses, gels, fluoride-containing dentifrices, and mouth rinses subsequently have played important roles in the use of F^- in caries prevention (Gron, 1977).

2.15.6.7 Cariostatic Mechanisms of Fluoride

The biochemical basis for the effectiveness of F^- in caries prevention is not completely understood. Plaque microbes initiate formation of dental caries by producing acid as a result of fermentation of sugars (such as glucose, fructose, sucrose, and lactose). Caries result from the acid-mediated dimineralization of tooth enamel. F^-, whether systematically available from diet or drinking water or topically applied through other treatments, inhibits this process, apparently by a combination of physical and biochemical mechanisms (Harper and Loesche, 1986). Cariostatic action has been attributed to systemic effects on preeruptive tooth development (Aasenden and Peebles, 1974) and protection of enamel acid-mediated demineralization by formation of less soluble fluoro-hydroxyapatite (Moreno et al., 1977). The uptake of F^- into dental tissue is most rapid during periods of growth and calcification, although there is debate over the timing and mechanism of F^- incorporation into developing enamel (Bawden et al., 1986). Topically applied fluorides can act by facilitation of remineralization of enamel surface (ten Cate and Arends, 1977) and possibly by interference with bacterial adherence and plaque fermentation on enamel surface (Kilian et al., 1979).

2.15.6.8 Action of Fluoride on Oral Bacteria

There is considerable evidence for a direct bacteriostatic action of F^- on plaque microbes (Hamilton, 1977; Jenkins and Edgar, 1977). In fact, inhibition of growth of oral bacteria was suspected as the cause of the cariostatic effects of F^- when these effects were first recognized. Although other mechanistic interpretations have emerged, continuing studies on the effect of F^- on microorganisms of dental plaque indicate that it is likely that inhibition of growth of acid-producing bacteria contributes to cariostatic action (Hamilton, 1977; Germaine and Tellefson, 1986b). For example, F^- has been shown to inhibit the phosphoenolpyruvate-dependent transport of sugars across the cell membrane of such enamel-colonizing bacteria as Streptococcus mutans (Germaine and Tellefson, 1986a).

Despite well-documented inhibition of bacterial growth *in vitro* by F^-, its low *in vivo* oral concentration has led to questions as to whether a similar inhibition could play a role in its cariostatic activity. At least two factors, both related to pH, may amplify the effectiveness of F^- in this capacity. The production of acid by oral bacteria lowers salivary pH, with two important consequences. The resulting acid-stimulated onset of demineralization of enamel will release F^- from fluorapatite, thus increasing the availability of F^- for bacteriostatic action (Harper and Loesche, 1986). A lower pH also results in more efficient passive uptake of F^- (as HF) into the cell. The intracellular pH of the oral bacteria is significantly higher, and F^- (as F^-) becomes concentrated by this pH gradient (Germaine and Tellefson, 1986a).

2.16 Biochemistry of Inorganic Fluoride—Summary

Fluorine is classified as a trace element, and possibly an essential trace element (Messer, 1984). F^-, in many instances because of its electronic and steric resemblance to hydroxide, can affect a host of biochemical processes. However, these effects frequently are manifested only at relatively high concentrations. An overview of the interactions of inorganic F^- with enzymes, cellular functions, and mineralized tissue has been given in this chapter. Recent developments in this biochemistry are noteworthy. For example, the role of AlF_4^- in the stimulation of various guanine-binding regulatory proteins by F^- (possibly as a mimic of phosphate) may have wider biochemical and physiological implications. For example, Al^{3+} has been implicated as a possible contributing factor to several neurological and skeletal disorders, including Alzheimer's disease (Macdonald and Martin, 1988). While the substitution of Al^{3+} for Mg^{2+} at critical regulatory and enzymatic sites has been suggested as the primary mechanism for this toxicity, the complex interactions of F^- and aluminum of G-proteins, and possibly at other biological targets, may also contribute to Al^{3+}-induced toxicity (Miller *et al.*, 1989). An additional medically relevant area remains the effects of F^- on mineralized tissue. Thus, recent developments suggest that F^- treatment indeed may provide a method for management of osteoporosis.

References

Aasenden, R., and Peebles, T. C., 1974. Effects of fluoride supplementation from birth on human deciduous and permanent teeth, *Arch. Oral Biol.* 19:321–326.

Baglioni, C., 1972. Heterogeneity of the small ribosomal subunit and mechanism of chain initiation in eukaryotes, *Biochim. Biophys. Acta* 287:189–193.

Baglioni, C., Vesco, C., and Jacobs-Lorena, M., 1969. The role of ribosomal subunits in mammalian cells, *Cold Spring Harbor Symp. Quant. Biol.* 34:555–565.

Baikov, A. A., Smirnova, I. N., and Volk, S. E., 1984. Membrane inorganic pyrophosphatase. "Syncatalytic" inactivation by the fluoride ion, *Biochemistry (Biokhimiya)* 49:696–702.

Bailey, K., and Webb, E. C., 1944. Purification and properties of yeast pyrophosphatase, *Biochem. J.* 38:394–398.

Bakuleva, N. P., Nazarova, T. I., Baykov, A. A., and Avaeva, S. M., 1981. The phosphorylation of yeast inorganic pyrophosphatase and formation of stoichiometric amounts of enzyme-bound pyrophosphate, *FEBS Lett.* 124:245–247.

Bang, S., 1978. Effects of fluoride on the chemical composition of inorganic bone structure, in *Fluoride and Bone* (B. Courvoisier, A. Donath, and C. A. Baud, eds.), Hans Huber Publishers, Bern, Switzerland, pp. 56–61.

Banks, R. E., 1986. Isolation of fluorine by Moissan: Setting the scene, *J. Fluorine Chem.* 33:3–26.

Banks, R. E., and Goldwhite, H., 1966. Fluorine chemistry, in *Handbook of Experimental Pharmacology, XX/1. Pharmacology of Fluorides, Part 1* (O. Eichler, A. Farah, H. Herken, and A. D. Welch, eds.), Springer-Verlag, Berlin, pp. 1–52.

Barnes, D., 1987. Close encounters with an osteoclast, *Science* 236:914–916.

Bassett, C. A. L., 1968. Biological significance of piezoelectricity, *Calcif. Tissue Res.* 1:252–272.

Bawden, J. W., McLean, P., and Deaton, T. G., 1986. Fluoride uptake at various stages of rat molar development, *J. Dent. Res.* 65:34–38.

Baykov, A. A., and Avaeva, S. M., 1974. Yeast inorganic pyrophosphatase: Studies on metal binding, *Eur. J. Biochem.* 47:57–66.

Baylink, D. J., and Berstein, D. S., 1967. The effects of fluoride on metabolic bone disease: A histological study, *Clin. Orthop.* 55:51–85.

Baylink, D., Wergedal, J., Stauffer, M., and Rich, C., 1970. Effects of fluoride on bone formation, mineralization, and resorption in the rat, in *Fluoride in Medicine* (T. L. Vischer, ed.), Hans Huber Publishers, Bern, Switzerland, pp. 37–69.

Baylink, D. J., Duane, P. B., Farley, S. M., and Farley, J. R., 1983. Monofluorophosphate physiology: The effects of fluoride on bone, *Caries Res.* 17 (Suppl. 1):56–76.

Berridge, M. J., 1987. Inositol triphosphate and diacylglycerol: Two interacting second messengers, *Annu. Rev. Biochem.* 56:159–193.

Berridge, M. J., and Irvine, R. F., 1984. Inositol trisphosphate, a novel second messenger in cellular signal transduction, *Nature* 312:315–321.

Bigay, J., Deterre, P., Pfister, C., and Chabre, M., 1985. Fluoroaluminates activate transducin-GDP by mimicking the γ-phosphate of GTP in its binding state, *FEBS Lett.* 191:181–185.

Bigay, J., Deterre, P., Pfister, C., and Chabre, M., 1987. Fluoride complexes of aluminum or beryllium act on G-proteins as reversibly bound analogues of γ phosphate of GTP, *EMBO J.* 6:2907–2913.

Blackmore, P. F., Bocckino, S. B., Waynick, L. E., and Exton, J. H., 1985. Role of guanine nucleotide-binding regulatory protein in the hydrolysis of hepatocyte phosphatidylinositol 4,5-biphosphate by calcium-mobilizing hormones and the control of cell calcium. Studies utilizing aluminum fluoride, *J. Biol. Chem.* 260:14477–14483.

Boeckaert, J., Deterre, P., Pfister, C., Guillon, G., and Chabre, M., 1985. Inhibition of hormonally regulated adenylate cyclase by the $\beta\gamma$ subunit of transducin, *EMBO J.* 4:1413–1417.

Bokoch, G. M., Katada, T., Northup, J. K., Hewlett, E. L., and Gilman, A. G., 1983. Identification of the predominant substrate for ADP-ribosylation by islet activating protein, *J. Biol. Chem.* 258:2072–2075.

Bokoch, G. M., Katada, T., Northup, J. K., Ui, M., and Gilman, A. G., 1984. Purification and properties of the inhibitory guanine nucleotide-binding regulatory component of adenylate cyclase, *J. Biol. Chem.* 259:3560–3577.

Borei, H., 1945. Inhibition of cellular oxidation by fluoride, *Ark. Kim., Mineral. Geol.* 20A:1–125.

Brandt, D. R., and Ross, E. M., 1986. Effect of Al^{3+} plus F^- on the catecholamine-stimulated GTPase activity of purified and reconstituted G_s, *Biochemistry* 25:7036–7041.

Brewer, J. M., 1981. Yeast enolase: Mechanism of activation by metal ions, *CRC Crit. Rev. Biochem.* 1981:209–254.

Brewer, J. M., and Ellis, P. D., 1983. ^{31}P-nmr studies of the effect of various metals on substrate binding to yeast enolase, *J. Inorg. Biochem.* 18:71–82.

Brewer, J. M., Carreira, K. M., Collins, K. M., Duvall, M. C., Cohen, C., and DerVartanian, D. V., 1983. Studies of activation and nonactivating metal ion binding to yeast enolase, *J. Inorg. Biochem.* 19:255–267.

Bunick, F. J., and Kashket, S., 1982. Binding of fluoride by yeast enolase, *Biochemistry* 21:4285–4290.

Burch, W. M., Hamner, G., and Wuthier, R. E., 1985. Phosphotyrosine phosphatase activity of alkaline phosphatase in mineralizing cartilage, *Metabolism* 34:169–175.

Cimasoni, G., 1966. Inhibition of cholinesterase by fluoride *in vitro*, *Biochem. J.* 99:133–137.

Cimasoni, G., 1972. The inhibition of enolase by fluoride *in vitro*, *Caries Res.* 6:93–102.

Cockcroft, S., and Gomperts, B. D., 1985. Role of guanine nucleotide binding protein in the activation of polyphosphoinositide phosphodiesterase, *Nature* 314:534–536.

Cockcroft, S., and Taylor, J. A., 1987. Fluoroaluminates mimic guanosine 5′[γ-thio]triphosphate in activating the polyphosphoinositide phosphodiesterase of hepatocyte membranes, *Biochem. J.* 241:409–414.

Colombo, B., Vesco, C., and Baglioni, C., 1968. Role of ribosomal subunits in protein synthesis in mammalian cells, *Proc. Natl. Acad. Sci. USA* 61:651–658.

Curnutte, J. T., Babior, B. M., and Karnovsky, M. L., 1979. Fluoride-mediated activation of the respiratory burst in human neutrophils, *J. Clin. Invest.* 63:637–747.

Dandona, P., Gill, D. S., and Khokher, M. A., 1989. Fluoride and osteoblasts (letter), *Lancet* 1989:449–450.

Dreisbuch, R. H., 1980. Fluorine, hydrogen fluoride and derivatives, in *Handbook of Poisoning*, Lange Medical Publishers, Los Altos, California, pp. 210–213.

Duncan, R. F., and Hershey, J. W., 1987. Initiation factor protein modifications and inhibition of protein synthesis, *Mol. Cell. Biol.* 7:1293–1295.

Eager, J. M., 1901. Denti di Chiaie (chiaie teeth), *Public Health Rep.* 16:2576.

Eanes, E. D., 1982. Effect of fluoride on mineralization of teeth and bones, Proceedings of the International Fluoride Symposium, Logan, Utah, pp. 195–198.

Eanes, E. D., and Reddi, A. H., 1979. The effect of fluoride on bone mineral apatite, *Metab. Bone Dis. Relat. Res.* 2:3–10.

Ericsson, Y., 1983. Monofluorophosphate physiology: General considerations, *Caries Res.* 17 (Suppl. 1):46–55.

Eriksen, E. F., Hodgson, S. F., and Riggs, B. L., 1988. Treatment of osteoporosis with sodium fluoride, in *Osteoporosis. Etiology, Diagnosis, and Management* (B. L. Riggs and L. J. Melton, eds.), Raven Press, New York, pp. 415–432.

Farley, J. R., Wergedal, J. E., and Baylink, D. J., 1983. Fluoride directly stimulates proliferation and alkaline phoshatase activity of bone-forming cells, *Science* 222:330–332.

Farley, J. R., Tarbaux, N. M., Lau, K.-H. W., and Baylink, D. J., 1987. Monofluorophosphate is hydrolyzed by alkaline phosphatase and mimics the actions of NaF on skeletal tissues, *in vitro*, *Calcif. Tissue Int.* 40:35–42.

Froede, H. C., and Wilson, I. B., 1985. The slow rate of inhibition of acetylcholinesterase by fluoride, *Mol. Pharmacol.* 27:630–633.

Fung, B. K.-K., 1983. Characterization of transducin from bovine retinal rod outer segments. I. Separation and reconstitution of the subunits, *J. Biol. Chem.* 258:10495–10502.

Fung, B. K.-K., and Stryer, L., 1980. Photolyzed rhodopsin catalyzes the exchange of GTP for bound GDP in retinal rol outer segments, *Proc. Natl. Acad. Sci. USA* 77:2500–2504.

Fung, B. K.-K., Hurley, J. B., and Stryer, L., 1981. Flow of information in the light-triggered cyclic nucleotide cascase of vision, *Proc. Natl. Acad. Sci. USA* 78:152–156.

Gabler, W. L., and Hunter, N., 1987. Inhibition of human neutrophil phagocytosis and intracellular killing of yeast cells by fluoride, *Arch. Oral Biol.* 32:363–366.

Gabler, W. L., and Leong, P. A., 1979. Fluoride inhibition of polymorphonuclear leukocytes, *J. Dent. Res.* 58:1933–1939.

Germaine, G. R., and Tellefson, L. M., 1986a. Role of the cell membrane in pH-dependent fluoride inhibition of glucose uptake by *Streptococcus mutans, Antimicrob. Agents Chemother,* 29:58–61.

Germaine, G. R., and Tellefson, L. M., 1986b. Effect of endogenous phosphoenolpyruvate on fluoride inhibition of glucose by *Streptococcus mutans, Infect. Immun.* 51:119–124.

Gilman, A. G., 1984a. Guanine nucleotide-binding regulatory proteins and dual control of adenylate cyclase, *J. Clin. Invest.* 73:1–4.

Gilman, A. G., 1984b. G proteins and dual control of adenylate cyclase, *Cell* 36:577–579.

Gilman, A. G., 1987. G proteins: Transducers of receptor-generated signals, *Annu. Rev. Biochem.* 56:615–649.

Godchaux, W., III and Atwood, K. C., IV, 1976. Structure and function of initiation complexes which accumulate during inhibition of protein synthesis by fluoride *ion, J. Biol. Chem.* 251:292–301.

Greenspan, C. M., and Wilson, I. B., 1970. The effect of fluoride on the reaction of acetylcholinesterase with carbamates, *Mol. Pharmacol.* 6:266–272.

Gron, P., 1977. Chemistry of topical fluorides, *Caries Res.* 11 (Suppl. 1):172–204.

Guy, W. S., Taves, D. R., and Brey, W. S., Jr., 1976. Organic fluorocompounds in human plasma: Prevalence and characterization, in *Biochemistry Involving Carbon–Fluorine Bonds* (R. Filler, ed.), ACS Symposium Series 28, American Chemical Society, Washington, D. C., pp. 117–134.

Hamilton, I. R., 1977. Effects of fluoride on enzymatic regulation of bacterial carbohydrate metabolism, *Caries Res.* 11 (Suppl. 1):262–278.

Haromy, T. P., Knight, W. B., Dunaway-Mariano, D., and Sundaralingam, M., 1982. X-ray crystallographic and nuclear magnetic resonance spectral studies of the products from yeast inorganic pyrophosphatase-Co(NH_3)$_4$ PP reaction. Investigation of the pyrophosphatase reaction mechanism, *Biochemistry* 21:6950–6956.

Harper, D. S., and Loesche, W. J., 1986. Inhibition of acid production from oral bacteria by fluorapatite-derived fluoride, *J. Dent. Res.* 65:30–33.

Haugen, D. A., and Suttie, J. W., 1974. Fluoride inhibition of rat liver microsomal esterases, *J. Biol. Chem.* 249:2723–2731.

Higashijima, T., Ferguson, K. M., Sternweis, P. C., Ross, E. M., Smigel, M. D., and Gilman, A. G., 1987. The effect of activating ligands on the intrinsic fluorescence of guanine nucleotide-binding regulatory proteins, *J. Biol. Chem.* 262:752–756.

Hildebrandt, J. D., Codina, J., Risinger, R., and Birnbaumer, L., 1984. Identification of a γ-subunit associated with the adenyl cyclase regulatory proteins N_s and N_i, *J. Biol. Chem.* 259:2039–2042.

Hodge, H. C., and Smith, F. A., 1981. Fluoride, *Mineral Metab.* I:439–483.

Hodge, H. C., and Smith, F. A., 1965a. Biological properties of inorganic fluorides, in *Fluorine*

Chemistry, Volume IV (J. H. Simon, ed.), Academic Press, New York, London, pp. 1–375.

Hodge, H. C., and Smith, F. A., 1965b. Effects of fluoride on bones and teeth, in *Fluorine Chemistry, Volume IV* (J. H. Simon, ed.), Academic Press, New York, London, pp. 376–691.

Holland, R. I., 1979a. Fluoride inhibition of DNA synthesis in cells in vitro, *Acta Pharmacol. Toxicol.* 45:96–101.

Holland, R. I., 1979b. Fluoride inhibition of protein synthesis, *Cell Biol. Int. Rep.* 3:701–705.

Hongslo, C. F., Hongslo, J. K., and Holland, R. I., 1980. Fluoride sensitivity of cells from different organs, *Acta Pharmacol. Toxicol.* 46:73–77.

Howlett, A. C., Sternweis, P. C., Macik, B. A., Van Arsdale, P. M., and Gilman, A. G., 1979. Reconstitution of catecholamine-sensitive adenylate cyclase. Association of a regulatory component of the enzyme with membranes containing the protein and β-adrenergic receptors, *J. Biol. Chem.* 254:2287–2295.

Humphreys, W. G., and Macdonald, T. L., 1988. The effects of tubulin polymerization and associated guanosine triphosphate hydrolysis of aluminum ion, fluoride ion, and fluoroaluminate species, *Biochem. Biophys. Res. Commun.* 151:1025–1032.

Jenkins, G. N., and Edgar, W. M., 1977. Distribution and forms of F in saliva and plaque, *Caries Res.* 11 (Suppl. 1):226–242.

Jowsey, J., and Kelley, P. J., 1968. Effect of fluoride treatment in a patient with osteoporosis, *Mayo Clinic Proc.* 43:435–443.

Jowsey, J., Riggs, B. L., Kelly, P. J., and Hoffman, D. L., 1972. Effect of combined therapy with sodium fluoride, vitamin D and calcium in osteoporosis, *Amer. J. Med.* 53:43–59.

Kanaho, Y., Moss, J., and Vaughan, M., 1985. Mechanism of inhibition of transducin GTPase activity by fluoride and aluminum, *J. Biol. Chem.* 260:11493-11497.

Katada, T., Bokoch, G. M., Northup, J. K., Ui, M., and Gilman, A. G., 1984a. The inhibitory guanine nucleotide-binding regulatory component of adenylate cyclase. Properties and functions of the purified protein, *J. Biol. Chem.* 259:3568–3577.

Katada, T., Northup, J. K., Bokoch, G. M., Ui, M., and Gilman, A. G., 1984b. The inhibitory guanine nucleotide-binding regulatory component of adenylate cyclase. Subunit dissociation and guanine nucleotide-dependent hormonal inhibition, *J. Biol. Chem.* 259:3578–3585.

Kienast, J., Arnout, J., Pliegler, G., Deckmyn, H., Hoet, B., and Vermylen, J., 1987. Sodium fluoride mimics effects of both agonists and antagonists on intact human platelets by simultaneous modulation of phospholipase C and adenylate cyclase activity, *Blood* 69:859–866.

Kilian, M., Larsen, M. J., Fejerskov, O., and Thylstrup, A., 1979. Effects of fluoride on the initial colonization of teeth *in vivo*, *Caries Res.* 13:319–329.

Kleiner, H. S., and Allmann, D. W., 1982. The effects of fluoridated water on rat urine and tissue cAMP levels, *Arch. Oral Biol.* 27:107–112.

Knight, W. B., Fitts, S. W., and Dunaway-Mariano, D., 1981. Investigation of the catalytic mechanism of yeast inorganic pyrophosphate, *Biochemistry* 20:4079–4086.

Knight, W. B., Ting, S.-J., Chuang, S., Dunaway-Mariano, D., Harmony, T., and Sundaralingam, M., 1983. Yeast inorganic pyrophosphatase substrate recognition, *Arch. Biochem. Biophys.* 227:302–309.

Kuhn, H., 1980. Light- and GTP-regulated interaction of GTPase and other proteins with bovine photoreceptor membranes, *Nature* 283:587–589.

Kuza, M., and Kazimierczak, W., 1982. On the mechanism of histamine release from sodium fluoride-activated mouse mast cells, *Agents Actions* 12:289–294.

Lange, A. J., Arion, W. J., Burchell, A., and Burchell, B., 1986. Aluminum ions are required for stabilization and inhibition of hepatic microsomal glucose-6-phosphatase by sodium fluoride, *J. Biol. Chem.* 261:101–107.

Lefkowitz, R. J., Caron, M. G., and Stiles, G. L., 1984. Mechanisms of membrane-receptor regulation. Biochemical, physiological, and clinical insights derived from studies of the adrenergic receptors, *N. Engl. J. Med.* 310:1570–1579.

Lin, I., Knight, W. B., Hsueh, A., and Dunaway-Mariano, D., 1986. Investigation of the regiospecificity and stereospecificity of proton transfer in the yeast inorganic pyrophosphatase catalyzed reaction, *Biochemistry* 25:4688–4692.

Litosch, I., and Fain, J. N., 1986. Regulation of phosphoinositide breakdown by guanine nucleotides, *Life Sci.* 39:187–194.

Lochrie, M. A., and Simon, M. I., 1988. G Protein multiplicity in eukaryotic signal transduction systems, *Biochemistry* 27:4957–4965.

Lundy, M. W., Farley, J. R., and Baylink, D. J., 1986. Characterization of a rapidly responding animal model for fluoride-stimulated bone formation, *Bone* 7:289–293.

Macdonald, T. L., and Martin, R. B., 1988. Aluminum ion in biological systems, *Trends Biochem. Sci.* 13:15–19.

Marks, P. A., Burka, E. R., Conconi, F. M., Perl, W., and Rifkind, R. A., 1965. Polyribosome dissociation and formation in intact reticulocytes with conservation of messenger ribonucleic acid, *Proc. Natl. Acad. Sci. USA* 53:1437–1443.

Matthews, 1970. Changes in cell function due to inorganic fluoride, in *Handbook of Experimental Pharmacology, XX/2. Pharmacology of Fluorides, Part 2* (A. F. Smith, ed.), Springer-Verlag, Berlin, pp. 98–143.

Maurer, P. J., and Nowak, T., 1981. Fluoride inhibition of yeast enolase. 1. Formation of the ligand complexes, *Biochemistry* 20:6894–6900.

McGrath, J. P., Capon, D. J., Goeddel, D. V., and Levinson, A. D., 1984. Comparative biochemical properties of normal and activated human *ras* p21 protein, *Nature* 310:644–649.

McIvor, M. E., Cummings, C. C., Mower, M. M., Baltazar, R. F., Wenk, R. E., Lustgarten, J. A., and Salomon, J., 1985. The manipulation of postassium efflux during fluoride intoxication: Implications for therapy, *Toxicology* 37:233–239.

McPhail, L. C., Shirley, P. S., Clayton, C. C., and Snyderman, R., 1985. Activation of the respiratory burst enzyme from human neutrophils in a cell-free system. Evidence for a soluble cofactor, *J. Clin. Invest.* 75:1735–1739.

Messer, H. H., 1984. Fluorine, in *Biochemistry of the Essential Ultratrace Elements* (E. Frieden, ed.), Plenum Press, New York, pp. 55–87.

Miki, N., Keirns, J. J., Marcus, F. R., Freeman, J., and Bitensky, M. W., 1973. Regulation of cyclic nucleotide concentrations in photoreceptors: An ATP-dependent stimulation of cyclic nucleotide phosphodiesterase by light, *Proc. Natl. Acad. Sci. USA* 70:3820–3824.

Miller, J. L., Hubbard, C. M., Litman, B. J., and Macdonald, T. L., 1989. Inhibition of transducin activation and guanosine triphosphatase activity by aluminum ion, *J. Biol. Chem.* 264:243–250.

Moreno, E. C., Kresak, M., and Zahradnik, R. T., 1977. Physicochemical aspects of fluoride-apatite systems relevant to the study of caries, *Caries Res.* 11 (Suppl. 1):142–171.

Mundy, G. R., 1989. Identifying mechanisms for increasing bone mass, *J. NIH Res.* 1:65–68.

Neuman, W. F., DiStefano, V., and Mulryan, B. J., 1951. The surface chemistry of bone. III. Observations on the role of phosphatase. *J. Biol. Chem.* 17:286–293.

Newbold, R., 1984. Mutant *ras* proteins and cell transformation, *Nature* 310:628–629.

Northup, J. K., Sternweis, P. C., and Gilman, A. G., 1983a. The subunits of the stimulatory regulatory component of adenylate cyclase. Resolution, activity, and properties of the 35,000-Da (β) subunit, *J. Biol. Chem.* 258:11361–11368.

Northup, J. K., Smigel, M. D., Sternweis, P. C., and Gilman, A. G., 1983b. The subunits of the stimulatory regulatory component of adenylate cyclase. Resolution of the activated 45,000-Dalton (α) subunit, *J. Biol. Chem.* 258:11369–11376.

Nowak, T., and Maurer, P. J., 1981. Fluoride inhibition of enolase 2. Structural and kinetic properties of the ligand complexes determined by nuclear relaxation rate studies, *Biochemistry* 20:6901–6911.

Nowak, T., Mildvan, A. S., and Kenyon, G. L., 1973. Nuclear relaxation and kinetic studies of the role of Mn^{2+} in the mechanism of enolase, *Biochemistry* 12:1690–1701.

Page, J. D., and Wilson, I. B., 1983. The inhibition of acetylcholinesterase by arsenite and fluoride, *Arch. Biochem. Biophys.* 226:492–497.

Page, J. D., Wilson, I. B., and Silman, I., 1985. Butyrylcholinesterase: Inhibition by arsenite, fluoride, and other ligands, cooperativity in binding, *Mol. Pharmacol.* 27:437–443.

Pain, V. M., 1986. Inhibition of protein synthesis in mammalian cells, *Biochem. J.* 235:625–637.

Pak, C. Y. C., Sakhaee, K., Zerwekh, J. E., Parcel, C., Peterson, R., and Johnson, K., 1989. Safe and effective treatment of osteoporosis with intermittent slow release sodium fluoride: Augmentation of vertebral bone mass and inhibition of fractures, *J. Clin. Invest. Metab.* 68:150–159.

Peck, W. A., and Woods, W. L., 1988. The cells of bone, in *Osteoporosis: Etiology, Diagnosis, and Management* (B. L. Riggs and J. L. Melton, eds.), Raven Press, New York, pp. 1–44.

Peters, R. A., Shorthouse, M., and Murray, L. R., 1964. Enolase and fluorophosphate, *Nature* 202:1331.

Poll, C., Kyrle, P., and Westwick, J., 1986. Activation of protein kinase C inhibits sodium fluoride-induced elevation of human platelet cytosolic free calcium and thromboxane B_2 generation, *Biochem. Biophys. Res. Commun.* 136:381–389.

Prince, R. C., and Gunson, D. E., 1987. Superoxide production by neutrophils, *Trends Biochem. Sci.* 12:86–87.

Reinbold, W.-D., Genant, H. K., Reiser, U. J., Harris, S. T., and Ettinger, B., 1986. Bone mineral content in early-postmenopausal and postmenopausal women: Comparison of measurement method, *Radiology* 160:469–478.

Revell, P. A., 1986. *Pathology of Bone*, Springer-Verlag, Berlin.

Rich, C., and Ensinck, J., 1961. Effect of sodium fluoride on calcium metabolism of human beings, *Nature* 191:184–185.

Rich, C., Ensinck, J., and Ivanovich, P., 1964. The effects of sodium fluoride on calcium metabolism of subjects with metabolic bone disease, *J. Clin. Invest.* 43:545–556.

Riggs, B. L., 1984. Treatment of osteoporosis with sodium fluoride: An appraisal, in *Bone and Mineral Research, Annual 2* (W. A. Peck, ed.), Elsevier, Amsterdam, pp. 366–393.

Riggs, B. L., and Melton, L. J., III (eds.), 1988. *Osteoporosis: Etiology, Diagnosis, and Management*, Raven Press, New York.

Riggs, B. L., Seeman, E., Hodgson, S. F., Taves, D. R., and O'Fallon, W. M., 1982. Effect of the fluoride/calcium regimen on vertebral fracture occurrence in postmenopausal osteoporosis: Comparison with conventional therapy, *N. Engl. J. Med.* 306:446–450.

Robinson, J. D., 1975. Functionally distinct classes of K^+ sites on the $(Na^+ + K^+)$-dependent ATPase, *Biochim. Biophys. Acta* 384:250–264.

Robinson, J. D., Davis, R. L., and Steinberg, M., 1986. Fluoride and beryllium interact with the $(Na + K)$-dependent ATPase as analogs of phosphate, *J. Bioenerg. Biomembr.* 18:521–531.

Robinson, R., 1923. The possible significance of hexose phosphoric esters in ossification, *Biochem. J.* 17:286–293.

Rodbell, M., 1980. The role of hormone receptors and GTP-regulatory proteins in membrane transduction, *Nature* 284:17–22.

Schelling, D. L., and Cohen, P., 1987. MgATP-Dependent inactivation of rat liver aminoacyl-tRNA synthetases is explained by pyrophosphate generation and not by phosphorylation, *Biochem. Soc. Proc.* 15:271–272.

Skorecki, K. L., Verkman, A. S., Jung, C. Y., and Ausiello, D. A., 1986. Evidence for vasopressin activation of adenylate cyclase by subunit dissociation, *Am. J. Physiol.* 250:C103–C123.

Smirnova, I. N., and Baikov, A. A., 1983. Two-step mechanism of inhibition of inorganic pyrophosphatase by fluoride, *Biochemistry (Biokhimiya)* 48:1414–1423.

Smith, F. A., 1966. Metabolism of inorganic fluoride, in *Handbook of Experimental Pharmacology, XX/I. Pharmacology of Fluorides, Part 1* (O. Eichler, A. Farah, H. Herken, and A. D. Welch, eds.), Springer-Verlag, Berlin, pp. 53–140.

Spencer, S. G., and Brewer, J. G., 1982. Substrate-dependent inhibition of yeast enolase by fluoride, *Biochem. Biophys. Res. Commun.* 106:553–558.

Springs, B., Welsh, K. M., and Cooperman, B. S., 1981. Thermodynamics, kinetics, and mechanism in yeast inorganic pyrophosphatase catalysis of inorganic pyrophosphate: Inorganic phosphate equilibration, *Biochemistry* 20:6384–6391.

Stein, P. J., Halliday, K. R., and Rasenick, M. M., 1985. Photoreceptor protein mediates fluoride activation of phosphodiesterase, *J. Biol. Chem.* 260:9081–9084.

Sternweis, P. C., and Gilman, A. G., 1982. Aluminum: A requirement for activation of the regulatory component of adenylate cyclase by fluoride, *Proc. Natl. Acad. Sci. USA* 79:4888–4891.

Stiles, G. L., Caron, M. C., and Lefkowitz, R. J., 1984. β-Adrenergic receptors: Biochemical mechanisms of physiological regulation, *Physiol. Rev.* 64:661–742.

Strnad, C. F., and Wong, K., 1985. Calcium mobilization in fluoride activated human neutrophils, *Biochem. Biophys. Res. Commun.* 133:161–167.

Susheela, A., K., and Singh, M., 1982. Adenyl cyclase activity following fluoride ingestion, *Toxicol. Lett.* 10:209–212.

Sutherland, E. W., and Rall, T. W., 1958. Fractionation and characterization of a cyclic adenine ribonucleotide formed by tissue particles, *J. Biol. Chem.* 232:1077–1091.

Swarup, G., Cohen, S., and Garbers, D. L., 1981. Selective dephosphorylation of proteins containing phosphotyrosine by alkaline phosphatases, *J. Biol. Chem.* 256:8197–8201.

ten Cate, J. M., and Arends, J., 1977. Remineralization of artificial enamel lesions *in vitro*, *Caries Res.* 11:277–286.

Vaughan, J., 1981. *The Physiology of Bone*, Clarendon Press, Oxford.

Verkman, A. S., Ausiello, D. A., Jung, C. Y., and Skorecki, K. L., 1986. Mechanisms of non-hormonal activation of adenylate cyclase based on target analysis, *Biochemistry* 25:4566–4572.

Vittur, F., and deBernard, B., 1973. Alkaline phosphatase activity associated to a calcium-binding glycoprotein from calf scapula cartilage, *FEBS Lett.* 38:87–90.

Wang, T., and Himoe, A., 1973. Kinetics of the rabbit muscle enolase-catalyzed dehydration of 2-phosphoglycerate: Fluoride and phosphate inhibition, *J. Biol. Chem.* 249:3895–3902.

Warburg, O., and Christian, W., 1942. Isolierung und Kristallisation des Gärungsferments Enolase, *Biochem. Z.* 310:384–421.

Weatherell, J. A., 1969. Fluoride and the skeletal and dental tissues, in *Handbook of Experimental Pharmacology XX/1. Pharmacology of Fluorides, Part 1* (O. Farah, H. Herken, and A. D. Welch, eds.), Springer-Verlag, Berlin, pp. 141–172.

Whyte, M. P., and Vrabel, L. A., 1987. Infantile hypophosphatasia fibroblasts proliferate normally in culture: Evidence against a role for alkaline phosphatase (tissue nonspecific isoenzyme) in the regulation of cell growth and differentiation, *Calcif. Tissue Int.* 40:1–7.

Wilson, I. B., 1985. Cholinesterase: Two surprising inhibitors, in *Molecular Basis of Nerve Activity* (J. P. Changeux, ed.), Walter de Gruyter & Co., Berlin, pp. 667–678.

Wiseman, A., 1970. Effect of inorganic fluoride on enzymes, *Handbook of Experimental Pharmacology XX/2. Pharmacology of Fluorides, Part 2* (F. A. Smith, ed.), Springer-Verlag, Berlin, pp. 48–97.

Wuthier, R. E., and Register, T. C., 1984. Role of alkaline phosphatase, a polyfunctional enzyme, in mineralizing tissues, in *The Chemistry and Biology of Mineralized Tissues* (W. T. Butler, ed.), Proceedings of the Second International Conference on the Chemistry and Biology of Mineralized Tissues, Gulf Shores, Alabama, September 9–14, 1984, Ebsco Media, Birmingham.

Biochemistry of Inorganic Chloride

3

3.1 Introduction

The issues involved with the biochemistry of inorganic chloride (Cl^-) differ considerably from those considered for the biochemistry of inorganic fluoride (F^-). Whereas F^- is present in trace amounts in the body, Cl^- is a normal and substantial constituent of biological fluids. Whereas extracellular fluid—the composition of which resembles that of pre-Cambrian era seawater—has a high concentration of Na^+ and Cl^- and a low concentration of K^+, intracellular fluid has a large quantity of K^+ and phosphate, primarily organically bound, but little Cl^-. The maintenance of proper compositions of these fluids, vital to the well-being of the cell, depends on the proper availability and identity of nutrients in the extracellular fluid, cellular metabolism, and transport properties of cellular membranes. Because of the position of chloride as the most abundant anion in the extracellular medium, membrane transport of chloride has assumed an important role in many processes, including absorption, secretion, and control of osmotic pressure, cell volume, fluid pH, and electrolyte balance. Cl^- is also a common counterion in proteins, especially basic ones.

In this chapter, the role of Cl^- transport mechanisms in such important functions as carbon dioxide transport, renal function, and gastric secretion will be reviewed. The implication of defective Cl^- transport in cystic fibrosis will be discussed, as will the function of Cl^- transport in GABA-dependent neural function. Cl^- interacts with several macromolecular systems as a natural substrate and has important roles in regulation of several enzymes. The role of Cl^- in the photosynthetic generation of oxygen will also be discussed, and examples of the direct interaction of Cl^- with several enzymes will be given.

An examination of the literature related to the biochemistry of Cl^- rapidly reveals that this anion is involved in a multitude of biological processes. In this chapter, the attempt has been made to choose examples such that the diversity of Cl^- biochemistry will be represented—an exhaustive review would be impossible and inappropriate for this volume.

3.2 Transport of Chloride Through Cell Membranes

3.2.1 Transmembrane Transport Mechanisms

Transport of material through the lipid bilayer making up the cell membranes occurs either by diffusion (passive transport) or by active transport (against an energy gradient). Simple diffusion through the lipid bilayer is related directly to the lipid solubility of the substance and is quite rapid for such nonpolar molecules as oxygen and carbon dioxide. On the other hand, ions cannot penetrate the lipid bilayer. Instead, diffusion of ions (and other substances) can occur through aqueous channels of transmembrane proteins. These channels may be selective for substances and are often "gated." Facilitated diffusion requires the intervention of a carrier protein and, unlike simple diffusion, can be saturated by high substrate concentrations. The net rate of diffusion of a substance through the cell membrane is dependent on the sum of the factors making up the electrochemical gradient across the membrane (for example, concentration, pressure, and electrical gradients), and no material can diffuse against this gradient. Active transport, on the other hand, is mediated by carrier proteins coupled to an energy-producing process—the enzymatic hydrolysis of membrane-bound ATP—which leads to transport against an electrochemical gradient. A schematic diagram of these transport processes is given in Fig. 3-1 [for a further discussion of transport mechanisms, see, e.g., Guyton, 1986 (pp. 88–100)].

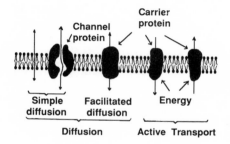

Figure 3-1. Schematic diagram of transport pathways through the cell membrane. From A. C. Guyton, *Textbook of Medical Physiology*, 7th ed., W. B. Saunders Co., Philadelphia (1986), p. 147, with permission.

3.2.2 Classification of Chloride Transport Mechanisms

Traditionally, the transport of Cl$^-$ and other anions had been assumed to be a secondary process, following electrochemical gradients established by a more physiologically relevant transport of cations. However, mediated translocations of anions across cell membranes have more recently become well recognized. Furthermore, anion transport has several important physiological functions, and the study of these processes has become a major field of membrane biology. In this section, the various types of Cl$^-$ transport that have been identified will be reviewed (for a more comprehensive review, see Hoffmann, 1986). In following sections, examples of these transport mechanisms, as related to specific physiological functions, will be given.

3.2.2.1 Electrodiffusion of Chloride

Electrodiffusion of Cl$^-$ through the cell membrane [(d) in Fig. 3-2 (Hoffmann, 1986)] occurs in response to anion gradients caused by active cation transport. Since the dissipation of these gradients is counter to the purpose of metabolically driven active transport, it is not surprising that, in many cells, this Cl$^-$ leak by anion diffusion is small under normal circumstances. On the other hand, gated Cl$^-$ channels have several important functions. For example, upon stimulation by inhibitory neurotransmitters, Cl$^-$ channels mediate the influx of Cl$^-$ into central nerve cells along the electrochemical gradient and cause hyperpolarization of these cells (Section 3.9). The role of a defect in adrenergically activated Cl$^-$ channels in the pathogenesis of cystic fibrosis (Section 3.7) demonstrates the importance of hormonally controlled Cl$^-$ channels in secretory processes.

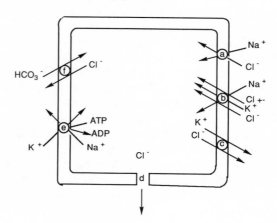

Figure 3-2. Schematic diagram of anion transport pathways through the cell membrane of an isolated vertebrate cell (see text for descriptions of pathways). After Hoffmann (1986), with permission.

3.2.2.2 Electroneutral Exchange of Chloride

Electroneutral exchange of choride [(f) in Fig. 3-2], exemplified by Cl^-/HCO_3^- exchange in red blood cells, is a type of facilitated transport that consists of a tightly coupled anion influx and efflux and thus causes no net change in electric charge across the cell membrane. This transport mechanism will be discussed more thoroughly in Section 3.4.

3.2.2.3 Cotransport of Chloride and Cations

Electroneutral Cl^- transport tightly coupled with transport of Na^+, K^+, and $Na^+ + K^+$ is shown schematically in Fig. 3-2 [(a), (c), and (b), respectively)]. Proof of true cotransport requires the demonstration that an observed dependence of anion transport on cation transport is not an indirect result of gradients established by cation transport. Mechanisms (a) and (b) are able to accumulate Cl^- inside the cell against an electrochemical gradient, by virtue of the downhill movement of Na^+ along its electrochemical gradient, and, accordingly, transport of Cl^- requires the removal of Na^+ from the cell by the action of the Na^+/K^+ pump driven by ATPase [(e) in Fig. 3-2]. The net transport of Cl^- against an electrochemical gradient is classified as a "secondary active" process since it is coupled to maintenance of the Na^+ gradient by the primary active transport system (Hoffmann, 1986). These cotransport processes are important for many cellular functions, including cell volume control and transepithelial Cl^- flow, and will be considered further in coming sections.

3.2.2.4 Chloride-Stimulated ATPase

In addition to the three widely accepted mechanisms of Cl^- transport described above, there have been efforts to document the presence of active Cl^- transport associated with stimulated ATPase (a Cl^- pump). Cl^--stimulated ATPase is present in nearly all animal cells but appears to be largely confined to the mitochondria. While a role for mitochondrial Cl^--stimulated ATPase in Cl^- transport across the plasma membrane is unlikely, it may be involved in intracellular transport. However, there is considerable controversy over the existence of nonmitochondrial Cl^--stimulated ATPase. Microsomal activity reported by some researchers has been attributed to mitochondrial contamination by others. This issue was debated, but not resolved, in a 1986 conference (Gerencser et al., 1988).

3.3 Chloride Transport and Body Fluid Homeostasis

As the most abundant anion in extracellular fluid, Cl^- transport is important in the control of body fluid volume, pH, electrolyte balance, and other factors. Concise discussions of these issues are found in textbooks on medical physiology [e.g., Guyton, 1986 (pp. 2–10, 382–392); Porth, 1986], and these topics will be considered only briefly. Roos and Boron (1981) have provided a detailed review of mechanisms by which intracellular pH is controlled, including anion transport mechanisms. Kregenow (1981) has reviewed salt-transporting mechanisms involved in the control of intracellular volume.

3.4 Chloride/Bicarbonate Cotransport in Erythrocyte Membrane

The transport of HCO_3^- and Cl^- across the red blood cell membrane (the Hamburger cycle) has the important function of facilitating the elimination of carbon dioxide (CO_2) from the body. Metabolically produced CO_2 diffuses from tissue cells, enters the capillary lumen by simple diffusion, and then diffuses into the red blood cell, where it is converted rapidly to HCO_3^- by carbonic anhydrase. The increased concentration of HCO_3^- provides a driving force for the one-for-one exchange of intracellular HCO_3^- for plasma Cl^-. Through this process, capillary CO_2 is converted into plasma HCO_3^- and carried to the lungs predominantly as such. In the pulmonary capillaries, the reverse process releases CO_2 into alveolar air [Fig. 3-3; Guyton, 1986 (pp. 500–503)]. This process also modulates pH gradients associated with CO_2 transport.

The erythrocyte Cl^-/HCO_3^- exchange has been studied extensively, and over the last two decades many qualitative and quantitative details have been elucidated. Many reviews of the huge volume of literature in this field are available, for example, Jennings (1985), Passow (1986), Lowe and Lambert 1983), Knauf (1979), and Cabantchik et al. (1978).

3.4.1 General Properties of Erythrocyte Anion Transport

Anion transport by the red cell membrane is mediated by a 97,000-Da transmembrane dimeric glycoprotein known as "band 3" (Steck, 1978) because of its position on sodium dodecyl sulfate (SDS) gels. Band 3 is present as approximately 550,000 noncovalent dimers on the human red cell and makes up approximately 25% of the red cell membrane protein. The predominance of band 3 in red cell membrane suggests additional

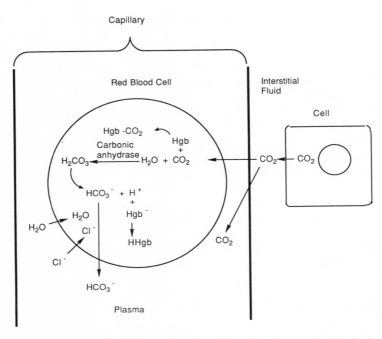

Figure 3-3. Transport of CO_2 in the blood. Adapted from A. C. Guyton, *Textbook of Medical Physiology*, 7th ed., W. B. Saunders Co., Philadelphia (1986), p. 500, with permission.

structural roles for the protein. Rapid self-exchange of small inorganic ions as well as heteroexchange between several anions such as Cl^-, Br^-, I^-, SO_4^{2-}, and PO_4^{3-} is also accomplished by this anion exchange system. The transport system is characterized by its efficiency. Thus, the half-time for Cl^-/HCO_3^- under physiological conditions is about 100 ms, permitting up to 90% exchange during usual capillary transit time. While the binding sites are relatively nonselective for transported anions ($I^- > HCO_3^- > Br^- > Cl^- > F^-$) (Dalmark, 1976), the rate of transport varies greatly, with Cl^- and HCO_3^- transported most rapidly (reviewed by Jennings, 1985).

Wright and Diamond (1977) have reviewed quantitative and qualitative information on anion selectivity in biological systems and have developed theoretical patterns to describe observed selectivities. Of relevance to this volume is the fact that of the many halide sequences (rank order of activity) possible (24), only a few occur in biological systems, and these have been predicted theoretically. In general, anion binding sites provided by quaternary ammonium groups (M^+) have higher anion selectivity than do those made up of protonated amines (MH^+).

3.4.2 Kinetics of Transport

Band 3-catalyzed anion transport occurs by an obligatory one-for-one exchange, and two anions thus are required for a complete catalytic cycle. Two kinetic schemes could, in principe, account for this catalysis. In a ping-pong mechanism, an anion binds to an available transport site and is transported to the opposite side of the membrane and released, making available the binding site to anions on this side of the membrane. Thus, two conformations of the carrier protein are required, one "inward facing" and one "outward facing." In a simultaneous mechanism, intracellular and extracellular anions can bind to the receptor at the same time and are then transported simultaneously in a process that requires the presence of both anions. In this mechanism, no conformational asymmetry is required. Schematic diagrams of ping-pong and simultaneous transport are given in Fig. 3-4.

There has accumulated much evidence to support asymmetric anion transport by band 3 protein. For example, resealed red cell ghosts containing $^{36}Cl^-$ suspended in a medium devoid of rapidly transported anions suffer a rapid loss of about 10^6 Cl^- per ghost—corresponding to one Cl^- per band 3 protein—followed by a much slower efflux. Thus, the partial catalytic cycle—binding at the intracellular site, transport, and release to the outside—can occur in the absence of influx of anion (Jennings, 1982). The identity and concentration of a *trans* ion also influence the flux of a *cis* ion, supporting the ping-pong mechanism (Gunn

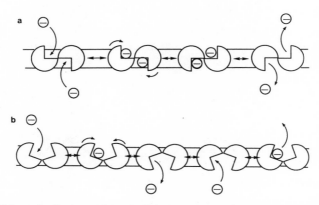

Figure 3-4. Schematic diagrams of (a) simultaneous and (b) ping-pong mechanisms for one-for-one exchange of Cl^- (see text). Adapted from W. Furuya, T. Tarshis, F.-Y. Law, and P. A. Knauf, *J. Gen. Physiol.* 83:657–681 (1984), by copyright permission of the Rockefeller University Press.

and Frölich, 1979; Jenings, 1985). Similarly, binding of a nonpermeating transport inhibitor, dihydro-4,4'-diisothiocyanostilbene-2,2'-disulfonate (H_2DIDS; Fig. 3-5), to band 3 (see below) increases as intracellular Cl^- [Cl_i] is increased, consistent with an increased number of outward-facing sites caused by higher [Cl_i] (Furuya *et al.*, 1984). Similar experiments with another inhibitor of transport, niflumic acid, have shown that when [Cl_i] = [Cl_0], about 15 times more sites face the inside than the outside of the cell (Knauf and Mann, 1984).

Extensive research has been carried out to define the kinetic properties of band 3 transport. Reviews by Passow (1986), Jennings (1985), and Gunn and Frölich (1982) contain thorough descriptions of these studies.

3.4.3 Molecular Probes for the Erythrocyte Cl^-/HCO_3^- Transport System

The isolated band 3 protein has no endogenous marker related to its function (such as ATPase associated with cation transport systems), and since substrate affinities are quite low, identification of the protein responsible for transport is difficult. In this regard, the isolation and characterization of band 3 have been facilitated greatly by the availability of specific inhibitors that bind to functional components of the membrane protein. Stilbene sulfonates, one class of such inhibitors (Fig. 3-5), are membrane-impermeant compounds that competitively inhibit anion transport from the external side of the membrane.

3.4.3.1 Identification of Band 3 as the Cl^- Transport Protein

Stilbene sulfonates react specifically with a small population of protonated amines on band 3, certain of which are required for anion

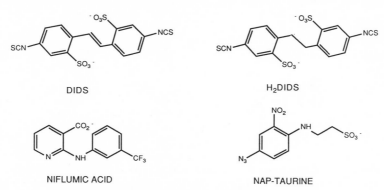

Figure 3-5. Structures of inhibitors of Cl^- transport.

transport, but apparently cannot reach sites that are involved with cation permeability. Binding of stilbene sulfonates such as 4,4'-diisothio-cyanostilbene-2,2'-disulfonate (DIDS, Fig. 3-5) to band 3 and the association of this binding with inhibition of transport are among many observations that convincingly demonstrated band 3 to be the transport protein (e.g., Cabantchik and Rothstein, 1974). Nonetheless, the stilbene binding site and the anion binding site appear not to be identical (reviewed by Passow *et al.*, 1980).

3.4.3.2 Probes for Transport and Modifier Sites on Band 3

The anion transport size of band 3 is half-saturated by Cl^- at 65 mM, and self-inhibition occurs at relatively high concentrations of Cl^- and HCO_3^- (Dalmark, 1976; reviewed by Lowe and Lambert, 1983). These and other data (Knauf *et al.*, 1978) support the existence of a "modifier site," occupation of which leads to inhibition of transport. Based on self-inhibition of anions and the behavior of transport inhibitors, a model for anion transport was developed (Macara and Cantley, 1981). This model is representative of a class of "knock-on" models (see below). Thus, as shown in Fig. 3-6, two positive charges function as transport sites, access to which is controlled by an anionic gate. At the outward-facing gate, binding Cl^- from the outside weakens the electrostatic attraction between the anionic gate and the positive charge, and the gate swings to the other positive charge, releasing Cl^- to the cytoplasm with formation of the inward-facing gate. In this model, self-inhibition would result from binding of anions to both positive charges spontaneously, and the "modifier site" thus is part of a transporting gate (Macara and Cantley, 1981). Knauf and Mann (1986) have recently provided evidence that the modifier site may reside on the inside of the cell membrane.

The above are examples of the use of inhibitors to probe binding sites and the band 3 transport mechanism. Many other studies could be cited, and extensive reviews that discuss this work are available. The relationship between inhibitor binding sites, transport sites, modifier sites, and conformational changes associated with interactions of these sites is still under active investigation (for further examples, see Dix *et al.*, 1986; Verkman *et al.*, 1983; Frölich, 1982).

3.4.4 The Structure of Band 3 and Models for Anion Transport

The work discussed above represents only a very small fraction of the extensive research done on the structure and mechanism of band 3, research that has included—in addition to kinetic studies and research

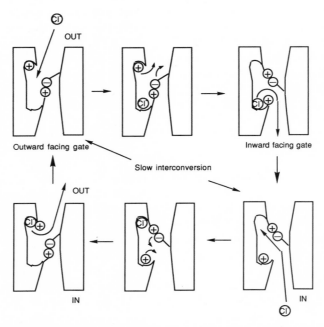

Figure 3-6. Proposed mechanism for anion exchange by band 3 protein, an example of a "knock-on" model (see text). Adapted, with permission, from I. G. Macara and L. C. Cantley, Mechanism of anion exchange across the red cell membrane by band 3: Interactions between stilbenedisulfonate and NAP-taurine binding sites, *Biochemistry* 20:5695–5701. Copyright 1981 American Chemical Society.

using transport inhibitors—reconstitution studies, proteolytic digestion of membrane-bound enzyme and peptide mapping of the obtained fragments, chemical modification of specific amino acid residues, and other approaches (Passow, 1986). As in many such cases, the techniques of molecular biology have proven quite effective. Thus, the complete amino acid sequence of murine band 3 protein has recently been deduced from the nucleotide sequence of a complementary DNA clone (Kopita and Lodish, 1985). The two main structural domains revealed are consistent with dual functions of band 3 as the anchor for the red cell cytoskeleton and as the mediator of the membrane anion exchange. The latter function is associated with a hydrophobic domain that crosses the membrane at least 12 times. Thus, as discussed by Passow (1986), several essentially parallel peptide segments apparently form a channel across the membrane. A gate near the outer surface of the membrane would consist of amino acid residues—some of which would be the stilbene binding site—from different adjacent segments. Anion accumulation at this gate would signal allosteric and conformational changes that would lead to binding and translocation. "Knock-on" and "lock-in" models currently under consideration incorporate the assumption that substrate molecules are accumulated at the channel orifice by a cluster of positive charges. An example of a "knock-on" model, proposed by Macara and Cantley (1981), was given in Section 3.4.3.2. Passow (1986) discusses these and other models in greater detail.

3.4.5 Band 3 and Chloride Transport—Summary

The physiological importance of the Cl^-/HCO_3^- exchange and the ready availability of band 3 protein have generated an enormous amount of research. A brief review of this work cannot adequately cover many important studies. The attempt has been made to highlight certain themes and strategies. Similar Cl^-/HCO_3^- exchange mechanisms are present in many other tissues.

3.5 Chloride/Cation Cotransport and Cell Volume Control

The presence of any appreciable transmembrane osmotic pressure differential will rupture the fragile vertebrate cell membrane. Thus, when duck red cells are challenged with an osmotic pressure differential, an initial shrinking or swelling ensues to compensate for this pressure. This is followed by rapid return to the original volume (Kregenow, 1981), leading to the proposal that these cells contain a volume-controlling mechanism (VCM) that senses cell size and transport salts to adjust that size. Subsequent research demonstrated that swelling of duck red cells in hypotonic media stimulates a Cl^--dependent, Na^+-independent, K^+-dependent transport whereas shrinking them in hypertonic media stimulates a $Na^+ + K^+ + 2Cl^-$ cotransport. The inhibition of $Na^+ + K^+ + 2Cl^-$ cotransport by furosemide and other loop diuretics (see below) suggested that erythrocyte volume control and transepithelial NaCl transport in the kidney are mechanistically related. The diuretic-sensitive $Na^+ + K^+ + 2Cl^-$ cotransport system has subsequently been identified in a wide variety of cell types and has been studied extensively (reviewed by O'Grady et al., 1987; Geck and Pfeiffer, 1985; Epstein and Silva, 1985; Kinne et al., 1985).

The relationship between the Na^+/K^+ pump and $Na^+ + K^+ + 2Cl^-$ cotransport during regulatory volume increase (RVI) is shown in Fig. 3-7. Activation of the Na^+/K^+ pump causes influx of two K^+ and efflux of three Na^+ for each molecule of ATP hydrolyzed. Influx of three Na^+ by the cotransport system, accompanied by three K^+ and six Cl^-, results in

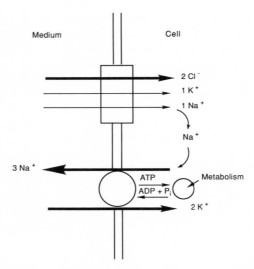

Figure 3-7. The Na^+/K^+ pump and $Na^+ + K^+ + 2Cl^-$ cotransport. Adapted, with permission, from P. Geck and B. Pfeiffer, 1985, $Na^+ + K^+ + 2Cl^-$ cotransport in animal cells—its role in volume regulation, *Ann. N.Y. Acad. Sci.* 456:166–182 (1985).

the net uptake of five KCl, thus increasing the content of osmotically active ions without changing drastically the makeup of intracellular fluid. This transmembrane recycling of Na^+ also facilitates transepithelial transport by doubling the number of ions transported for each ATP hydrolyzed (see below). During regulatory volume decrease (RVD), loss of KCl is accomplished by the activation of specific K^+ and Cl^- channels (Geck and Pfeiffer, 1985).

An increase in intracellular Ca^{2+} [Ca_i] has recently been described as a necessary and sufficient condition for a volume decrease in acinar (salivary gland) cells, and this decrease was linked to secretion. This volume decrease apparently reflects a loss of KCl from the cell due to agonist-enhanced K^+ and Cl^- plasma membrane conductances (Foskett and Melvin, 1989).

3.6 Chloride and Transepithelial Transport

3.6.1 Transepithelial Transport

Epithelia of various kinds that line inner surfaces and cover outer surfaces throughout the body consist of polarized cells in mono- or multicellular layers. Epithelial cells carry out transport of solutes through cellular layers and may be classified as absorptive or secretory epithelia, these functions being determined by the asymmetric distribution of transport mechanisms in the various types. The cells of the basolateral membrane of epithelia are in contact with each other or with connective tissue and basement membrane whereas the cells of the apical membrane face the free or luminal surface. Examples of absorptive epithelia are the epithelium of renal tubules and small intestine, while examples of secretory epithelia include the epithelium of salivary and sweat glands, exocrine glands, and trachea. The cellular processes that mediate transepithelial transport consist of active transport on one side of the cellular sheet (transport that increases the intracellular electrochemical gradient) and simple or facilitated diffusion on the other side (Rechkemmer, 1988; Griepp and Robbins, 1988). Examples of chloride absorption and secretion across epithelial membrane will be given in the next sections.

3.6.2 Chloride Transport and Renal Function

To illustrate processes by which Cl^- is absorbed through epithelia, the role of Cl^- transport in renal function will be considered. The kidneys together contain approximately 2,400,000 nephrons, each of which can produce urine. The nephron is composed of a glomerulus (a capillary network from which fluid is filtered) and a long tubule in which the glomerular filtrate is converted to urine. This occurs by a process of reabsorption of solutes as the filtrate passes through the several sections of the tubule [proximal tubule, loop of Henle (each with a descending and an ascending limb), and distal tubule] to the collecting tubule and duct. A schematic diagram of the functional nephron is shown in Fig. 3-8 [Guyton, 1986 (pp. 393–409)]. Since Cl^- constitutes nearly two-thirds of the anions in plasma and other extracellular fluids, this is the major anion filtered by the glomerulus and reabsorbed in renal tubules.

3.6.2.1 Chloride Transport in the Proximal Tubule

Approximately 70% of the Cl^- present in the glomerular filtrate is reabsorbed in the proximal tubule (reviewed by Schild et al., 1988). A Na^+/H^+ antiport system promotes HCO_3^- absorption, which, in turn, increases Cl^- concentration in the filtrate progressively along the proximal tubule. The increasing electrochemical gradient causes reabsorption of Cl^- by means of a passive flux through a paracellular shunt pathway—the presence of which is indicated by a high chloride conductance. This

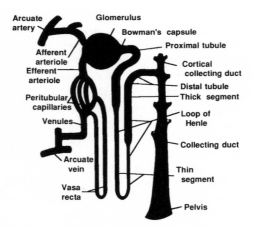

Figure 3-8. A schematic diagram of the functional nephron. From A. C. Guyton, *Textbook of Medical Physiology*, 7th ed., W. B. Saunders Co., Philadelphia (1986) p. 395, with permission.

accounts for a substantial portion of the total Cl^- reabsorption in the proximal tubule, although controversy exists concerning the exact proportion that passive transport contributes. Electroneutral transport of Cl^- at the luminal membrane appears to occur by a Cl^-/HCO_3^- exchanger, similar to that in the red blood cell. There is little evidence for neutral cotransport mechanisms at this surface. A model that would link Na^+/H^+ exchange with Cl^-/formate (HCO_2^-) exchange has been proposed (Karniski and Aronson, 1985) (Fig. 3-9).

Verkman *et al.* (1989) have developed a new procedure for the study of Cl^- transport based on quantitative fluorescence of the Cl^--sensitive fluorophore 6-methoxy-*N*-(3-sulfopropyl)quinolinium (SPQ; Fig. 3-10) entrapped in membrane vesicles. With the use of this procedure, Cl^- transport was studied in apical (brush border) membrane vesicles isolated from

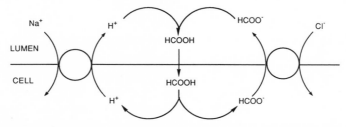

Figure 3-9. Proposed model for the role of Cl^-/formate exchange in mediating the Na^+-coupled active transport of Cl^- across the luminal membrane of the proximal tubular cell. After Karniski and Aronson (1985), with permission.

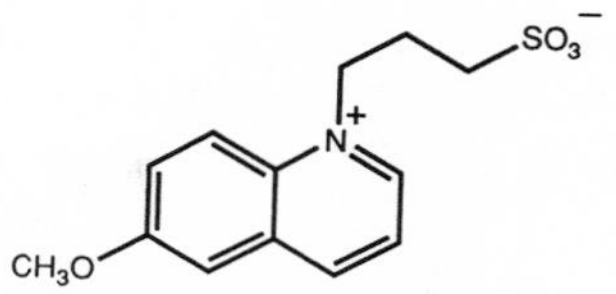

Figure 3-10. The structure of SPQ, a Cl^- sensitive fluorophore.

rabbit renal cortex. Evidence was found to support the existence of stilbene-sensitive Cl^-/OH^- and Cl^-/HCO_2^- exchange and for Cl^- conductance. No $Na^+ + Cl^-$ or $Na^+ + K^+ + 2Cl^-$ cotransport was found (Chen et al., 1988).

Considerable controversy exists over the mechanism of extrusion of Cl^- at the basolateral membrane. Cl^- conductance, Cl^-/anion exchange, K^+/Cl^- cotransport, and $Na^+/Cl^-/HCO_3^-$ coupled transport have been suggested (Chen and Verkman, 1988; Schild et al., 1988). Using fluorescence quenching of SPQ to address this issue, Chen and Verkman (1988) obtained evidence to support the presence of a stilbene-sensitive $Cl^-/OH^-(HCO_3^-)$ exchange and a $Na^+/Cl^-/HCO_3^-$ coupled transport, but no K^+/Cl^- cotransport or Cl^- conductance was detected.

3.6.2.2 Chloride Transport in Distal Nephron Segments

Unlike the situation in the proximal tubule, most Cl^- absorption in the distal nephron segments occurs through the cell. Much progress has been made in understanding transport mechanisms in the thick ascending limb of the loop of Henle, known as the "diluting" segment, and these mechanisms will be summarized. The reviews by Greger (1988), de Rouffignac and Elalouf (1988), and Hebert and Andreoli (1984) describe these processes more thoroughly, including discussions of biochemical regulation of transport as well as descriptions of transport in other distal tubule segments.

The transport of Cl^- through the thick ascending limb is shown schematically in Fig. 3-11. The Na^+/K^+-ATPase-mediated efflux of Na^+ at the basolateral membrane generates an inside negative potential and lowers the intracellular electrochemical gradient for Na^+. This activates the electroneutral $Na^+ + K^+ + 2Cl^-$ cotransport system at the luminal membrane. The Na^+ delivered by this transporter is extruded by the Na^+/K^+ pump, Cl^- exits through a basolateral conductance channel and by KCl cotransport, and K^+ is recycled through a K^+ channel in the luminal

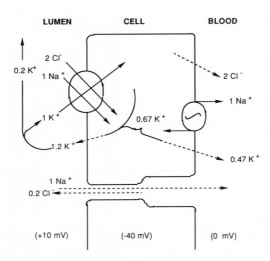

Figure 3-11. NaCl absorption in the mouse medullary thick ascending limb of the loop of Henle. Conductive pathways are denoted by dashed arrows. See text for further details. After Hebert and Andreoli (1984), with permission.

membrane. The presence of the $Na^+ + K^+ + 2Cl^-$ cotransporter in this nephron segment, absent in other distal nephron segments, provides an extremely efficient mechanism for NaCl transport, with as many as six NaCl transported for each ATP hydrolyzed (Greger, 1985; de Rouffignac and Elalouf, 1988).

3.6.2.3 Inhibitors of Renal Cl^- Transport

A class of potent diuretics, exemplified by bumetanide and furosemide (Fig. 3-12), act primarily in the thick ascending limb of the loop of Henle and thus are termed "loop" diuretics. By blocking solute reabsorption in this nephron segment, a more highly concentrated solution is delivered to the distal portions of the nephrons. Here, the osmotic activity of the solutes prevents water reabsorption, and urine volume is increased. These sulfamoyl benzoic acids, active only at the tubular lumen, are relatively specific inhibitors of $Na^+ + K^+ + 2Cl^-$ cotransport. An anionic moiety is a structural requirement for the inhibitor, suggesting that these compounds bind to a site normally occupied by Cl^-. Current research directed toward the identification and purification of the $Na^+ + K^+ + 2Cl^-$ cotransporter is based on binding studies and affinity labeling using these compounds (reviewed by Greger, 1988; O'Grady et al., 1987).

A series of aminobenzoic acids (Fig. 3-12) block Cl^- channels on the

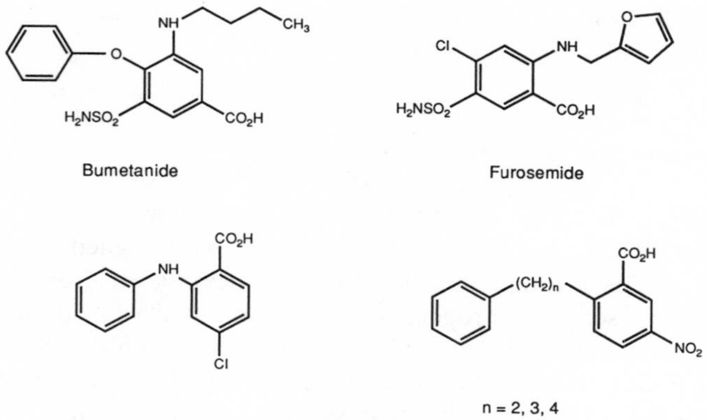

Figure 3-12. Examples of diuretics.

basolateral membrane from the outside. Wangemann *et al.* (1986) have published a recent extensive study of the structure–activity relationship of these drugs. An anionic group again is necessary, as is a secondary amine. These compounds are lipophilic whereas loop diuretics are relatively hydrophilic, a difference which may contribute to their different sites of action.

3.7 Epithelial Chloride Secretion and Cystic Fibrosis

3.7.1 Altered Electrolyte Transport in Cystic Fibrosis

Cystic fibrosis (CF), the most common lethal genetic disease of Caucasians (1 in about 2000 live births), is characterized by an accumulation of airway mucus that impairs the pulmonary function of victims. CF patients also have an abnormally high NaCl content in sweat, an abnormality long recognized as a diagnostic feature of CF. Quinton (1983) demonstrated that this salt elevation is caused by low Cl^- permeability in the reabsorptive duct isolated from CF sweat glands and suggested that a similar impairment could account for abnormal electrolyte transport in respiratory epithelium. Knowles *et al.* (1983) obtained similar results in nasal mucosa, and Widdicombe *et al.* (1985) reported that Cl^- permeability of the apical membrane of monolayers of cultured cells from CF tracheal epithelium was impaired. These and other studies of electrolyte transport in CF exocrine and epithelial cells have confirmed that the primary defect in CF is impaired permeability of Cl^- at the apical membrane in various secretory organs.

3.7.2 Mechanism of Secretion in Airway Epithelia

The identification of impaired Cl^- transport as an underlying mechanism of CF has spurred research efforts designed to define the molecular events leading to this defect. A schematic diagram of a model developed to describe electrogenic Cl^- secretion in airway epithelia is given in Fig. 3-13. In a fashion similar to the renal epithelial absorption described above, Na^+/K^+-ATPase associated with the basolateral membrane provides energy for the process and maintains a low intracellular Na^+ concentration. This provides a driving force for a $Na^+ + K^+ + 2Cl^-$ cotransport, and K^+ is recycled through the basolateral membrane through K^+ channels. These basolateral transport systems lead to intracellular accumulation of Cl^- above electrochemical equilibrium, and opening of the apical membrane channel permits Cl^- to exit down the electrochemical gradient. Paracellular transport of Na^+, driven by the transepithelial electrical gradient arising from Cl^- exit into the mucosal solution, results in net NaCl secretion. Absorptive processes are also important in controlling the quantity and composition of airway fluids. For a further discussion of these mechanisms, see Frizzel (1988) and Rechkemmer (1988).

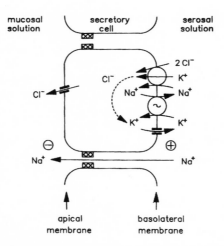

Figure 3-13. Model for electrogenic Cl^- secretion in epithelia (see text) (Rechkemmer, 1988; used with permission from the American Review of Respiratory Diseases).

3.7.3 Regulation of the Apical Chloride Channel and Cystic Fibrosis

3.7.3.1 Protein Kinase Activation of the Chloride Channel

The rate-limiting step in airway epithelial secretion is the Cl^- permeability at the apical membrane, permeability that is dependent on activation of the apical Cl^- channel. This channel, apparently unique to Cl^--secreting epithelia (Welsh, 1986), is regulated by β-adrenergic agonists and other agents that produce cAMP in response to receptor binding (Cotton et al., 1987). Thus, as analyzed by Rechkemmer (1988), defective Cl^- transport could, in principe, result from a lack of β-adrenergic receptors, defective coupling of the receptors to adenylate cyclase, absence of cAMP-dependent protein kinase A (A-kinase) activity, defective phosphorylation of the Cl^- channels or of a regulatory protein, or an absence of Cl^- channels. There is much evidence from a number of studies that CF epithelial cells contain the same apical Cl^- channel as do normal cells, that binding of β-adrenergic agonists to the basolateral membrane receptor is normal, and that cAMP accumulates normally as a consequence of this binding (Frizzell et al., 1986; Welsh and Liedtke, 1986; reviewed by McPherson et al., 1988). Indeed, the demonstration that A-kinase opens apical Cl^- channels in normal but not in CF airway epithelial cells localizes the defect between A-kinase and channel activation. Thus, a model has been proposed in which a protein, phosphorylation of which is required for channel opening, is altered such that it becomes an inappropriate substrate for A-kinase (Schoumacher et al., 1987). Brautigan (1988) has reviewed cycles of protein phosphorylation and has proposed sites of mutations on channel regulatory proteins that could be important in CF pathogenesis.

Epithelial Cl^- secretion is also regulated by cAMP-independent mechanisms. Hwang et al. (1989) and Li et al. (1989) have recently reported that cAMP-independent protein kinase C (C-kinase) activates normal airway cell Cl^- channels but does not activate cells from CF patients. Li et al. also reported that, under certain conditions, C-kinase down-regulates the Cl^- channel in both normal and CF cells. These results suggest further that the CF defect lies in an inability of the apical Cl^- to be phosphorylated or in a failure of the phosphorylated channel to become activated.

3.7.3.2 Calcium Metabolism and Chloride Secretion

The intracellular protein calmodulin is a "second messenger" that is activated by Ca^{2+} influx into the cell. Formation of Ca^{2+}/calmodulin leads to multiple biochemical effects, including regulation of protein and ion transport. Significant alterations in Ca^{2+} metabolism in CF cells have suggested that defects in Ca^{2+} regulation of Cl^- transport could be important in CF pathogenesis. Activation of cAMP phosphodiesterase (PDE) by Ca^{2+}/calmodulin represents an example of interaction between Ca^{2+}/calmodulin and cAMP-mediated processes. Increased activation of PDE in CF cells—an increase that would decrease the response to β-adrenergic agonists—and an alteration of a specific calmodulin-binding protein have been investigated as possible molecular lesions in CF cells (reviewed by McPherson et al., 1988). In this regard, Willumsen and Boucher (1989) have presented evidence that a Ca^{2+}-dependent path for Cl^- secretion is preserved in CF epithelia and have suggested that agonists targeted to changes in cytosolic Ca^{2+} might be considered as therapeutic agents.

3.7.4 Chloride Secretion and Cystic Fibrosis—Summary

Regulation of the apical Cl^- channel in airway epithelia is a complex process, but it is also evident that a much clearer understanding of this regulation is emerging rapidly (Levitan, 1989). New approaches to attacking this problem are also emerging. Isolation of the apical Cl^- channel protein from both normal and CF cells would be a major step in elucidating structural defects in the latter. To this end, Landry et al. (1989) have isolated from kidney and trachea several proteins that, when reconstituted into artificial phospholipid bilayer membranes, exhibit Cl^- channel activity. In another important advance, Jetten et al. (1989) have established a stable transformed epithelial cell line that maintains the defect in the apical Cl^- channel. This recent rapid progress made in understanding CF suggests that a detailed understanding of CF, at the molecular level, soon may be at hand. Indeed, the recent identification of the CF gene (Rommens et al., 1989), the cloning and characterization of the complementary DNA (Riordan et al., 1989), and genetic analysis of this gene (Kerem et al., 1989) will have a major impact on understanding of the disease as well as possible new strategies for treatment of the disease. For example, approximately 70% of CF patients have base pair deletions in this gene that correspond to loss of a phenylalanine residue in the putative gene product, a protein which has properties consistent with membrane association.

3.8 Hydrochloric Acid Secretion in the Stomach

Primarily in response to the presence of food, secretory glands throughout the gastrointestinal tract provide digestive enzymes, mucus, and electrolytes needed for digestion. The secretion of HCl in the stomach, part of this overall function, provides an excellent example of how integrated mechanisms combine to carry out an exquisitely regulated secretory function. The parietal (oxyntic) cells of the stomach oxyntic (acid-forming) glands secrete an approximately 180 mM solution of hydrochloric acid. The H^+ concentration of this solution, having a pH of about 0.8, is some 3×10^7 fold higher than that of arterial blood, and to achieve this concentration, approximately 1500 calories per liter of gastric juice is required. H^+/K^+-ATPase, an energy source for HCl secretion, is confined to cytoplasmic tubulovesicles while the cell is not secreting acid. On stimulation by altered cAMP or intracellular Ca^{2+} concentrations [Ca_i], the tubulovesicles migrate to and fuse with the apical membrane of the secretory canaliculus of the cell. Active secretion of protons energized by ATP hydrolysis is accompanied by Cl^- transport through a Cl^- conductance channel to produce HCl in the secretory canaliculus. K^+ actively transported into the cell also exits through a K^+ conductance channel (Fig. 3-14) (Perez *et al.*, 1989; Muallem *et al.*, 1988; Sachs *et al.*, 1988; and references therein).

Compensation for a continuing loss of H^+ and Cl^-, as well as a lesser amount of K^+ caused by incomplete recycling of K^+ by the H^+/K^+

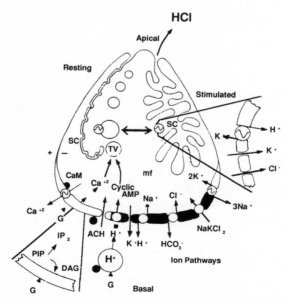

Figure 3-14. A composite model of the parietal cell; tv, tubulovesicles; sc, secretory canaliculus (see text and Sachs *et al.*, 1988, for details) (Sachs, 1986; used with permission).

pump, is accomplished by several basolateral surface transporters. As shown in Fig. 3-14, Na^+/K^+-ATPase, Na^+/H^+ exchange, and Cl^-/HCO_3^- exchange are known to be present. CO_2, entering the cell by diffusion or formed by metabolism, is converted to H_2CO_3 by the action of carbonic anhydrase. Reaction of H_2CO_3 with OH^- produced by H^+ secretion gives HCO_3^-, which is exchanged at the basolateral membrane for Cl^-, a sequence that serves both to control pH and to supply Cl^- for HCl production (Muallem *et al.*, 1988).

Muallem *et al.* (1988) have provided evidence that an initial increase in intracellular pH [pH_i] caused by stimulation of the parietal cell—an increase that *precedes* acid secretion—is associated with a cAMP and [Ca_i] stimulation of basolateral Na^+/H^+ exchange, an exchange that provides Na^+ for Na^+/K^+-ATPase. Activation of this enzyme is necessary to maintain intracellular K^+ since stimulation of the cell activates basolateral and apical channels for K^+ efflux. The increase in HCO_3^- resulting from Na^+/K^+ exchange also stimulates the Cl^-/HCO_3^- exchanger, providing Cl^- for acid secretion at the apical membrane. Thus, activation and regulation of these several basolateral exchange mechanisms are critical to the initiation and maintenance of apical HCl secretion.

3.9 Glycine- and GABA-Gated Chloride Channels

3.9.1 Chloride Channels and Neural Hyperpolarization

In the resting neuron of the central nervous system (CNS), the Na^+/K^+ pump and K^+/Na^+ leak channels combine to provide an intracellular potential of about $-65\,mV$. The nerve action potential is initiated by a rapid increase in Na^+ permeability, an increase that can be triggered by such stimulatory factors as chemical transmitters or electric charges. The resulting influx of Na^+ depolarizes the neuron. At about $-45\,mV$, voltage-gated Na^+ channels are activated, and an extremely rapid Na^+ influx ensues to produce the action potential of about $+35\,mV$. Inactivation gates rapidly block further Na^+ influx, and opening of K^+ channels permits rapid repolarization of the neuron [Guyton, 1986 (pp. 546–561)].

From the above very brief description of neural activation, it is evident that any process that makes the neuron more negative, or hyperpolarized, will inhibit the achievement of the action potential. In fact, the less negative potential ($-65\,mV$) of central neurons relative to large peripheral nerve fibers and skeletal muscle fibers (-80 to $-90\,mV$) permits both positive and negative control of neural excitability, and negative control is now recognized as an important aspect of central neural networks, serving to prevent overexcitability of neurons and to "fine tune" the responsiveness of excitatory networks. A major mechanism for this control in the CNS relies on gated Cl^- channels, the opening of which causes an influx of Cl^- along its electrochemical gradient, thus making the interior of the neuron more negative [Guyton, 1986 (pp. 546–561)]. Activation of these Cl^- channels is mediated primarily by glycine in the spinal cord and brain stem and by γ-aminobutyric acid (GABA) in other parts of the brain, acting on postsynaptic glycine and $GABA_A$ receptors, respectively.

Neurons that produce GABA (GABAergic neurons) are widespread in the CNS and, in certain areas of the brain, can make up to 20–40% of all neurons present (reviewed by Krnjević, 1974). Almost without exception, neurons in every part of the CNS are sensitive to the inhibitory action of GABA. The following discussion will concentrate primarily on the GABA-gated channel.

3.9.2 Identification of a GABA/Benzodiazepine Receptor–Chloride Ionophore Complex

The results of extensive electrophysiological studies that identified GABA as the primary inhibitory neurotransmitter in the CNS were confirmed by parallel biochemical studies stimulated by the intimate pharmacological connection between GABA and benzodiazepines, barbiturates, and related compounds (Costa and Guidotti, 1979). Important progress came from studies on the mechanism of action of benzodiazepines, such as diazepam (Valium; Fig. 3-15), which were used extensively in the 1960s to induce sleep and for relief of anxiety, pain, and muscle spasms. Two independent groups demonstrated the presence of high-affinity benzodiazepine binding sites in several species, including humans (Squires and Braestrup, 1977; Mohler and Okada, 1977). A neurochemical link between GABA and benzodiazepine recognition sites was established when Tallman et al. (1978) showed that GABA and the GABAmimetic muscimol (Fig. 3-15) increased the affinity of [^{3}H]diazepam for tissue benzodiazepine receptors. Shortly thereafter, Costa et al. (1979) reported that Br^-, I^-, and Cl^- also increased the apparent affinity of [^{3}H]diazepam. Barbiturates such as phenobarbital (Fig. 3-15) also enhance both benzodiazepine and GABA binding, and this enhancement requires the presence of Cl^-. Further, GABAmimetics and barbiturates increase the efflux and uptake of $^{36}Cl^-$ in synaptoneurosomes, a clear demonstration of the functional link between the GABA/benzodiazepine receptor complex and the chloride ionophore (Schwartz et al., 1986). These and several other studies provided evidence

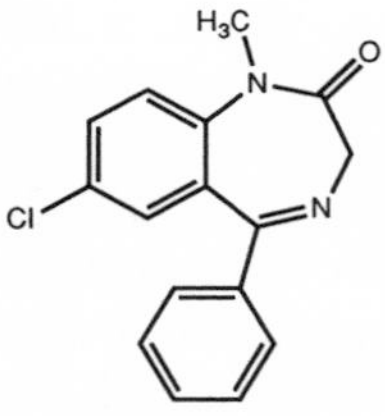

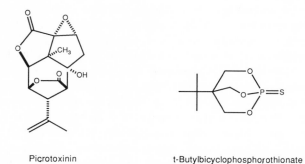

Diazepam

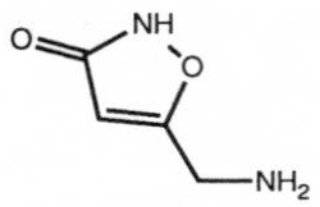

Muscimol

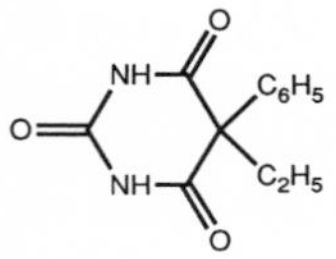

Phenobarbital

Figure 3-15. Structures of diazepam (Valium), muscimol, and phenobarbital.

that recognition sites for benzodiazepines and GABA are on the same molecule and are linked functionally to a Cl^- ionophore (reviewed by Skolnick *et al.*, 1987).

Cage convulsants, such as picrotoxinin and *t*-butylbicyclophosphorothionate (TBS) (Fig. 3-16), exert their action by binding to anion recognition sites on the Cl^- ionophore, thus blocking hyperpolarization of the neuron. Binding is dependent on the presence of Cl^- or other permeable anion, and Havoundjian *et al.* (1986) found close correlation between the ability of an anion to enhance [^{35}S]TBS binding and its

Picrotoxinin   t-Butylbicyclophosphorothionate

Figure 3-16. Structures of cage convulsants picrotoxinin and *t*-butylbicyclophosphorothionate (TBS).

permeability through the channel. The toxic effects of certain insecticides (e.g., lindane) occur through a similar mechanism (see Chapter 8, Section 8.2.2.2). Binding of barbiturates also occurs near the luminal surface of the ionophore (reviewed by Skolnick and Paul, 1988). Thus, four types of receptors have been demonstrated on the GABA receptor complex—(1) the $GABA_A$ receptor, (2) receptors for picrotoxinin and other cage convulsants, (3) a receptor for barbiturates, and (4) the anion recognition site. Ligands for these receptors allosterically modulate each other's binding, a fact that has helped unravel the molecular mechanisms of GABAergic activity and which has prompted intense interest in further development of useful medicinal agents based on modulation of GABAergic activity (for a recent review of pharmacology of the GABA/benzodiazepine/receptor–Cl^- ionophore complex, a discussion of which is beyond the scope of this chapter, see Bruun-Meyer, 1987). A comparison of the binding characteristics of [^3H]strychnine and [^3H]TBS binding to glycine and GABA-gated chloride channels, respectively, revealed several similarities, suggesting that these ligands bind to similar sites on the receptor complexes (Marvizón and Skolnick, 1988).

3.9.3 Structure of the GABA/Benzodiazepine Receptor–Chloride Ionophore Complex

Binding of GABA to the $GABA_A$ receptor opens the Cl^- channel, an effect that is enhanced by occupation of a benzodiazepine receptor located in close proximity to the $GABA_A$ receptor. Isolation and structural studies have led to the proposal of an α_2, β_2 tetrameric structure for the GABA/benzodiazepine receptor–Cl^- ionophore "supramolecular complex" (Nielsen et al., 1985; Schwartz et al., 1985a). Photoaffinity labeling has shown that the benzodiazepine binding site is on the α subunit and the $GABA_A$ receptor is on the β subunit. Several models of this complex have been constructed that schematically depict the four types of binding sites on an α_2, β_2 tetrameric structure (e.g., Olsen, 1981; Skolnick and Paul, 1983; Guidotti et al., 1983; Bruun-Meyer, 1987) (Fig. 3-17). Skolnick et al. (1987) have described the interplay of these components of the supramolecular complex as a "wiring diagram."

The GABA receptor–Cl^- ionophore protein complex has been isolated, antibodies have been raised against it, and cDNA clones have been obtained (Schofield et al., 1987; Olsen et al., 1988; and references therein). Amino acid sequences of both subunits, deduced from cloned cDNAs, revealed four hydrophobic domains (designated M1–M4). In a schematic model for the topology of the $GABA_A$ receptor, these

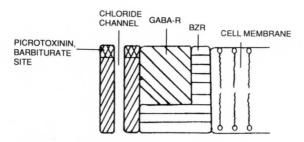

Figure 3-17. Model for the GABA/benzodiazepine receptor–chloride ionophore complex. Reprinted with permission from S. Bruun-Meyer, The GABA/benzodiazepine receptor–chloride ionophore complex: Nature and modulation, *Prog. Neuro-Psychopharmacol. Biol. Psychiat.* 11:365–387. Copyright 1987 Pergamon Press plc.

hydrophobic domains span the membrane and are suggested to stabilize the walls of the channel (Fig. 3-18). Clusters of positively charged side chains found at the ends of the membrane-spanning domains are presumed to form the channel mouth. According to this model, ion flow in the channel is enhanced by the presence of threonine and serine side chains. A speculative gating mechanism was proposed based on the presence of a proline in the M1 hydrophobic domain of each subunit. This residue was postulated to produce a protrusion in the channel in the absence of GABA, thus keeping the channel closed. GABA binding would reposition the M1 domain, opening the channel, and also cause a conformational change that would expose positive-charged residues at the channel mouth, thereby causing ion flux. Such conformational changes would be subject to modulation by the binding of benzodiazepines or barbiturates (Schofield *et al.*, 1987). While such a model is speculative, the authors note that mutational analysis of cloned α- and β-subunit cDNAs will facilitate its evaluation.

Homology that has been shown between sequences in the α and β subunits of the nicotinic acetylcholine receptor, particularly in the extracellular ligand-binding domain, suggest the existence of a "superfamily" of chemically gated ion channel receptors, including both anionic and cationic channel types. The proline residue, implicated in the gating mechanism, is conserved (Schofield *et al.*, 1987).

3.9.4 Physiological Relevance of Modulation of the GABA-Gated Chloride Channel

The anxiolytic effects of benzodiazepines have prompted extensive efforts to define a physiological role for the benzodiazepine receptor,

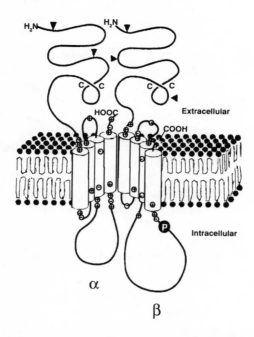

Figure 3-18. A schematic model for the topology of the GABA$_A$ receptor (Schofield *et al.*, 1987). Adapted, with permission, from *Nature* 328:221–227. Copyright 1987 Macmillan Magazines Limited.

although attempts to identify an endogenous ligand have been inconclusive. Skolnick *et al.* (1987) have demonstrated that, in the rat, acute stress causes a rapid and substantial increase in Cl$^-$ conductance through the GABA-gated Cl$^-$ channel, an increase that may be designed to compensate for the effects of stress.

3.10 Stimulation and Inhibition of Enzymes by Chloride

The ability of Cl$^-$ to bind to macromolecular components of cell membranes is apparent from the above discussions. Cl$^-$ can also bind to and affect the activity of several enzymes. Binding occurs at electrophilic centers such as protonated amines or metal centers, causing such consequences as conformational changes or changes in ionization of neighboring acids or bases.

3.10.1 Chloride and the Photosynthetic Formation of Oxygen

Perhaps no other function of Cl^- is more important to life on the planet than its role as an indispensible cofactor in the photosynthetic evolution of oxygen (O_2). An O_2-evolving complex is contained in photosystem II (PS II), one of the two photosystems (PS I and PS II) that perform complementary functions in the series of finely orchestrated events that make up the photosynthetic process. In PS II, light is absorbed by chlorophyll molecules, and electronic energy derived from this process is funneled to a reaction center chlorophyll called P680. Photoexcited P680 (P680*) transfers an electron to a bound porphyrin called pheophytin (ph) to form $P680^+$ and reduced ph and ultimately reduced plastoquinol, a weak reductant. $P680^+$, a strong oxidant, is the electron acceptor during the oxidation of water to O_2. A cluster of four manganese ions in the reaction center is oxidized by $P680^+$ through a series of five oxidation states ($S_0 \rightarrow S_4$) in a sequence characterized by addition of water molecules to the cluster and ultimately by a splitting of O_2 from the complex ($S_4 \rightarrow S_0$) (reviewed by Stryer, 1988).

Cl^- depletion leads to a complete and reversible inhibition of O_2 evolution. Certain anions, especially Br^- and, to a lesser extent, NO_3^-, can substitute for Cl^-, but smaller anions, including OH^- and F^-, competitively inhibit O_2 production in the presence of Cl^-. Recent research into the basis of the Cl^- requirement for O_2 evolution has provided convincing evidence that Cl^- depletion interferes with the manganese oxidation cycle (reviewed by Homann, 1987; Dismukes, 1986). Results from a variety of spectroscopic and other physicochemical studies have shown that oxidation to the S_2 state proceeds normally, but transitions beyond the S_2 state are blocked in the absence of Cl^- (Ono et al., 1987, and references therein). Two models have been advanced that are consistent with much of the available data. According to the first, binding of Cl^- is required to neutralize positive charge accumulation during extraction of electrons in the S state transitions. This neutralization is proposed as a requisite step in the attainment of essential conformations required for further activity (e.g., Johnson et al., 1983; Homann, 1985). In the second model, binding of Cl^- as a ligand—perhaps as a bridging ligand—to the functional pool of manganese is proposed. A bridging ligand function would be consistent with the fact that large, soft nucleophiles such as Cl^-, Br^-, and I^- function as activators, while small, hard nucleophiles such as F^-, OH^-, and amines are inhibitors of O_2 evolution (Sandusky and Yocum, 1986, and references therein).

Consistent with the requirement for Cl^- in O_2 evolution, chloroplasts possess a highly regulated and very efficient Cl^--concentration mechanism

(Robinson and Downton, 1984). A clearer understanding of the Cl^- effect in O_2 evolution undoubtedly will be forthcoming in light of the very active research currently directed to that end.

3.10.2 Activation of α-Amylase

The activation of mammalian α-amylase by Cl^-, known for well over a century, is historicaly the first example of enzyme activation by a modulatory effector ligand. Cl^- binding, associated with suppression of exchange of 26 protons, produces a subtle conformational change that results in a 240-fold increase in amylase–Ca^{2+} binding. It is this conformational change, probably confined to the catalytic center, that produces the 30-fold increase in activity toward starch (Levitzki and Steer, 1974, and references therein). An ε-amino group of lysine, within the substrate binding site, has been implicated as an integral part of the Cl^- binding site (Lifshitz and Levitzki, 1976). Other anions, including Br^- and I^-, can substitute for Cl^-, but their effectiveness decreases as anionic radius increases (Levitzki and Steer, 1974). This order of effectiveness is similar to that seen in activation of the PS II O_2-evolving complex, despite the absence of a redox metal associated with α-amylase. On the other hand, while F^- inhibits O_2 formation, it has no effect on α-amylase.

3.10.3 Inhibition of [Glu¹]Plasminogen Activation

Human [Glu¹]plasminogen ([Glu¹]Pg), a 791-amino acid glycoprotein, is the precursor of the fibrinogenolytic and fibrinolytic enzyme plasmin. [Glu¹]Pg is activated by cleavage of the Arg^{561}–Val^{562} bond, mediated by a variety of proteins such as urokinase (u-PA), streptokinase (SK), and tissue plasminogen activator (t-PA). t-PA, considered the major plasminogen activator in human plasma, has recently been advanced as a clinical agent for clot dissolution (Urano et al., 1988, and references therein).

Activation of plasminogen by u-PA, SK, and t-PA is inhibited by physiological levels of Cl^-. Other anions are also effective, with the effectiveness of inhibition of initial rate of activation of t-PA by various anions showing the rank order $I^- > Cl^- > HCO_3^- > F^- > OAc^-$. The observation that activation by different proteins (u-PA, SK, and t-PA) is inhibited to a comparable degree by Cl^- has led to the proposal that the anion binding site is on $[Glu^1]Pg$ and that this binding leads to a $[Glu^1]Pg$ conformation that has lower affinity for the activating protein. This proposal is supported by several experimental results, including detection of large conformational changes in $[Glu^1]Pg$ induced by Cl^- binding (Urano *et al.*, 1988, and references therein). The demonstration of Cl^- inhibition and fibrin stimulation of $[Glu^1]Pg$ activation in intact plasma emphasizes the importance of consideration of the potential influence that these natural plasma constituents may have during the use of $[Glu^1]Pg$ in thrombolytic therapy (Gaffney *et al.*, 1988).

3.10.4 Stimulation of Angiotensin Converting Enzyme (ACE)

Angiotensin converting enzyme (ACE) is a zinc metalloexopeptidase that has the important function of catalyzing the hydrolysis of the decapeptide angiotensin I to the potent vasoconstricting octapeptide angiotensin II. ACE also catalyzes the inactivation of the vasodilating nonapeptide bradykinin. While the association of lung tissue ACE with blood pressure control is well established, physiological functions of ACE found in a large number of other tissues (for example, eye, brain, pancreas, testis, kidney, stomach, and intestine) have not been determined (Ehlers and Riordan, 1989; Ehlers and Kirsch, 1988; Shapiro *et al.*, 1983; and references therein).

The kinetics and mechanism of ACE-catalyzed hydrolyses have been studied extensively since the initial characterization of the enzyme in 1954, research spurred in part by the recognition that inhibitors of ACE have potential as antihypertensive drugs (for a related approach to hypotensive agents based on inhibition of the angiotensinogen to angiotensin I conversion, see Chapter 2 in vol. 9B of the series). The strong activation of ACE by Cl^-, a characteristic of the enzyme that has been the subject of many investigations, is a complex process dependent on pH, substrate, and species from which the enzyme is obtained. For example, whereas rabbit lung enzyme has an absolute requirement for Cl^-, with many substrates (Shapiro *et al.*, 1983), Cl^- is a nonessential activator in the hydrolysis of angiotensin I by the human enzyme (Ehlers and Kirsch, 1988). Effects of pH on enzyme activity and on Cl^- activation are consistent with the presence of a critical lysine residue at the active site, a residue that appears to be a component of the Cl^--binding site (Shapiro and Riordan, 1983). Ehlers and Kirsch (1988) note that in certain tissues, such as intestinal microvilli, ion fluxes might regulate activity, whereas in other tissues, such an human vascular and renal proximal tubular epithelium, pH and Cl^- concentration are relatively constant and are optimal for ACE activity.

3.10.5 Regulation of GTP-Dependent Regulatory Proteins by Chloride

In view of the attention given in Chapter 2 to the stimulation of GTP-dependent regulatory proteins by F^-, the interaction of Cl^- with this class of protein deserves comment. Cl^- (at physiological concentrations) and Br^- affect the activities of GTP-dependent proteins by increasing the affinity of the α subunit of the protein for guanine nucleotides. Thus, the mode of action is different from that of F^-, and a binding site for anions on the α subunit is implied (Higashijima *et al.*, 1987). The fact that physiological concentrations of Cl^- are effective suggests that this regulatory role may have *in vivo* significance. Indeed, the excitatory amino acid-induced accumulation of cAMP has recently been shown to be absolutely dependent on the presence of Cl^- (or Br^-) in the incubation medium (Baba *et al.*, 1988).

3.11 Chloride and Neutrophil Function

3.11.1 Chloride Transport and Neutrophil Activation

Although the precise mechanism of leukocyte activation is unknown, this activation is accompanied by a variety of ionic translocations. For example, intracellular pH has been proposed to have a regulatory role in several neutrophil functions. Extensive studies of anion transport in human neutrophils have revealed three mechanisms, including a stilbene-insensitive anion exchange mechanism that is responsible for most of the anion movement, a transport dependent on ATPase [no direct evidence for direct Cl^--stimulated ATPase (a Cl^- pump) is given], and a small electro-diffusive flux (Simchowitz, 1988, and references therein). Whereas recovery of neutrophils from acid conditions is dependent on a Na^+/H^+ exchange, the Cl^-/HCO_3^- exchange appears to mediate recovery from imposed alkalinization (Simchowitz and Roos, 1985). Thus, Cl^- may also play additional regulatory roles in neutrophil activation.

3.11.2 Role of Chloride in the Microbicidal Action of Neutrophils

As discussed in Chapter 2 (Section 2.13.2), neutrophils provide an important mechanism against bacterial infections. Bacteria that encounter these activated leukocytes lose their ability to divide within minutes, well before lysosomal enzymes effect cellular digestion. As part of the diverse set of potentially lethal processes accompanying phagocytosis, phagocytosing neutrophils discharge the heme enzyme myeloperoxidase (MPO) into the phagocytic vacuole or into the extracellular medium. MPO catalyzes the oxidation by hydrogen peroxide (H_2O_2) of Cl^-, Br^-, and I^- (but not F^-) to hypohalous acids (HOX), products that have been implicated in the microbicidal action of neutrophils (Fig. 3-19) (Weiss, 1983; Albrich et al., 1981; and references therein). The role of MPO and other peroxidases in biohalogenation of organic substrates will be considered in Chapter 7.

Because Cl^- is present in much higher concentrations than Br^- (20–100 μM) or I^- (0.1–0.5 μM), HOCl has been considered a likely important microbicidal product, although, as will be seen in Chapter 4, oxidation of Br^- by eosinophils has recently been assigned an important defensive role. Albrich et al. (1981) have examined the reactions of several biological compounds with HOCl in order to define more precisely the molecular event(s) that lead to the death of the microorganism. Several critical biological components were affected by HOCl under conditions that mimicked phagosome oxidation, including, in decreasing order of sensitivity, iron–sulfur proteins, carotenes, nucleotides, porphyrins, and hemoproteins. Furthermore, exposure of bacteria to exogenous HOCl leads to irreversible oxidation of cytochromes, nucleotides, and carotene pigments, accompanied by loss of aerobic respiration. Based on these data, HOCl was proposed as the primary microbicidal agent produced by activated neutrophils. Toxicity would result from interruption of electron transport chains and destruction of adenine nucleotide pools.

3.11.3 Chloride and the Pathogenesis of Inflammation

In addition to playing important roles in host defense mechanisms against invading microbes, neutrophils also can contribute to tissue injury under certain pathological conditions. Collagenase, a lysosomal collagenolytic enzyme, is released in an inactive form by triggered neutrophils that accumulate at areas of acute inflammation. Weiss et al. (1985) have shown that activation of collagenase is mediated by HOCl, providing a mechanism by which neutrophils can harness HOCl reactivity through coupling with a degradative enzyme. The process of HOCl-mediated collagenase activation is not known.

Another latent collagenolytic enzyme, gelatinase, is also released independently by triggered neutrophils. However, activation of gelatinase is

$$H_2O_2 + Cl^- + H^+ \longrightarrow HOCl + H_2O$$

Figure 3-19. Biosynthesis of HOCl.

accomplished by two processes, one mediated by HOCl and the other by respiratory burst reactions. Roles for gelatinase have been proposed in several processes that include, in addition to inflammation, diapedesis (a process by which monocytes and neutrophils squeeze through pores of blood vessels), chemotaxis, and wound repair (Peppin and Weiss, 1986).

3.12 Biochemistry of Chloride—Summary

The list of physiological functions in which Cl^- is important obviously is quite long. The examples of Cl^- transport discussed in this chapter were chosen to illustrate various transport mechanisms and functional roles of this transport, but many other examples could have been selected. Research on anion transport in plasma membranes of several vertebrate cells— including lymphocytes, glial cells, Ehrlich ascites tumor cells, hepatocytes, and mammalian muscles—has been reviewed by Hoffmann (1986). Many additional examples of secretory and absorptive functions of epithelia involving Cl^- transport could be given. For example, a recent report describes a 4,4'-diisothiocyanostilbene-2,2'-disulfonic acid (DIDS)-sensitive Cl^-/HCO_3^- exchanger in human placental epithelial cells, possibly responsible for mediating transplacental transfer of CO_2 from the fetus to the mother (Grassl, 1989). The recent description of a histamine-gated Cl^- channel involved with neurotransmission in a photoreceptor (blowfly) synapse provides an additional example of a synaptic receptor functioning as a ligand-gated Cl^- channel (Hardie, 1989).

The abundance of Cl^- in vertebrate extracellular fluid, a reflection of the composition of seawater at the time animals with closed circulations evolved, has mandated that organisms develop mechanisms for transport and metabolism of this ion. Indeed, it is clear that many of these mechanisms are exquisitely fitted to fulfill specific and vital functions.

References

Albrich, J. M., McCarthy, C. A., and Hurst, J. K., 1981. Biological reactivity of hypochlorous acid: Implications for microbicidal mechanisms of leukocyte myeloperoxidase, *Proc. Natl. Acad. Sci. USA* 78:210–214.

Baba, A., Nishiuchi, Y., Uemura, A., and Iwata, H., 1988. Mechanism of excitatory amino acid-induced accumulation of cyclic AMP in hippocampal slices: Role of extracellular chloride, *J. Pharmacol. Exp. Ther.* 245:299–304.

Brautigan, D. L., 1988. Molecular defects in ion channel regulation in cystic fibrosis predicted from analysis of protein phosphorylation/dephosphorylation, *Int. J. Biochem.* 20:745–752.

Bruun-Meyer, S., 1987. The GABA/benzodiazepine receptor–chloride ionophore complex: Nature and modulation, *Prog. Neuro-Psychopharmacol. Biol. Psychiat.* 11:365–387.

Cabantchik, Z. I., and Rothstein, A., 1974. Membrane proteins related to anion permeability of human red blood cells. I. Localization of disulfonic stilbene binding sites in proteins involved in permeation, *J. Membr. Biol.* 15:207–226.

Cabantchik, Z. I., Knauf, P. A., and Rothstein, A., 1978. The anion transport system of the red blood cell. The role of membrane protein evaluated by the use of "probes," *Biochim. Biophys. Acta* 515:239–302.

Chen, P.-Y., and Verkman, A. S., 1988. Sodium-dependent chloride transport in basolateral membrane vesicles isolated from rabbit proximal tubule, *Biochemistry* 27:655–660.

Chen, P.-Y., Illsley, N. P., and Verkman, A. S., 1988. Renal brush-border chloride transport mechanisms characterized using a fluorescent indicator, *Am. J. Physiol.* 254:F114–F120.

Costa, E., and Guidotti, A., 1979. Molecular mechanisms in the receptor action of benzodiazepines, *Annu. Rev. Pharmacol. Toxicol.* 19:531–545.

Costa, T., Rodbard, D., and Pert, C. B., 1979. Is the benzodiazepine receptor coupled to a chloride ion channel? *Nature* 277:315–317.

Cotton, C. U., Stutts, M. J., Knowles, M. R., Gatzy, J. T., and Boucher, R. C., 1987. Abnormal apical cell membrane in cystic fibrosis respiratory epithelium: An in vitro electrophysiological analysis, *J. Clin. Invest.* 79:80–85.

Dalmark, M., 1976. Effects of halides and bicarbonate on chloride transport in human red blood cells, *J. Gen. Physiol.* 67:223–234.

de Rouffignac, C., and Elalouf, J.-M., 1988. Hormonal regulation of chloride transport in the proximal and distal nephron, *Annu. Rev. Physiol.* 50:123–140.

Dismukes, G. C., 1986. The metal centers of the photosynthetic oxygen-evolving complex, *Photochem. Photobiol.* 43:99–115.

Dix, J. A., Verkman, A. S., and Solomon, A. K., 1986. Binding of chloride and a disulfonic stilbene transport inhibitor to red cell band 3, *J. Membr. Biol.* 89:211–223.

Ehlers, M. R. W., and Kirsch, R. E., 1988. Catalysis of angiotensin I hydrolysis by human angiotensin-converting enzyme: Effect of chloride and pH, *Biochemistry* 27:5538–5544.

Ehlers, M. R. W., and Riordan, J. F., 1989. Angiotensin-converting enzyme: New concepts concerning its biological role, *Biochemistry* 28:5311–5318.

Epstein, F. H., and Silva, P., 1985. Na–K–Cl cotransport in chloride-transporting epithelia, *Ann. N.Y. Acad. Sci.* 456:187–197.

Foskett, J. K., and Melvin, J. E., 1989. Activation of salivary secretion: Coupling of cell volume and $[Ca^{2+}]_i$ in single cells, *Science* 244:1582–1585.

Frizzell, R. A., 1988. Role of absorptive and secretory processes in hydration of the airway surface, *Am. Rev. Respir. Dis.* 138:S3–S6.

Frizzell, R. A., Rechkemmer, G., and Shoemaker, R. L., 1986. Altered regulation of airway epithelial cell chloride channels in cystic fibrosis, *Science* 233:558–560.

Frölich, O., 1982. The external anion binding site of the human erythrocyte anion transporter: DNDS binding and competition with chloride, *J. Membr. Biol.* 65:111–123.

Furuya, W., Tarshis, T., Law, F.-Y., and Knauf, P. A., 1984. Transmembrane effects of intracellular chloride on the inhibitory potency of extracellular H_2DIDS. Evidence for two conformations of the transport site of the human erythrocyte anion exchange protein, *J. Gen. Physiol.* 83:657–681.

Gaffney, P. J., Urano, T., de Serrano, V. S., Mahmoud-Alexandroni, M., Metzger, A. R., and Castellino, F. J., 1988. Roles for chloride ion and fibrinogen in the activation of [Glu[1]]plasminogen in human plasma, *Proc. Natl. Acad. Sci. USA* 85:3595–3598.

Geck, P., and Pfeiffer, B., 1985. $Na^+ + K^+ + 2Cl^-$ cotransport in animal cells—its role in volume regulation, *Ann. N.Y. Acad. Sci.* 456:166–182.

Gerencser, G. A., White, J. F., Gradmann, D., and Bonting, S. L., 1988. Is there a Cl^- pump? *Am. J. Physiol.* 255:R677–R692.

Grassl, S. M., 1989. Cl/HCO$_3$ exchange in human placental brush border membrane vesicles, *J. Biol. Chem.* 264:11103–11106.

Greger, R., 1985. Ion transport mechanisms in the thick ascending limb of Henle's loop of the mammalian nephron, *Physiol. Rev.* 65:760–797.

Greger, R., 1988. Chloride transport in thick ascending limb, distal convolution, and collecting duct, *Annu. Rev. Physiol.* 50:111–122.

Griepp, E. B., and Robbins, E. S., 1988. Epithelium, in *Cell and Tissue Biology. A Textbook of Histology*, 6th ed. (L. Weiss, ed.), Urban and Schwartzenberg, Baltimore, pp. 115–153.

Guidotti, A., Corda, M. G., Wise, B. C., Vaccarino, F., and Costa, E., 1983. GABAergic synapses: Supramolecular organization and biochemical regulation, *Neuropharmacology* 22:1471–1479.

Gunn, R. B., and Frölich, O., 1979. Asymmetry in the mechanism for anion exchange in human red blood cell membranes. Evidence for reciprocating sites that react with one transported ion at a time, *J. Gen. Physiol.* 74:351–374.

Gunn, R. B., and Frölich, O., 1982. Arguments in support of a single transport site on each anion transporter in human red cells, in *Chloride Transport in Biological Membranes* (J. A. Zadunaisky, ed.), Academic Press, New York, pp. 33–59.

Guyton, A. C., 1986. *Textbook of Medical Physiology*, 7th ed., W. B. Saunders Co., Philadelphia.

Hardie, R. C., 1989. A histamine–activated chloride channel involved in neurotransmission at a photoreceptor synapse, *Nature* 339:704–706.

Havoundjian, H., Paul, S. M., and Skolnick, P., 1986. The permeability of γ-aminobutyric acid-gated chloride channels is described by the binding of a "cage" convulsant, *t*-butylbicyclophosphoro[^{35}S]thionate, *Proc. Natl. Acad. Sci. USA* 83:9241–9244.

Hebert, S. C., and Andreoli, T. E., 1984. Control of NaCl transport in the thick ascending limb, *Am. J. Physiol.* 246:F745–F756.

Higashijima, T., Ferguson, K. M., and Sternweis, P. C., 1987. Regulation of hormone-sensitive GTP-dependent regulatory proteins by chloride, *J. Biol. Chem.* 262:3597–3602.

Hoffmann, E. K., 1986. Anion transport systems in the plasma membrane of vertebrate cells, *Biochim. Biophys. Acta* 864:1–31.

Homann, P. H., 1985. The association of functional anions with the oxygen-evolving center in chloroplasts, *Biochim. Biophys. Acta* 809:311–319.

Homann, P. H., 1987. The relations between chloride, calcium, and polypeptide requirements of photosynthetic water oxidation, *J. Bioenerg. Biomembr.* 19:105–123.

Hwang, T.-C., Luo, L., Zeitlin, P. L., Greunert, D. C., Huganir, R., and Guggino, W. B., 1989. Cl$^-$ channels in CF: Lack of activation by protein kinase C and cAMP-dependent protein kinase, *Science* 244:1351–1353.

Jennings, M. L., 1982. Stoichiometry of a half-turnover of band 3, the chloride transport protein of human erythrocytes, *J. Gen. Physiol.* 79:169–185.

Jennings, M. L., 1985. Kinetics and mechanism of anion transport in red blood cells, *Annu. Rev. Physiol.* 47:519–533.

Jetten, A. M., Yankaskas, J. R., Stutts, M. J., Willumsen, N. J., and Boucher, R. C., 1989. Persistence of abnormal chloride conductance regulation in transformed cystic fibrosis epithelia, *Science* 244:1472–1475.

Johnson, J. D., Pfister, V. R., and Homann, P. J., 1983. Metastable proton pools in thylakoids and their importance for the stability of photosystem-II, *Biochim. Biophys. Acta* 723:256–265.

Karniski, L. P., and Aronson, P. S., 1985. Chloride/formate exchange with formic acid recycling: A mechanism of active chloride transport across epithelial membranes, *Proc. Natl. Acad. Sci. USA* 82:6362–6365.

Kerem, B., Rommens, J. M., Buchanan, J. A., Markiewicz, D., Cox, T. K., Chakravarti, A., Buchwald, M., and Tsui, L.-C., 1989. Identification of the cystic fibrosis gene: Genetic analysis, *Science* 245:1073–1080.

Kinne, R., Hannafin, J. A., and König, B., 1985. Role of the NaCl–KCl cotransport system in active chloride absorption and secretion, *Ann. N.Y. Acad. Sci.* 456:198–206.

Knauf, P. A., 1979. Erythrocyte anion exchange and the band 3 protein: Transport kinetics and molecular structure, *Curr. Top. Membr. Transp.* 12:249–363.

Knauf, P. A., and Mann, N. A., 1984. Use of niflumic acid to determine the nature of the asymmetry of the human erythrocyte anion exchange system, *J. Gen. Physiol.* 83:703–725.

Knauf, P. A., and Mann, N. A., 1986. Location of the chloride self-inhibitory site of the human erythrocyte anion exchange system, *Am. J. Physiol.* 251:C1–C9.

Knauf, P. A., Ship, S., Breuer, W., McCulloch, L., and Rothstein, A., 1978. Asymmetry of the red cell anion exchange system. Different mechanisms of reversible inhibition by *N*-(4-azido-2-nitrophenyl)-2-aminosulfonate (NAP-taurine) at the inside and outside of the membrane, *J. Gen. Physiol.* 72:607–630.

Knowles, M. R., Stutts, M. J., Spock, A., Fischer, N., Gatzy, T. J., and Boucher, R. C., 1983. Abnormal ion permeation through cystic fibrosis respiratory epithelium, *Science* 221:1067–1070.

Kopita, R. R., and Lodish, H. F., 1985. Primary structure and transmembrane orientation of the murine anion exchange protein, *Nature* 316:234–238.

Kregenow, F. M., 1981. Osmoregulatory salt transporting mechanisms: Control of cell volume in anisotonic media, *Annu. Rev. Physiol.* 43:493–505.

Krnjević, K., 1974. Chemical nature of synaptic transmission in vertebrates, *Physiol. Rev.* 54:418–540.

Landry, D. W., Akabas, M. H., Redhead, C., Edelman, A., Cragoe, E. J., Jr., and Al-Awqati, Q., 1989. Purification and reconstitution of chloride channels from kidney and trachea, *Science* 244:1469–1472.

Levitan, I. B., 1989. The basic defect in cystic fibrosis, *Science* 244:1423.

Levitzki, A., and Steer, M. L., 1974. The allosteric activation of mammalian α-amylase by chloride, *Eur. J. Biochem.* 41:171–180.

Li, M., McCann, J. D., Anderson, M. P., Clancy, J. P., Liedtke, C. M., Nairn, A. C., Greengard, P., and Welsh, M. J., 1989. Regulation of chloride channels by protein kinase C in normal and cystic fibrosis airway epithelia, *Science* 244:1353–1356.

Lifshitz, R., and Levitzki, A., 1976. Identity and properties of the chloride effector binding site in hog pancreatic α-amylase, *Biochemistry* 15:1987–1993.

Lowe, A. G., and Lambert, A., 1983. Chloride–bicarbonate exchange and related transport processes, *Biochim. Biophys. Acta* 694:353–374.

Macara, I. G., and Cantley, L. C., 1981. Mechanism of anion exchange across the red cell membrane by band 3: Interactions between stilbenedisulfonate and NAP-taurine binding sites, *Biochemistry* 20:5695–5701.

Marvizón, J. C. G., and Skolnick, P., 1988. Enhancement of *t*-[^{35}S]butylbicyclophosphorothionate and [^{3}H]strychnine binding by monovalent anions reveals similarities between γ-aminobutyric acid- and glycine-gated chloride channels, *J. Neurochem.* 50:1632–1639.

McPherson, M. A., Shori, D. K., and Dormer, R. L., 1988. Defective regulation of apical membrane chloride transport and exocytosis in cystic fibrosis, *Biosci. Rep.* 8:27–33.

Mohler, H., and Okada, T., 1977. Benzodiazepine receptor: Demonstration in the central nervous system, *Science* 198:849–851.

Muallem, S., Blissard, D., Cragoe, E. J., and Sachs, G., 1988. Activation of the Na^+/H^+ and Cl^-/HCO_3^- exchange by stimulation of acid secretion in the parietal cell, *J. Biol. Chem.* 263:14703–14711.

Nielsen, M., Honore, T., and Braestrup, C., 1985. Radiation inactivation of brain (^{36}S)t-butylbicyclophosphorothionate binding sites reveals complicated molecular arrangements of the GABA/benzodiazepine receptor chloride channel complex, *Biochem. Pharmacol.* 34:3633–3642.

O'Grady, S. M., Palfrey, H. S., and Field, M., 1987. Characteristic and functions of the Na–K–Cl cotransport in epithelial cells, *Am. J. Physiol.* 253:C177–C192.

Olsen, R. W., 1981. GABA-benzodiazepine–barbiturate receptor interactions, *J. Neurochem.* 37:1–13.

Olsen, R. W., Bureau, M., Ransom, R. W., Deng, L., Dilber, A., Smith, G., Krestchatisky, M., and Tobin, A. J., 1988. The GABA receptor–chloride ion channel protein complex, *Adv. Exp. Med. Biol.* 236:1–14.

Ono, T.-A., Nakayama, H., Gleiter, H., Inoue, Y., and Kawamori, A., 1987. Modification of the properties of S_2 state in photosynthetic O_2-evolving center by replacement of chloride with other anions, *Arch. Biochem. Biophys.* 256:618–624.

Passow, H., 1986. Molecular aspects of band 3 protein-mediated anion transport across the red blood cell membrane, *Rev. Physiol. Biochem. Pharmacol.* 103:61–203.

Passow, H., Fasold, H., Gartner, E. M., Legrum, B., Ruffing, W., and Zaki, L., 1980. Anion transport across the red blood cell membrane and the conformation of the protein in band 3, *Ann. N.Y. Acad. Sci.* 341:361–383.

Peppin, G. J., and Weiss, S. J., 1986. Activation of the endogenous metalloproteinase, gelatinase, by triggered human neutrophils, *Proc. Natl. Acad. Sci. USA* 83:4322–4326.

Perez, A., Blissard, D., Sachs, G., and Hersey, S. J., 1989. Evidence for a chloride conductance in secretory membrane of parietal cells, *Am. J. Physiol.* 256:G299–G305.

Porth, C. M., 1986. *Pathophysiology* (C. M. Porth, ed.), J. B. Lippincott Co., Philadelphia, pp. 411–439.

Quinton, P. M., 1983. Chloride impermeability in cystic fibrosis, *Nature* 301:421–422.

Rechkemmer, G. R., 1988. The molecular biology of chloride secretion in epithelia, *Am. Rev. Respir. Dis.* 138:S7–S9.

Riordan, J. R., Rommens, J. M., Kerem, B., Alon, N., Rozmahel, R., Grselczak, Z., Zielenski, J., Lok, S., Plavsic, N., Chou, J.-L., Drumm, M. L., Iannuzzi, M. C., Collins, F. S., and Tsui, L.-C., 1989. Identification of the cystic fibrosis gene: Cloning and characterization of complementary DNA, *Science* 245:1066–1073.

Robinson, S. P., and Downton, W. J. S., 1984. Potassium, sodium, and chloride content of isolated intact chloroplasts in relation to ionic compartmentation in leaves, *Arch. Biochem. Biophys.* 228:197–206.

Rommens, J. M., Iannuzzi, M. C., Kerem, B., Drumm, M. I., Melmer, G., Dean, M., Rozmahel, R., Cole, J. L., Kennedy, D., Hidaka, N., Zsiga, M., Buchwald, M., Riordan, J. R., Tsui, L.-C., and Collins, F. S., 1989. Identification of the cystic fibrosis gene: Chromosome walking and jumping, *Science* 245:1059–1065.

Roos, A., and Boron, W. F., 1981. Intracellular pH, *Physiol. Rev.* 61:296–434.

Sachs, G., 1986. The parietal cell as a therapeutic target, *Scand. J. Gastroenterol., Suppl.* 118:1–10.

Sachs, G., Muallem, S., and Hersey, S. J., 1988. Passive and active transport in the parietal cell, *Comp. Biochem. Physiol.* 90A:727–731.

Sandusky, P. O., and Yocum, C. F., 1986. The chloride requirement for photosynthetic oxygen evolution: Factors affecting nucleophilic displacement of chloride from the oxygen-evolving complex, *Biochim. Biophys. Acta* 849:85–93.

Schild, L., Giebish, G., and Green, R., 1988. Chloride transport in the proximal renal tubule, *Annu. Rev. Physiol.* 50:97–110.

Schofield, P. R., Darlison, M. G., Fujita, N., Burt, D. R., Stephenson, F. A., Rodriguez, H., Rhee, L. M., Ramachandran, J., Reale, V., Glencorse, T. A., Seeburg, P. H., and Barnard,

E. A., 1987. Sequence and functional expression of the GABA$_A$ receptor shows a ligand-gated receptor super-family, *Nature* 328:221–227.

Schoumacher, R. A., Shoemaker, R. L., Halm, D. R., Tallant, E. A., Wallace, R. W., and Frizzel, R. A., 1987. Phosphorylation fails to activate chloride channels from cystic fibrosis airway cells, *Nature* 330:752–754.

Schwartz, R. D., Thomas, J. W., Kempner, E. S., Skolnick, P., and Paul, S. M., 1985a. Radiation inactivation studies of the benzodiazepine/γ-aminobutyric acid/chloride ionophore complex, *J. Neurochem.* 45:108–115.

Schwartz, R., Skolnick, P., Seale, T., and Paul, S. M., 1986. Demonstration of GABA/barbiturate-receptor-mediated chloride transport in rat brain synaptoneurosomes: A functional assay of GABA receptor–effector coupling, *Adv. Biochem. Psychopharmacol.* 41:33–49.

Shapiro, R., and Riordan, J. F., 1983. Critical lysine residue at the chloride binding site of angiotensin converting enzyme, *Biochemistry* 22:5315–5321.

Shapiro, R., Holmquist, B., and Riordan, J. F., 1983. Anion activation of angiotensin converting enzyme: Dependence on nature of substrate, *Biochemistry* 22:3850–3857.

Simchowitz, L., 1988. Interactions of bromide, iodide, and fluoride with the pathways of chloride transport and diffusion in human neutrophils, *J. Gen. Physiol.* 91:835–860.

Simchowitz, L., and Roos, A., 1985. Regulation of intracellular pH in human neutrophils, *J. Gen. Physiol.* 85:443–470.

Skolnick, P., and Paul, S., 1983. New concepts in the neurobiology of anxiety, *J. Clin. Psychiat.* 44:12–19.

Skolnick, P., and Paul, S., 1988. The benzodiazepine/GABA receptor chloride channel complex, *ISI Atlas of Science* 2:19–22.

Skolnick, P., Havoundjian, H., and Paul, S. M., 1987. Modulation of the benzodiazepine-GABA receptor chloride ionophore complex by multiple allosteric sites: Evidence for a barbiturate "receptor," in *Clinical Pharmacology in Psychiatry* (Psychopharmacology Series 3) (S. G. Dahl, L. F. Gram, S. M. Paul, and W. Z. Potter, eds.), Springer-Verlag, Berlin, pp. 29–36.

Squires, R. F., and Braestrup, C., 1977. Benzodiazepine receptors in rat brain, *Nature* 266:732–734.

Steck, T. L., 1978. The band 3 protein of the human red cell membrane: A review, *J. Supramolec. Struct.* 8:311–324.

Stryer, L., 1988. *Biochemistry*, 3rd ed., W. H. Freeman and Co., pp. 517–540.

Tallman, J. F., Thomas, J. W., and Gallager, D. W., 1978. GABAergic modulation of benzodiazepine binding site sensitivity, *Nature* 274:383–385.

Urano, T., de Serrano, V. S., Gaffney, P. J., and Castellino, F. J., 1988. Effectors of the activation of human [Glu^1]plasminogen by human tissue plasminogen activator, *Biochemistry* 27:6522–6528.

Verkman, A. S., Dix, J. A., and Solomon, A. K., 1983. Anion transport inhibitor binding to band 3 in red blood cell membranes, *J. Gen. Physiol.* 81:421–449.

Verkman, A. S., Takla, R., Sefton, B., Basbaum, C., and Widdicombe, J. H., 1989. Quantitative fluorescence measurement of chloride transport mechanisms in phospholipid vesicles, *Biochemistry* 28:4240–4244.

Wangemann, P., Wittner, M., Di Stefano, A., Englert, H. C., Lang, H. J., Schlatter, E., and Greger, R., 1986. Cl$^-$ channel blockers in the thick ascending limb of the loop of Henle. Structure activity relationship, *Pflügers Arch. Eur. J. Phys.* 407(S2):S128–S141.

Weiss, S. J., 1983. Oxygen as a weapon in the phagocyte armamentarium, in *Handbook of Inflammation*, Vol. 4 (P. A. Ward, ed.), Elsevier, Amsterdam, pp. 37–87.

Weiss, S. J., Peppin, G., Ortiz, X., Ragsdale, C., and Test, S. T., 1985. Oxidative autoactivation of latent collagenase by human neutrophils, *Science* 227:747–749.

Welsh, M. J., 1986. An apical-membrane chloride channel in human tracheal epithelium, *Science* 232:1648–1650.

Welsh, M. J., and Liedtke, C. M., 1986. Chloride and potassium channels in cystic fibrosis airway epithelia, *Nature* 322:467–470.

Widdicombe, J. H., Welsh, M. J., and Finkbeiner, W. E., 1985. Cystic fibrosis decreases the apical membrane chloride permeability of monolayers cultured from cells of tracheal epithelium, *Proc. Natl. Acad. Sci. USA* 82:6167–6171.

Willumsen, N. J., and Boucher, R. C., 1989. Activation of an apical Cl^- conductance by Ca^{2+} ionophores in cystic fibrosis airway epithelia, *Am. J. Physiol.* 256:C226–C233.

Wright, E. M., and Diamond, J. M., 1977. Anion selectivity in biological systems, *Physiol. Rev.* 57:109–156.

Biochemistry of Inorganic Bromide

4.1 Introduction

Bromine is recognized as the most abundant and ubiquitous of trace elements. Despite this, essential roles in plants, microorganisms, or animals have been difficult to demonstrate (Nielsen, 1986). Early interest in the biochemistry of bromide (Br^-) stemmed from the use of bromides as sedatives and anticonvulsants, a use introduced in 1857. Toxicity associated with Br^- ingestion through use of over-the-counter bromine-containing drugs—a medical problem that, though rare, still persists—and the recognition of the presence of increased concentrations of Br^- in food and water due to the use of brominated pesticides and postharvest fumigants are among factors that have caused interest in the biochemistry, pharmacology, and toxicology of Br^- to be maintained.

The first identified physiological function of Br^- in humans appears to be a role in defense mechanisms against parasites mediated by the recently demonstrated preferential oxidation of Br^- by eosinophil peroxidase (Weiss *et al.*, 1986) (Section 4.6). Nevertheless, Br^- can replace Cl^- as a substrate for many biochemical processes, including transport mechanisms and enzyme activation or inhibition, often with equal or better avidity. Accordingly, the literature contains numerous biochemical studies in which the effectiveness of Br^-, and other anions, as a replacement for Cl^- has been examined. Although such competition must also occur *in vivo*, the much lower physiological concentration of Br^- relative to Cl^-—for example, 20–100 μM versus ~ 100 mM in human extracellular fluid—renders physiological relevance unlikely.

In this chapter, the uptake, biodistribution, and metabolism of Br^- will be reviewed briefly. Mechanisms for sedative actions and toxicity of Br^- will be discussed. A limited number of examples of the interaction of Br^- with enzymes and anion transport mechanisms will be given to illustrate the information that can be obtained based on selectivities of anion recognition sites. The role of Br^- in eosinophil-mediated responses to parasitic infections will be discussed. Peroxidase-catalyzed brominations of organic substrates will be discussed in Chapter 7.

4.2 Occurrence and Biodistribution of Bromide

Inorganic salts of Br^- are widely distributed in nature, particularly in seawater (65–67 mg/kg)—reported to contain as much as 99% of the world supply of Br^-—and in salt lakes. Soil and rocks normally contain 2.5–10 mg/kg, but levels vary over a wider range depending on the origin of the mineral (reviewed by van Leeuwen and Sangster, 1987).

Human dietary intake of Br^- is substantial. For example, 24 mg/day of Br^- was found in U.S. diets, averaged over two years, while an average intake of 8.4 mg/day was determined for English diets (reviewed by Nielsen, 1986). Ingested Br^- is readily absorbed, becomes widely distributed in body tissue (Bowen, 1966), and crosses the blood–brain barrier (reviewed by Rauws, 1983). The similar biodistribution of Br^- and Cl^- has led to the use of radiolabeled Br^- for the determination of extracellular body water. In this procedure, administration of a known concentration of Br^- permits ready calculation of fluid volume (Cheek, 1961), a determination aided also by the relatively long biological half-life of 10–12 days (Hellerstein *et al.*, 1960; Mason, 1936). This long biological half-life is explained by the fact that Br^-, which is excreted primarily in the urine, is reabsorbed preferentially to Cl^- by renal tubular cells. The concentration of Br^- in the serum of healthy individuals is about 6.0 μg/ml (Versieck and Cornelis, 1989).

4.3 Pharmacology and Toxicology of Bromide

4.3.1 Historical Background

Bromides were introduced initially into medicine for the treatment of syphilis and epilepsy (reviewed by van Leeuwen and Sangster, 1987; Mendelsohn, 1980). NaBr was also used in the first part of this century for treatment of opium addiction ("bromide sleep" treatment) (Kolb and Himmelsbach, 1938, and references therein). While Br^- indeed has sedative and anticonvulsant activity, and produced a substantial improvement in the treatment of epileptic patients when introduced in 1857, the margin of safety in the use of Br^- increasingly was recognized as being quite narrow. Thus, treatments led in many cases to toxic manifestations associated with

Br^- overload, and these Br^--based drugs now have been replaced with other more selective anxiolytics such as barbiturates (phenobarbital was introduced for the treatment of seizures in 1912). Despite this, the effects of Br^- in the central nervous system (CNS) have led to abuse of bromine-containing drugs.

4.3.2 "Bromism"—Chronic Bromide Toxicity

In his classic study of Br^- toxicity, Mason (1936) recognized that an increased ingestion of Br^- caused a lower physiological Cl^- concentration, such that the molar sum of Cl^- and Br^- remained essentially constant at 110 mmol/liter. As will be seen below, Br^- is an effective replacement for Cl^- in many transport systems and has similar effects on a number of enzymes. Nonetheless, the several effects of excess Br^- make it clear that Br^- cannot effectively replace Cl^- in many of its important physiological functions.

Bromism is a term used to describe symptoms of chronic Br^- toxicity. The sedative effects of Br^- at elevated but low concentrations are replaced in chronic Br^- intoxication by CNS symptoms that include apathy, a decrease in coordination, memory loss, drowsiness, and emotional disturbances. Tremors, hallucinations, and agitation may be present, and stupor and coma may also occur. A minority of patients also exhibit dermatological problems, including acne and rash. A no-effect level of 4 mg/kg body weight has been determined in normal volunteers, suggesting an acceptable daily intake of Br^- of 0.4 mg/kg, or 24 mg for a 60-kg individual. Thus, normal dietary amounts (see above) leave little leeway for additional intake (van Leeuwen and Sangster, 1987).

4.3.3 Effects of Bromide on the Central Nervous System

The clinical use of Br^- for the treatment of neurological and psychic disorders, together with the symptoms of chronic Br^- toxicity, makes it clear that Br^- can affect CNS function. However, the mechanisms of the CNS effects of Br^- have yet to be elucidated clearly. The role of GABA-gated Cl^- channels in the functioning of inhibitory neurotransmitters has suggested that Br^- might be more efficient than Cl^- in causing neural hyperpolarization. Indeed, Montoya and Riker (1982) found that extracellular Br^- hyperpolarizes nerve cells (frog sympathetic ganglion), similarly to other anticonvulsants. However, their data provided presumptive evidence that this hyperpolarization results from Br^- inhibition of Na^+ influx into the resting cell rather than from a greater Br^- permeability. Because of the transient nature of this hyperpolarization, they questioned its relevance to anticonvulsant activity.

Inhibition by Br^- of carbonic anhydrase and HCO_3^-/ATPase, two enzymes related to Cl^- transport, was proposed as a site of action, since this inhibition might be expected to inferfere with Cl^- transport and thus promote hyperpolarization of central neurons. Hyperpolarization caused by a more rapid passive influx into glial cells of Br^- relative to Cl^- also was suggested (Woodbury and Pippenger, 1982).

4.3.4 Bromide and Thyroid Function

In view of the importance of I^- in thyroid function (Chapters 5 and 6) and the similar behavior of Br^- and I^- in many processes (see below and Chapters 5 and 6), an effect of Br^- overload on thyroid function would not be surprising. A striking physiological consequence of NaBr contained in the diet of rats for 90 days (>1200 mg/kg diet) was an increased activity of the thyroid gland, accompanied by an increase in weight and histopathological changes in the thyroid (van Logten *et al.*, 1974). A decrease in serum thyroxine was also observed (van Leeuwen *et al.*, 1983). Van Leeuwen *et al.* (1988) recently studied the effects of Br^- on active uptake of I^- into the thyroid, effects on oxidation of I^- (required for the iodination of tyrosyl residues; see chapter 6), and other parameters. Their data revealed that I^- uptake is inhibited as is the oxidation of I^- to I_2, resulting in lower incorporation of I_2 into tyrosine residues. The coupling of tyrosine residues to thyronine (Chapter 6) is also inhibited. In the pituitary, increased activity of thyroid-stimulating hormone was also observed, apparently reflecting an attempt to compensate for the decrease in serum thyroid hormones (reviewed by van Leeuwen and Sangster, 1987).

At a Br^- dose of 4–9 mg/day for 12 weeks, healthy female, but not male, human volunteers showed an increase in serum thyroxine and triiodothyronine, though a later, larger study placed these results in question. No increase in thyroid-stimulating hormone (TSH) or other parameters of endocrine function was observed in either sex. Thus, there appears to be little effect of Br^- on thyroid function at the concentrations used in human studies. In fact, Br^- is a poor inhibitor of I^- transport and—on a molar basis—is poorly concentrated in the thyroid. The effects seen with rats presumably reflect the larger doses used (van Leeuwen and Sangster, 1987, and references therein).

4.4 Bromide and Anion Transport Mechanisms

4.4.1 Theoretical Basis for Anion Selectivity

In Chapter 3, reference was made to the existence of a limited number of halide sequences related to selectivities of biological systems. If the energy associated with the binding of an anion to a cationic site is considered simply as the difference between the hydration energy of the anion and the energy of attraction of the anion with the binding site, seven sequences can be predicted: (1) $I^- > Br^- > Cl^- > F^-$; (2) $Br^- > I^- > Cl^- > F^-$; (3) $Br^- > Cl^- > I^- > F^-$; (4) $Cl^- > Br^- > I^- > F^-$; (5) $Cl^- > Br^- > F^- > I^-$; (6) $Cl^- > F^- > Br^- > I^-$; (7) $F^- > Cl^- > Br^- > I^-$ (Wright and Diamond, 1977). All but sequences 6 and 7 have been observed. Sequence 7 is a limiting case in which a very high positive charge is associated with the binding site, such that the hydration energy becomes unimportant and the sequence becomes the inverse order of the ionic radii—the smaller anion having the greater columbic attraction to the positive site. Conversely, sequence 1 corresponds to the case of a weak positive charge associated with the binding site, and anions having weak hydration energies are selectively bound. Differences in the radius, coordination, and charge at the cationic site thus are suggested to contribute to the observed differences in selectivity of natural systems (Wright and Diamond, 1977). It follows that observed selectivities can reveal information on the nature of the anion binding site. Examples will be given below to illustrate this point.

4.4.2 Examples of Selectivity in Halide Transport

The comparable biodistributions of Br^- and Cl^- suggest that most transport systems do not discriminate effectively between these two halides, consistent with the selectivity sequences shown in the previous section. For example, as discussed in Chapter 3, anion binding sites on erythrocyte band 3 protein are relatively nonselective, giving halide sequence 1 ($I^- > Br^- > Cl^- > F^-$), indicative of weak charges at this site. The modifier site shows the same sequence. On the other hand, the order of halide flux at $0\ °C$ is $Cl^- > F^- > Br^- > I^-$, showing that binding to the anion binding site does not determine the rate of transport (reviewed by Cabantchik et al., 1978).

The Na^+–K^+–$2Cl^-$ cotransporter also accepts Br^- as a substrate, but I^- does not support loop diuretic-sensitive anion influx (O'Grady et al., 1987). Thus, in erythrocytes (Palfrey and Greengard, 1981) and in canine kidney cells (McRoberts *et al.*, 1982), the Na^+–Cl^- cotransport systems show a selectivity for halides according to the sequence $Cl^- > Br^- \gg I^-$.

Two apical Cl^- channels of tracheal epithelia having differing halide selectivities—I^- (1.6) > Br^- (1.3) > Cl^- (1) and Cl^- (1) > Br^- (0.4) = I^- (0.4)—have been reported (Frizzell, 1987). On the other hand, apical membrane vesicles prepared from bovine tracheal epithelial cells showed virtually no discrimination among these three halides, perhaps reflecting an average value for the two types of channels described by Frizzell (Fong *et al.*, 1988). The apical membrane Cl^- channels of secretory colonic epithelial cells discriminate among halide anions according to sequence 1 [I^- (1.78) > Br^- (1.40) > Cl^- (1) > F^- (0.37)], indicative of weak interactions between the site and the anion (Halm *et al.*, 1988a, b). In contrast, transepithelial movement has been shown in other secretory epithelial cells to follow the sequence $Cl^- > Br^- \gg I^-$ (canine tracheal epithelium) and $Cl^- > Br^-$ (shark rectal gland). The ability of the secretory cells to discriminate for Cl^- apparently reflects the selectivity of Na^+-coupled uptake for Cl^-, discussed above, operating at the basolateral membrane. Thus, Br^- and I^- are not accumulated in the cell and are not available for exit through the nonselective apical channels.

Other transport mechanisms also show a preference for weakly hydrated halides. Glycine- and GABA-gated anion channels have identical permeability sequences for halides: $I^- > Br^- > Cl^- > F^-$ (Marvizón and Skolnick, 1988). On the other hand, whereas isolated gastric glands use NaCl and NaBr with equal facility, NaI is a significantly poorer substrate for acid production (20%) (Berglindh, 1977).

From these few examples, the ability of Br^- to serve as a substrate for anion transport mechanisms can be seen. In processes in which hydration energy differences among anionic substrates are important, as appears to be the case in many transport systems, Br^- and I^- are the preferred substrates [the bases for anion selectivities encompass other more complex and subtle factors—the review by Wright and Diamond (1977) should be consulted by interested readers].

4.5 Inhibition and Stimulation of Enzymes by Bromide

In Chapter 3, several examples of activation and inhibition of enzymes by Cl^- were given. In most of these examples, Br^- is as effective as, or more effective than, Cl^- in producing the response. Thus, Br^- is only slightly less effective than Cl^- in reactivation of Cl^--depleted photosystem II (PS II) O_2 evolution. This is consistent with the requirement for binding of a univalent anion with optimum charge density to the PS II anion binding site for O_2 production (Damoder *et al.*, 1986). Both F^- and I^- inhibit O_2 production, the former possibly by disconnecting Mn from the PS II centers, and the latter by efficiently donating electrons to PS II and by modifying PS II proteins (Ono *et al.*, 1987).

In another example, the effectiveness of halides in activating α-amylase follows the order $Cl^- \geqslant Br^- > I^- > F^-$ (sequence 4), suggestive of significant charge density at the anion binding site (a protonated ε-amino group of lysine) (Levitzki and Steer, 1974). Likewise, Br^- and Cl^- are equally effective in increasing the affinity of the α subunit of G_0 for GTPγS, while I^- and F^- are without effect (Higashijima *et al.*, 1987). In hippocampal slices, Br^- is also as effective as Cl^- in stimulation of excitatory amino acid-induced cAMP formation (Baba *et al.*, 1988).

A final example will be given in which sequences of halide inhibition have been used to study active-site accessibility. Tyrosinase is a copper-containing mixed-function oxidase that catalyzes the hydroxylation of monophenols (for example, formation of dihydroxyphenylalanine from tyrosine) and also catalyzes the oxidation of catechols to quinones. Halide inhibition of tyrosinases from three different sources revealed markedly different orders of inhibition. Frog epidermis tyrosinase gave, in order of decreasing inhibitory activity, the sequence $I^- > Br^- > Cl^- > F^-$, mushroom tyrosinase gave the sequence $F^- > I^- > Cl^- > Br^-$, while mouse melanoma tyrosinase gave $F^- > Cl^- \gg Br^- > I^-$. These results were interpreted in terms of accessibility of the active site to ligands. Since the strength of the reaction between halides and copper follows the sequence $I^- > Br^- > Cl^- > F^-$, the copper at the active site of frog epidermis tyrosinase may be readily accessible, since this order of inhibition is observed with no apparent effects related to ionic size. In contrast, inhibition of mouse melanoma tyrosinase follows a halide sequence related directly to ionic size, suggesting poor access to the active site by ligands. Mushroom tyrosinase would represent an intermediate situation (Martinez *et al.*, 1986).

4.6 Bromide and Eosinophil Function

4.6.1 Eosinophil Function and Host Defense

Eosinophils, which make up 2–3% of human leukocytes, have an important specialized role in host defense against parasitic infections. As with neutrophils, they can also contribute to host tissue destruction in inflammatory diseases. The cellular mechanisms also are similar in that parasite destruction or host tissue degradation is mediated in part by the ability of the cell to generate reactive oxygen metabolites. Thus, eosinophil peroxidase (EPO) catalyzes the oxidation of many substances, including halides, by H_2O_2, and the combination of EPO, H_2O_2, and halide is toxic to many organisms, including schistosomes, toxoplasmas, trypanosomes, mast cells, and tumor cells (reviewed by Gleich and Loegering, 1984).

4.6.2 Brominating Oxidants from Human Eosinophils

Recent research has revealed that, in contrast to neutrophils, eosinophils produce reactive oxidizing intermediates through EPO-catalyzed selective oxidation of Br^-, despite a concentration of Cl^- at least 1000-fold higher than that of Br^-. For example, human eosinophils triggered in the presence of 100 mM Cl^- and 20–100 μM Br^- oxidized Br^- preferentially to produce an oxidizing species tentatively identified as HOBr. Under the same conditions, neutrophils used myeloperoxidase to generate HOCl, but little, if any, HOBr. HOBr is a powerful oxidant, reactive toward thiols, amines, aromatic compounds, amino acids, and other biomolecules, and thus should be toxic to invading organisms and to host tissue in inflammatory responses. Eosinophils contain large amounts of EPO and coat targets with this protein, indicating the importance of this enzyme in the functioning of these leukocytes. Together, the role of eosinophils in host defense against parasites and the demonstration that EPO has high selectivity for Br^- as a substrate suggest strongly a physiological role for Br^- *in vivo* (Weiss *et al.*, 1986).

4.6.3 Bromide and the Biological Production of Singlet Oxygen

Singlet oxygen $[O_2(^1\Delta_g)]$ is highly reactive toward a number of organic substrates, including structural moieties important in biological molecules. The possibility that toxicity associated with peroxidase/H_2O_2/halide systems might be mediated, in part, by generation of $O_2(^1\Delta_g)$ (Fig. 4-1) has been the subject of much interest for nearly two decades.

$$H_2O_2 + Br^- + H^+ \longrightarrow HOBr + H_2O$$

$$H_2O_2 + HOBr \longrightarrow H_2O + H^+ + Br^- + O_2(^1\Delta_g)$$

Figure 4-1. Mechanism for the eosinophil peroxidase-catalyzed production of singlet oxygen.

Allen *et al.* (1972) first suggested that $O_2(^1\Delta_g)$ might contribute to the bactericidal activity mediated by MPO in neutrophils and reported preliminary evidence for the generation of excited oxygen in neutrophils, based on detection of chemical luminescence associated with phagocytosis. However, attempts to confirm the presence of $O_2(^1\Delta_g)$ in this and later studies were complicated by interference from secondary chemical luminescence and by a lack of specificity of $O_2(^1\Delta_g)$ traps. For these reasons, the quantitative significance of $O_2(^1\Delta_g)$ as a product of the respiratory burst of neutrophils was placed in doubt (Harrison *et al.*, 1978; Foote *et al.*, 1981; Kanofsky and Tauber, 1983). Kanofsky *et al.* (1984) subsequently made a careful analysis of the biochemical requirements for $O_2(^1\Delta_g)$ production by purified human MPO and concluded that the $MPO/H_2O_2/Cl^-$ system indeed could produce $O_2(^1\Delta_g)$, but the yield was low even under optimal, nonphysiological conditions. The low reactivity of HOCl toward H_2O_2 relative to its reactivity toward other biological targets was proposed as an explanation for the apparent inability of neutrophils to produce $O_2(^1\Delta_g)$.

In contrast to the case of neutrophils, Weiss *et al.* (1986) noted that several factors might favor production of $O_2(^1\Delta_g)$ by human eosinophils in physiologically significant quantities. Thus, the selective oxidation of Br^- by EPO produces HOBr, a species that reacts faster with H_2O_2 than does HOCl. Furthermore, other peroxidases had been shown to produce $O_2(^1\Delta_g)$ in the presence of supraphysiological concentrations of Br^-. Using a sensitive detection procedure—measurement of specific $O_2(^1\Delta_g)$ chemiluminescence at 1268 nm—Kanofsky *et al.* (1988) recently demonstrated the efficient production of $O_2(^1\Delta_g)$ both by purified EPO and by intact eosinophils, using physiological concentrations of Br^-. For example, at pH 7.0 in the presence of 100 μM Br^-, EPO generated 20% of the theoretical yield of $O_2(^1\Delta_g)$. These data suggest that toxic effects of $O_2(^1\Delta_g)$ indeed may contribute to eosinophil-mediated host defense mechanisms.

4.7 Biochemistry of Bromide—Summary

In the study of the biochemistry of the halides, Br^- has received the least attention of all the halides with the exception of astatine. The abundance and many biological functions of Cl^-, the special behavior of F^- (incorporation into mineralized tissue, for example), and the specialized role of I^- in thyroid function have assured intense scrutiny of the biochemistry of these elements. In contrast, with the exception of eosinophil function, Br^- appears to have no role in mammalian physiology (biobromination by marine organisms will be discussed in Chapter 7). Nonetheless, interest in the toxic properties of Br$^-$ revealed in therapeutic schemes based on Br$^-$-containing drugs has produced a significant amount of research. An imperfect mimicking of the functions of Cl$^-$ and I$^-$ appears to be the basis for the anticonvulsant activity and also the toxicity of supraphysiological concentrations of Br$^-$. The relevance of these studies may increase as local environmental levels of Br$^-$ increase through pesticide use.

References

Allen, R. C., Stjernholm, R. L., and Steele, R. H., 1972. Evidence for the generation of an electronic excitation state(s) in human polymorphonuclear leukocytes and its participation in bactericidal activity, *Biochim. Biophys. Res. Commun.* 47:679–684.

Baba, A., Nishiuchi, Y., Uemura, A., and Iwata, H., 1988. Mechanism of excitatory amino acid-induced accumulation of cyclic AMP in hippocampal slices: Role of extracellular chloride, *J. Pharmacol. Exp. Ther.* 245:299–304.

Berglindh, T., 1977. Absolute dependence on chloride for acid secretion in isolated gastric glands, *Gastroenterology* 73:874–880.

Bowen, H. J. M., 1966. *Trace Elements in Biochemistry*, Academic Press, London, p. 76.

Cabantchik, Z. I., Knauf, P. A., and Rothstein, A., 1978. The anion transport system of the red blood cell. The role of membrane protein evaluated by the use of "probes," *Biochim. Biophys. Acta* 515:239–302.

Cheek, D. B., 1961. Extracellular volume; its structure and measurement and the influence of age and disease, *J. Pediatr.* 58:103–125.

Damoder, R., Klimov, V. V., and Dimukes, G. C., 1986. The effect of Cl$^-$ depletion and X$^-$ reconstitution on the oxygen-evolution rate, the yield of the multiline manganese EPR signal and EPR signal II in isolated photosystem-II complex, *Biochim. Biophys. Acta* 848:378–391.

Fong, P., Illsley, N. P., Widdecombe, J. H., and Verkman, A. S., 1988. Chloride transport in apical membrane vesicles from bovine tracheal epithelium: Characterization using a fluorescent indicator, *J. Membr. Biol.* 104:233–239.

Foote, C. S., Abakerli, R. B., Clough, R. L., and Lehrer, R. I., 1981. On the question of singlet oxygen production in polymorphonuclear leukocytes, in *Bioluminescence and Chemiluminescence* (M. A. DeLuca and W. D. McElroy, eds.), Academic Press, New York, pp. 81–88.

Frizzell, R. A., 1987. Cystic fibrosis: A disease of ion channels? *Trends Neurosci.* 10:190–193.

Gleich, G. J., and Loegering, D. A., 1984. Immunology of eosinophils, *Annu. Rev. Immunol.* 2:429–459.

Halm, D. R., Rechkemmer, G., Schoumacher, R. A., and Frizzell, R. A., 1988a. Apical membrane chloride channels in a colonic cell line activated by secretory agonists, *Am. J. Physiol.* 254:C505–C511.

Halm, D. R., Rechkemmer, G. R., Schoumacher, R. A., and Frizzell, R. A., 1988b. Biophysical properties of a chloride channel in the apical membrane of a secretory epithelial cell, *Comp. Biochem. Physiol.* 90A:597–601.

Harrison, J. E., Watson, B. D., and Schultz, J., 1978. Myeloperoxidase and singlet oxygen: A reappraisal, *FEBS Lett.* 92:327–332.

Hellerstein, S., Kaiser, C., Des Darrow, D., and Darrow, D. C., 1960. The distribution of bromide and chloride in the body, *J. Clin. Invest.* 39:282–287.

Higashijima, T., Ferguson, K. M., and Sternweis, P. C., 1987. Regulation of hormone-sensitive GTP-dependent regulatory proteins by chloride, *J. Biol. Chem.* 262:3597–3602.

Kanofsky, J. R., and Tauber, A. I., 1983. Non-physiologic production of singlet oxygen by human neutrophils and by myeloperoxidase-H_2O_2-halide system, *Blood* 62:82a.

Kanofsky, J. R., Wright, J., Miles-Richardson, G. E., and Tauber, A. I., 1984. Biochemical requirements for singlet oxygen production by purified human myeloperoxidase, *J. Clin. Invest.* 74:1489–1495.

Kanofsky, J. R., Hoogland, H., Wever, R., and Weis, S. J., 1988. Singlet oxygen production by human eosinophils, *J. Biol. Chem.* 263:9692–9696.

Kolb, L., and Himmelsbach, C. K., 1938. Clinical studies of drug addiction, III. A critical review of the withdrawal treatments with method of evaluating abstinence syndromes, *Am. J. Psych.* 94:759–799.

Levitzki, A., and Steer, M. L., 1974. The allosteric activation of mammalian α-amylase by chloride, *Eur. J. Biochem.* 41:171–180.

Martinez, J. H., Solano, F., Peñafiel, R., Galindo, J. D., Iborra, J. L., and Lozano, J. A., 1986. Comparative study of tyrosinases from different sources: Relationship between halide inhibition and the enzyme active site, *Comp. Biochem. Physiol.* 83B:633–636.

Marvizón, J. C. G., and Skolnick, P., 1988. Enhancement of t-[^{35}S]butylbicyclophosphorothionate and [^{3}H]strychnine binding by monovalent anions reveals similarities between γ-aminobutyric acid- and glycine-gated chloride channels, *J. Neurochem.* 50:1632–1639.

Mason, M. F., 1936. Halide distribution in body fluids in chronic bromide intoxication, *J. Biol. Chem.* 113:61–74.

McRoberts, J. A., Erlinger, S., Rindler, M. J., and Saier, M. H., Jr., 1982. Furosemide-sensitive salt transport in the Madin–Darby canine kidney cell line. Evidence for the cotransport of Na^+, K^+, and Cl^-, *J. Biol. Chem.* 257:2260–2266.

Mendelsohn, W. B., 1980. *The Use and Misues of Sleeping Pills. A Clinical Guide*, Plenum, New York, p. 156.

Montoya, G. A., and Riker, W. K., 1982. A study of the actions of bromide ion on frog sympathetic ganglion, *Neuropharmacology* 21:581–585.

Nielsen, F. H., 1986. Other elements: Sb, Ba, B, Br, Cs, Ge, Rb, Ag, Sr, Sn, Ti, Zr, Be, Bi, Ga, Au, In, Nb, Sc, Te, Tl, W, in *Trace Elements in Human and Animal Nutrition*, Vol. II (W. Mertz, ed.), 5th ed., Academic Press, Orlando, Florida, pp. 415–463.

O'Grady, S. M., Palfrey, H. C., and Field, M., 1987. Characteristics and functions of Na–K–Cl cotransport in epithelial tissues, *Am. J. Physiol.* 253:C177–C192.

Ono, T.-A., Nakayama, H., Gleiter, H., Inoue, Y., and Kawamori, A., 1987. Modification of the properties of S_2 state in photosynthetic O_2-evolving center by replacement of chloride with other anions, *Arch. Biochem. Biophys.* 256:618–624.

Palfrey, H. C., and Greengard, P., 1981. Hormone-sensitive ion transport systems in erythrocytes as models for epithelial ion pathways, *Ann. N.Y. Acad. Sci.* 372:291–308.

Rauws, A. G., 1983. Pharmacokinetics of bromide ion—an overview, *Food Chem. Toxicol.* 21:379–382.

van Leeuwen, F. X. R., and Sangster, B., 1987. The toxicology of bromide ion, *CRC Crit. Rev. Toxicol.* 18:189–213.

van Leeuwen, F. X. R., den Tonkelaar, E. M., and van Logten, M. J., 1983. Toxicity of sodium bromide in rats: Effect on endocrine system and reproduction, *Food Chem. Toxicol.* 21:383–389.

van Leeuwen, F. X. R., Hanemaaijer, R., and Loeber, J. G., 1988. The effect of sodium bromide on thyroid function, *Toxicology*, Suppl. 12:93–97.

van Logten, M. J., Wolthuis, M., Rauws, A. G., Kroes, R., den Tonkelaar, E. M., Berkvens, H., and van Esch, G. J., 1974. Semichronic toxicity study of sodium bromide in rats, *Toxicology* 2:257–267.

Versieck, J., and Cornelis, R., 1989. *Trace Elements in Human Plasma or Serum*, CRC Press, Boca Raton, Florida, pp. 76–78, 168–169.

Weiss, S. J., Test, S. T., Eckmann, C. M., Roos, D., and Regiani, S., 1986. Brominating oxidants generated by human eosinophils, *Science* 234:200–203.

Woodbury, D. M., and Pippenger, C. E., 1982. Bromides, in *Antiepileptic Drugs* (D. M. Woodbury, J. K. Penry, and C. E. Pippenger, eds.), Raven Press, New York, pp. 791–801.

Wright, E. M., and Diamond, J. M., 1977. Anion selectivity in biological systems, *Phys. Rev.* 57:109–156.

Biochemistry of Inorganic Iodide 5

5.1 Introduction

As a constituent of thyroid hormones, iodine—the heaviest element metabolized in biological materials—plays a unique and essential role in mammalian biochemistry, and, accordingly, the vast majority of research dealing with inorganic iodide (I^-) has been related to this role. The biochemistry of thyroid hormones is reviewed in Chapter 6, which includes an overview of the process by which I^- is incorporated into thyroid hormones through bioiodination. In this chapter, the more general aspects of the biochemistry of I^- will be considered, including occurrence, uptake, biodistribution, and metabolism of I^-. Transport mechanisms, in particular, the mechanisms by which the thyroid gland concentrates I^-, will be reviewed as will the interaction of I^- with enzymes.

5.2 Occurrence, Uptake, and Biodistribution of Iodide

The occurrence, uptake, and biodistribution of iodide have been the subject of recent thorough reviews (e.g., Hetzel and Maberly, 1986; Alexander, 1984), and only a brief summary will be given here. Iodine, the 64th most abundant element, is widely distributed in nature but makes up only $10^{-5}\%$ of the earth's crust. The ocean is a primary source of I^-, containing about 50–60 μg/liter, comparable to the concentration in human serum, while the average concentration in earth is 300 μg/kg. An important commercial source of iodine is caliche, a nitrate rock that contains up to 0.2% iodine in the form of iodate salts. The amount of I^- available in food supplies is quite variable, being dependent on the amount of I^- in the soil, which, in turn, is influenced by such environmental factors as rainfall and flooding (which leaches I^- from the soil). Thus, crops grown in certain areas, such as the Ganges plains of India, are I^- deficient, and animal and human populations in these areas suffer from the same deficiency. A minimum I^- requirement of 100–200 μg/day has been suggested for the adult human diet (Alexander, 1984; Hetzel and Maberly, 1986).

Ingested I^- is absorbed rapidly from the gastrointestinal tract and, similarly to Cl^- and Br^-, is distributed throughout the extracellular space and in red blood cells. The body of a normal adult contains about 15–20 mg total iodine, 70–80% of which is present in the thyroid, most covalently linked to thyroglobulin. The total iodine pool consists of plasma inorganic I^- (PII), protein-bound plasma iodine (PBI), and intrathyroidal iodine. The kidney effectively competes with the thyroid for I^-—plasma I^- is cleared by the kidney at a rate of about 35 ml/min, compared to about 1 ml/min for Cl^-, and greater than 40 μg/day is excreted in urine. The half-life of I^- in the inorganic phase of the iodine pool is 6–7 h, while the half-life of iodine in equilibrium with the whole body pool is about 9.5 days (Alexander, 1984; Hetzel and Maberly, 1986).

5.3 Iodide Transport into Thyroid and Other Iodide-Concentrating Tissues

5.3.1 The Thyroid Gland

The thyroid gland is made up of closed follicles formed by thyroid epithelial cells (Fig. 5-1). These follicles contain a colloid made up primarily of thyroglobulin secreted by the boundary epithelial cells. Thyroglobulin is iodinated at the apical membrane of the follicular lumen and/or within the lumen (see below) and further processed to thyroxine and triiodothyronine, the hormones that are subsequently excreted to the

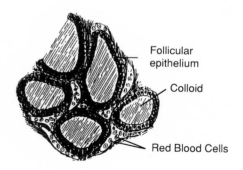

Figure 5-1. Microscopic appearance of the thyroid gland. From A. C. Guyton, *Textbook of Medical Physiology*, 7th ed., W. B. Saunders Co., Philadelphia (1986), p. 897; with permission.

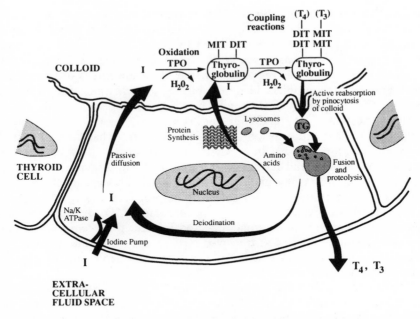

Figure 5-2. Diagram showing pathways of synthesis of thyroid hormones from iodine within the thyroid. After Hetzel and Maberly (1986), with permission.

plasma (Chapter 6) (Guyton, 1986). A diagram of the cellular mechanism for I^- transport and of thyroid hormone biosynthesis and release is shown in Fig. 5-2.

5.3.2 The Iodide Pump

The thyroid gland makes up about 0.03% of the weight of the adult human body yet contains 70–80% of the total body iodine, a fact that demonstrates clearly that efficient mechanisms are available to the thyroid for concentrating iodine (I^- trapping). I^- concentration requires (1) transport across the basolateral membrane of the thyroid epithelial cells against an electrochemical and chemical gradient, (2) accumulation within the cell, (3) diffusion along the electrochemical gradient across the apical membrane into the follicular lumen, and (4) storage (e.g., Weiss *et al.*, 1984a; for a recent review, see Wolff, 1983). The first step in this sequence, uptake across the basolateral membrane, must be accomplished against a large concentration gradient—estimated to be 100–500-fold under certain conditions (Nakamura *et al.*, 1988). This transport was shown early to be dependent on extracellular Na^+, to be stimulated by low extracellular concentrations of K^+, and to be inhibited by the ATPase inhibitor ouabain (for a review of early work, see Wolff, 1964). From these and other data, I^-/Na^+ cotransport was proposed whereby I^- transport is driven by the Na^+ gradient, this gradient being established by extrusion of intracellular Na^+ by the Na^+/K^+-ATPase pump (Fig. 5-3) [secondary active I^- transport (cf. Section 3.2.2.3)] (Wolff and Halmi, 1963). The association of Na^+/K^+-ATPase with transport of I^- into the thyroid has been confirmed by many additional studies in a variety of systems (e.g., Chow *et al.*, 1986; Weiss *et al.*, 1984a; and references therein). Thus, in a continuous line of cultured cells from rat thyroid, uptake of I^- is linearly dependent on external Na^+ concentration, is saturable with respect to I^- ($K_M = 31 \ \mu M$), is inhibited by thiocyanate (SCN^-), perchlorate (ClO_4^-), and ouabain, requires the presence of external K^+, and is affected by temperature and pH (Weiss *et al.*, 1984a). The transport mechanism, the iodide "pump," has been shown to be confined to the basolateral membrane of thyroid epithelia (Chambard *et al.*, 1983).

The effectiveness of large, lipophilic anions such as SCN^- and ClO_4^- as competitive inhibitors of I^- transport is related to the fact that the hydration enthalpies of these anions are similar to that of I^- (O'Neill *et al.*, 1987; Wolff, 1983). Selection of ions by the thyroid system has been viewed as an anion exchange process mediated by quaternary nitrogen exchangers (reviewed by Wolff, 1983). As discussed in Chapter 4 (Section 4.4.1), the selectivity of this system for large lipophilic anions indicates that a weak field is associated with these exchangers. Anion selectivities, the Hofmeister lyotropic series, and other "thyroid-like" systems are discussed further in

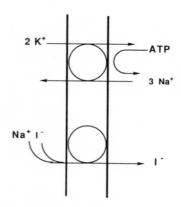

Figure 5-3. A schematic diagram of Na^+/K^+-ATPase and I^- transport.

Section 5.4. For a more thorough discussion of these principles, see the reviews by Wolff (1983) and Collins and Washabaugh (1985).

The identity of the molecular species responsible for thyroid iodide translocation has not been clearly established. While a phospholipid was proposed in early work as the likely transporter, recent research, using a transport model consisting of phospholipid vesicles made from porcine thyroid plasma membranes, has suggested that a protein or a lipid–protein complex is involved. Thus, the translocator is heat sensitive, is sensitive to trypsin, and appears to be inserted in the lipid bilayer (Saito *et al.*, 1984; reviewed by Wolff, 1983).

The Na^+/K^+-ATPase can establish a Na^+ gradient of 10–30-fold, significantly less than the magnitude of the I^- gradient achieved by the I^- transporter. Nakamura *et al.* (1988) examined kinetic properties of the transport process using phospholipid vesicles developed by Saito *et al.* The dependence of the rate of initial uptake on external Na^+ concentrations showed that transport requires initial binding of two Na^+ to the carrier, followed by binding of I^-. Evidence was found for different Na^+- and I^--binding modes on the outside and inside of the vesicles, suggesting that such differences may be important in I^- uptake against its concentration gradient.

Saito *et al.* (1989) have recently demonstrated the presence of Na^+-dependent counterflow mediated by the I^- carrier. This special case of transport, present in facilitated diffusion processes, results when addition of a substrate to one side of the membrane causes net movement of another substrate from the other side against a concentration gradient. In this case, preloading of thyroid plasma membrane-reconstituted phospholipid vesicles with unlabeled I^- or with SCN^- caused increased uptake of extravesicular radiolabeled I^-. This increased net uptake results from competitive inhibition of efflux of labeled I^- by the high concentration of intravesicular transport substrate. External SCN^- and ClO_4^-, or an absence of external Na^+, inhibited the counterflow, implicating the I^- carrier in this process. On the other hand, preloading with ClO_4^- did not lead to counterflow, suggesting that this anion can bind efficiently only to the outside of the membrane.

5.3.3 Regulation of Iodide Uptake: Thyroid-Stimulating Hormone (TSH)

As will be seen in Chapter 6, a precise level of thyroid hormones is required to maintain proper metabolic activity in the body. Thyroid-stimulating hormone (TSH, thyrotropin), a glycoprotein secreted by the anterior pituitary, increases the release of thyroid hormones by several mechanisms, including stimulation of the I^- pump. While the initial rapid increase in circulating hormones evoked by TSH is caused by an increased rate of proteolysis of iodinated thyroglobulin (Chapter 6), an increase in the activity of the I^- pump is not seen until 8–12 h. Early studies on the time course of TSH-stimulated I^- uptake (and other studies) using whole tissues such as thyroid slices or lobes were complicated by the tendency of such studies to measure the result of the entire sequence of events leading to hormone synthesis, rather than the isolated step in question. In addition, it was necessary to free tissues of endogenous TSH before measurements of exogenous TSH effects could be studied. The establishment of a continuous line of cultured thyroid cells has alleviated many of these problems and permitted closer examination of the effects of TSH on the I^- pump (Weiss et al., 1984a). In studies using this cell line, a biphasic time course of TSH stimulation of I^- uptake seen in earlier studies was confirmed. Thus, an initial decrease in intrathyroidal I^- concentration was followed several hours later by an enhancement of uptake. Several lines of evidence support previous suggestions that this enhancement may be related to cAMP-linked protein synthesis. Thus, dibutyryl-cAMP mimicked the effects of TSH, and TSH and dibutyryl-cAMP-stimulated increases in methionine incorporation could be blocked by cycloheximide and other inhibitors of protein synthesis. However, the effect of TSH appears to reflect an increase in general protein synthesis rather than induced synthesis of a single carrier protein (Weiss et al., 1984b).

The initial TSH-mediated depression of I^- uptake, described earlier by Halmi et al. (1960) and confirmed by Weiss et al. (1984a), could not be explained by a decrease in I^- influx. Weiss et al. (1984c), using the continuous line of cultured rat thyroid cells, confirmed earlier indications that this decrease in intrathyroidal I^- concentration actually reflects an increase of I^- efflux, possibly at the apical membrane. Moreover, this increased efflux is independent of I^- influx and is dissociated from the effect of TSH to stimulate cAMP. Comparable effects on efflux were observed after the addition of norepinephrine or the Ca^{2+} ionophore A23187 in the presence of external Ca^{2+}. An increased membrane permeability induced by intracellular Ca^{2+} or adrenergic stimulation was suggested by these results. Bidey and Tomlinson (1987) reported similar results, supporting the possible role of α-adrenergic agonists and intracellular Ca^{2+} in controlling the apical membrane permeation of I^- in vivo.

Certain substances, such as SCN^- and ClO_4^-, that interfere with I^- uptake into the thyroid function as goitrogens. In response to lowered thyroid hormone levels, the pituitary increases TSH output, and this in turn induces increased thyroid activity and hypertrophy of the gland.

5.3.4 Autoregulation of Iodide Uptake

While the autoregulation of I^- uptake by I^- has been recognized for many years, the mechanism for this autoregulation has not been determined precisely. Multiple factors appear to be involved, including possible effects of increased I^- concentrations on thyroid peroxidase (TPO), formation of an iodinated or oxidized inhibitor of I^- transport, and inhibition by I^- of TSH stimulation of adenylate cyclase (for reviews, see Ingbar, 1972; Wolff, 1969). Since most of the effects of TSH are mediated by adenylate cyclase, an important role in autoregulation has been implied for I^- effects on adenylate cyclase stimulation by TSH. Tseng et al. (1989) studied the effects of I^- on cAMP-dependent and cAMP-independent processes in dog thyroid slices and concluded that, in addition to cAMP-mediated mechanisms, other I^--sensitive processes, including glucose and phosphate transport, may be important to thyroid autoregulation. A possible action on protein kinase C was suggested by these results.

The paradoxical inhibition of formation of organic iodine by high levels of I^- ("iodide goiter," the Wolff–Chaikoff effect) is caused by I^- amounts in vast excess of the daily requirements for hormone synthesis. Daily intakes of $2000\ \mu g$/day have been suggested as being potentially harmful (Wolff, 1969). Such levels are not supplied normally by diets, but residents of certain areas—most notably the coastal regions of Hokkaido—where diets contain large amounts of seaweed can consume 50,000 to 80,000 μg/day, although incidence of goiter is low (reviewed by Hetzel and Maberly, 1986; Alexander, 1984).

While I^- toxicity has been studied extensively in humans, and in laboratory and farm animals, the mechanisms of this toxicity are unclear (Wolff, 1969; Hetzel and Maberly, 1986). Reaction of an active form of iodine with I^- to form I_3^- has been suggested (Alexander, 1984).

5.3.5 Additional Thyroid Ion Transport Mechanisms

Transport mechanisms not involved with active I^- transport have been demonstrated in thyroid tissue. For example, the loop diuretic furosemide, which has little effect on Na^+/K^+-ATPase but is a potent inhibitor of Na^+/Cl^- cotransport, reduced the concentrations of Na^+ and Cl^-, as well as of I^-, in turtle thyroid tissue and in cultured turtle thyroid cells. Similarly, amyloride, a specific inhibitor of Na^+ transport, decreased Na^+ and increased Cl^- concentrations in cultured turtle thyroid cells, presumably due to inhibition of Na^+/H^+ exchange. A combination of ouabain and furosemide did not block I^- transport completely, indicating that an additional I^- transport mechanism may be present (Chow *et al.*, 1986).

Stilbene sulfonate inhibitors of anion transport have been shown to increase the uptake of I^- into thyroid cells (e.g., Weiss, 1984a). I^- uptake in cultured thyroid cells has been shown to be inversely proportional to HCO_3^- concentration in the medium but unaffected by Cl^- concentration. Effects on intracellular pH have been proposed as an explanation of these observations. Thus, blockade of anion channels by inhibitors such as DIDS increases intracellular Cl^- and decreases intracellular HCO_3^-, thus lowering intracellular pH, leading to increased I^- accumulation (Chow *et al.*, 1987, 1984).

These data suggest that additional transport mechanisms may be present in thyroid cells—similar to those present in many other secretory epithelial cells (Fig. 5-4). Nonetheless, Na^+/K^+-ATPase-coupled Na^+/I^-

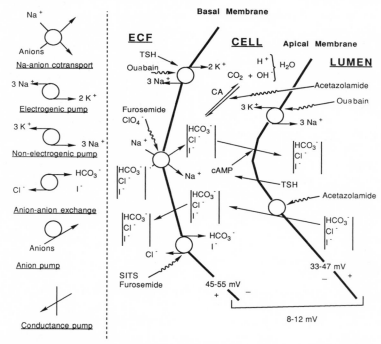

Figure 5-4. Schematic model of anion (I^-, HCO_3^-, Cl^-) and cation (Na^+ and K^+) transport across the basolateral and apical membranes of the turtle thyroid gland. The effects of various agents that inhibit (wavy lines) or stimulate (straight lines) the various pumps and conductance processes are also shown. (ECF = extracellular fluid) Adapted, with permission, from S. Y. Chow, J. W. Kemp, and D. M. Woodbury, Role of carbonic anhydrase in thyroidal iodide transport, *Ann. N.Y. Acad. Sci.* 429:604–606 (1984).

cotransport is overwhelmingly the most important mechanism for the extremely efficient concentration of I^- into the thyroid gland.

5.3.6 Iodide Transport and Thyroglobulin Iodination

Thyroglobulin, synthesized in thyroid follicle cells, is transported to the follicle lumen in exocytotic vesicles. Although these vesicles also contain thyroperoxidase and can accumulate I^- (Bastiani and Rognoni, 1985), no iodination of thyroglobulin occurs in the vesicles. Thyroperoxidase is a membrane-bound enzyme that is also found on the apical membrane of the follicle cell. Hydrogen peroxide, also required for iodination, is formed on the external surface of the apical membrane of the follicle cell, and, accordingly, iodination of thyroglobulin has been proposed to occur at this membrane surface (Ekholm, 1981). However, double-labeling experiments that permitted autoradiographic determination of both newly synthesized and newly iodinated protein revealed that iodine is incorporated not only into protein newly delivered to the follicle but also into molecules stored in the lumen (Ofverholm and Ericson, 1984). The mechanism of thyroglobulin iodination will be discussed further in Chapter 6.

5.3.7 Additional Iodine-Concentrating Organs

While most tissues do not concentrate I^- above the level found in plasma or surrounding medium, several tissues, in addition to thyroid, possess I^--concentrating ability (Wolff, 1964). Kidneys, salivary glands, and gastric glands compete effectively with the thyroid for the iodine pool. However, I^- cleared from the plasma by salivary and gastric glands is resecreted into the gastrointestinal tract and is recycled into the pool, thus providing a mechanism for endogenous replenishment of this pool (reviewed by Hetzel and Maberly, 1986). The ciliary body (responsible for formation of aqueous humor in the eye) and the choroid plexus (a secretor of cerebrospinal fluid in the brain) actively transport I^- toward the plasma. Thus, cerebrospinal fluid has a low concentration of I^- with respect to plasma. The I^- transport mechanism in these tissues appears to be similar to that in the thyroid (reviewed by Wolff, 1964).

5.4 Stimulation and Inhibition of Enzymes by Iodide

Anion selectivities related to transport and enzyme functions were discussed in Chapter 4. This discussion will be expanded in this section to include examples with special relevance to the biochemistry of I^-.

5.4.1 The Hofmeister Lyotropic Series and "Thyroid-Like" Systems

As seen in Chapter 4, several biological systems show a preference for I^- over other halides ($I^- > Br^- > Cl^- > F^-$). In their review of anion selectivities, Wright and Diamond (1977) called such systems "thyroid-like." While such organs as the choroid plexus, ciliary body of the eye, salivary glands, and other I^--concentrating tissues indeed are similar to the thyroid in that they appear to possess similar specific I^- transport mechanisms, other systems may be considered to have a preference for I^- based on a more general physicochemical phenomenon. Such systems include several enzymes that are sensitive to inhibition by anions, for example, acetoacetate decarboxylase and carbonic anydrase.

In his classic study of acetoacetate decarboxylase inhibition by anions, Fridovich (1963) observed a sequence of effectiveness of several anions that was similar to the sequence observed by Hofmeister in 1888 in research on the effects of anions on egg albumin solubilities. (A series of anions that cause an observable effect on a system that follows a sequence similar to that observed by Hofmeister is frequently called the Hofmeister, or lyotropic, series.) Thus, acetoacetate decarboxylase was inhibited by anions according to the sequence $HSO_3^- > SCN^- > ClO_4^- > I^- > NO_3^- > Br^- > Cl^- > BrO_3^- > F^- > IO_3^- > CCl_3CO_2^-$, and inhibition was consistent with the binding of each anion to a cationic site having a pK of about 5.8. Thermodynamic parameters of inhibition were investigated, and a relationship between anionic volume and entropy of inhibition was found. Large monovalent anions such as I^-, Br^-, and SCN^- do not have sufficiently strong external fields to permit strong hydration and thus may approach and interact with the cationic site without severe energy-requiring disruption of hydration spheres. [As discussed in Chapter 4, Wright and Diamond (1977) analyzed other data and extended the theoretical treatment of anion selectivities to other observed sequences.] Thus, the "preference" of acetoacetate decarboxylase for I^- relative to other halides does not involve a specific recognition site for I^-, but rather reflects the position of I^- in the Hofmeister series.

5.4.2 Carbonic Anhydrase Inhibition

Another enzyme that is very sensitive to inhibition by monovalent anions—inhibition which also corresponds to the Hofmeister series—is carbonic anhydrase (carbonate hydrolyase), discussed in Chapter 3 with respect to carbon dioxide transport and above with respect to thyroid function. The isozyme II is a very efficient catalyst for the hydration of CO_2, having a turnover of about $10^6 s^{-1}$ at ph 9. Two ionizing groups, zinc-bound water ionizing to OH^- and a histidine residue, are involved in the catalytic step. Inhibition of carbonic anhydrase by monovalent anions involves binding of the anion to the metal ion and can be considered as a competition between the anion and OH^- for the metal binding site (Tibell *et al.*, 1984; Pocker and Deits, 1982; and references therein). Carbonic anhydrase is "thyroid-like" in that the halide sequence for efficiency of inhibition is $I^- > Br^- > Cl^- > F^-$ (Wright and Diamond, 1977).

5.4.3 Inhibition of Thyroid Adenylate Cyclase Activity

The autoregulation of I^- transport into the thyroid by I^- has been discussed above. As noted, I^- inhibits adenylate cyclase activity in the thyroid, and a blockade of TSH-induced stimulation of cyclase activity has been proposed as the mechanism of this autoregulation. The mechanism of inhibition of thyroid adenylate cyclase activity by I^- has been studied extensively but remains unclear (Van Sande *et al.*, 1985; Cochaux *et al.*, 1987; and references therein). The inhibition requires an intact cell structure, an intact I^- transporter, and thyroperoxidase activity. I^- can block cyclase activity stimulated not only by TSH and prostaglandin E_1, hormones that act at receptor subunits of the cyclase system, but also activity stimulated by forskolin, fluoride, or cholera toxin, agents that function by interacting with regulatory or catalytic subunits. I^- depresses the maximum velocity of the system, rather than the coupling between the regulatory and catalytic subunits. From these and other observations, interaction of I^- with the catalytic and possibly also the regulatory subunit has been proposed (Cochaux *et al.*, 1987).

Additional examples of the interaction of I^- with enzyme systems were presented as part of the discussion of halide selectivity given in Chapter 4. While many other examples could be cited, this would not necessarily add to a clearer understanding of the principles that have been examined in this section.

5.5 Biochemistry of Inorganic Iodide–Summary

The position of I^- in the Hofmeister lyotropic series (as the least solvated and most polarizable of the halides) causes this anion to be an excellent substrate for many biological systems. Thus, I^- is accepted readily by anion recognition sites associated with many transport proteins and enzymes. However, as with Br^-, the low *in vivo* concentration of I^-

relegates these interactions to little, if any, biological significance. However, the ability of the thyroid gland to concentrate I^- efficiently, the initial step in the synthesis of thyroid hormones, is of profound significance. As will be seen in Chapter 6, these hormones, which mediate many of their actions through modulation of gene transcription, are essential to the growth and development of vertebrates. The mechanism of transport, and regulation of transport, of I^- into thyroid tissue is an important and very active area of research, and valuable new information continues to be produced. Examples of this research have been provided in this chapter; the likewise important and fascinating biochemistry of thyroid hormones is reviewed in Chapter 6.

References

Alexander, N. M., 1984. Iodine, in *Biochemistry of the Essential Ultratrace Elements* (E. Frieden, ed.), Plenum Press, New York, pp. 33–53.

Bastiani, P., and Rognoni, J. B., 1985. Iodide accumulation into thyroid golgi vesicles, *Cell Biochem. Function* 3:21–23.

Bidey, S. P., and Tomlinson, S., 1987. Differential modulation by Ca^{2+} of iodide transport processes in a cultured rat thyroid cell strain, *J. Endocrinol.* 112:51–56.

Chambard, M., Verrier, B., Gabrion, J., and Mauchamp, J., 1983. Polarization of thyroid cells in culture: Evidence for the basolateral localization of the iodide "pump" and of the thyroid-stimulating hormone receptor–adenyl cyclase complex, *J. Cell Biol.* 96:1172–1177.

Chow, S. Y., Kemp, J. W., and Woodbury, D. M., 1984. Role of carbonic anhydrase in thyroidal iodide transport, *Ann. N.Y. Acad. Sci.* 429:604–606.

Chow, S. Y., Yen-Chow, Y. C., White, H. S., and Woodbury, D. M., 1986. Effects of sodium on iodide transport in primary cultures of turtle thyroid cells, *Am. J. Physiol.* 250:E464–E469.

Chow, S. Y., Yen-Chow, Y. C., White, H. S., and Woodbury, D. M., 1987. Effects of 4,4'-di-isothiocyano-2,2'-stilbene disulphonate on iodide uptake by primary cultures of turtle thyroid follicular cells, *J. Endocrinol.* 113:403–412.

Cochaux, P., Van Sande, J., Swillens, S., and Dumont, J. E., 1987. Iodide-induced inhibition of adenylate cyclase activity in horse and dog thyroid, *Eur. J. Biochem.* 170:435–442.

Collins, K. D., and Washabaugh, M. W., 1985. The Hofmeister effect and the behaviour of water at interfaces, *Quart. Rev. Biophys.* 18:323–422.

Ekholm, R., 1981. Iodination of thyroglobulin. An intracellular or extracellular process? *Mol. Cell. Endocrinol.* 24:141–163.

Fridovich, I., 1963. Inhibition of acetoacetic decarboxylase by anions. The Hofmeister lyotropic series, *J. Biol. Chem.* 238:592–598.

Guyton, A. C., 1986. *Textbook of Medical Physiology*, 7th ed., W. B. Saunders Co., Philadelphia, pp. 897–908.

Halmi, N. S., Granner, D. K., Doughman, D. J., Peters, B. H., and Müller, G., 1960. Biphasic effect of TSH on thyroidal iodide collection in rats, *Endocrinology* 67:70–81.

Hetzel, B. S., and Maberly, G. F., 1986. Iodine, in *Trace Elements in Human and Animal Nutrition* (W. Mertz, ed.), 5th ed., Vol. 2, Academic Press, Orlando, Florida, pp. 139–208.

Ingbar, S. H., 1972. Autoregulation of the thyroid. Response to iodide excess and depletion, *Mayo Clinic Proc.* 47:814–823.

Nakamura, Y., Ohtaki, S., and Yamazaki, I., 1988. Molecular mechanism of iodide transport by thyroid plasmalemmal vesicles: Cooperative sodium activation and asymmetrical affinities for the ions on the outside and inside of vesicles, *J. Biochem.* 104:544–549.

Ofverholm, T., and Ericson, L. E., 1984. Intraluminal iodination of thyroglobulin, *Endocrinology* 114:827–835.

O'Neill, B., Magnolato, D., and Semenza, G., 1987. The electrogenic, Na^+-dependent I^- transport system in plasma membrane vesicles from thyroid glands, *Biochim. Biophys. Acta* 896:263–274.

Pocker, Y., and Deits, T. L., 1982. Effects of pH on the anionic inhibition of carbonic anhydrase activities, *J. Am. Chem. Soc.* 104:2424–2434.

Saito, K., Yamamoto, K., Takai, T., and Yoshida, S., 1984. Characteristics of the thyroid iodide translocator and of iodide-accumulating phospholipid vesicles, *Endocrinology* 114:868–872.

Saito, K., Yamamoto, K., Nagayama, I., Uemura, J., and Kuzuya, T., 1989. Effect of internally loaded iodide, thiocyanate, and perchlorate on sodium-dependent iodide uptake by phospholipid vesicles reconstituted with thyroid plasma membranes: Iodide counterflow mediated by the iodide transport carrier, *J. Biochem.* 105:790–793.

Tibell, L., Forsman, C., Simonsson, I., and Lindskog, S., 1984. Anion inhibition of CO_2 hydration catalyzed by human carbonic anhydrase II. Mechanistic considerations, *Biochim. Biophys. Acta* 789:302–310.

Tseng, F.-Y., Rani, C. S. S., and Field, J. B., 1989. Effect of iodide on glucose oxidation and ^{32}P incorporation into phospholipids stimulated by different agents in dog thyroid slices, *Endocrinology* 124:1450–1455.

Van Sande, J., Cochaux, P., and Dumont, J. E., 1985. Further characterization of the iodide inhibitory effect of the cyclic AMP system in dog thyroid slices, *Mol. Cell. Endocrinol.* 40:181–192.

Weiss, S. J., Philp, N. J., and Grollman, E. F., 1984a. Iodide transport in a continuous line of cultured cells from rat thyroid, *Endocrinology* 114:1090–1098.

Weiss, S. J., Philp, N. J., Ambesi-Impiombata, F. S., and Grollman, E. F., 1984b. Thyrotropin-stimulated iodide transport mediated by adenosine 3',5'-monophosphate and dependent on protein synthesis, *Endocrinology* 114:1099–1107.

Weiss, S. J., Philp, N. J., and Grollman, E. F., 1984c. Effect of thyrotropin on iodide efflux in FRTL-5 cells mediated by Ca^{2+}, *Endocrinology* 114:1108–1113.

Wolff, J., 1964. Transport of iodide and other anions in the thyroid glands, *Physiol. Rev.* 44:45–90.

Wolff, J., 1969. Iodide goiter and the pharmacological effects of excess iodide, *Am. J. Med.* 47:101–124.

Wolff, J., 1983. Congenital goiter with defective iodide transport, *Endocrine Rev.* 4:240–254.

Wolff, J., and Halmi, N. S., 1963. Thyroidal iodide transport. V. The role of Na^+-K^+-activated, ouabain-sensitive adenosinetriphosphatase activity, *J. Biol. Chem.* 238:847–851.

Wright, E. M., and Diamond, J. M., 1977. Anion selectivity in biological systems, *Phys. Rev.* 57:109–156.

Iodotyrosine, Iodothyronines, and Thyroid Function

6.1 Iodotyrosine as a Biological Tracer

Several methods are available for the facile radioiodination of tyrosine residues in peptides and proteins. Applications of radioiodinated macromolecules in such areas as receptor biochemistry, immunology, transport phenomena, and metabolic studies are now routine. References for methodology and applications are given in the review by Alexander (1984).

6.2 Iodothyronines and Thyroid Function

$3,5,3'5'$-Tetraiodothyronine (thyroxine, T_4) is the primary secretion product of the thyroid gland. $3,5,3'$-Triiodothyronine (T_3)—also secreted, but formed mainly by deiodination of T_4 outside the thyroid—now is thought to be responsible for most biological responses to thyroid hormones. Deiodination of T_4 produces $3,3',5'$-triiodothyronine (reverse T_3, rT_3) to a comparable extent as it does T_3, and a regulator role of rT_3 in thyroid function has been proposed recently (see Section 6.2.5.2). The thyroid hormones (Fig. 6-1) are unique to vertebrates, where they play important roles both in development and in metabolic regulation. The immense diversity of actions of these agents includes important roles in the complex processes of growth and differentiation, as well as regulation of basal metabolism, oxygen consumption, intermediary anabolic and catabolic metabolism, and receptor protein synthesis. The medical problems associated with thyroid dysfunction reflect the importance of thyroid hormones to human physiology.

The events that led to the isolation and identification of T_3 and T_4 make up a fascinating part of medical history. Subsequent research that has detailed most aspects of the biosynthesis of thyroid hormones, mechanisms of their biological transport, the scope of biological processes controlled by

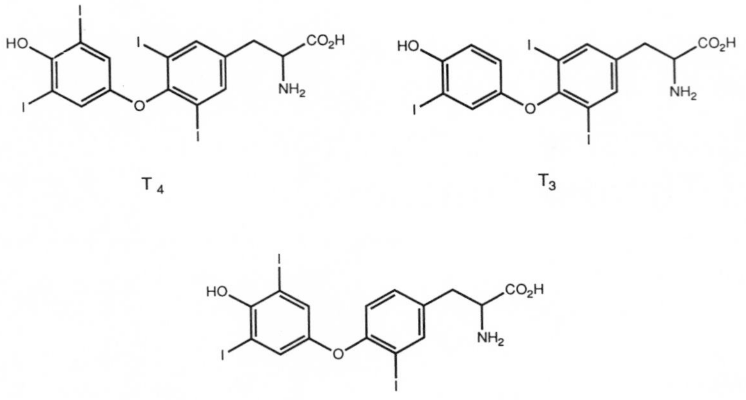

Figure 6-1. The thyroid hormones.

these hormones, and mechanisms by which they are elicited has been no less exciting and continues to make up an important area of biomedical research. Several reviews have provided valuable material used in the preparation of this section (e.g., Köhrle *et al.*, 1987; Alexander, 1984; Jorgensen, 1981; Oppenheimer, 1979). Thyroid research encompasses many disciplines, and any attempt to prepare a concise overview invariably will result in omissions of important material. For example, in this discussion, the many medical and physiological aspects of thyroid biology are dealt with only minimally.

6.2.1 A Brief Historical Perspective

Several recent reviews are available which contain excellent historical discussions of early thyroid hormone research (e.g., Frieden, 1981; Jorgensen, 1981), and this subject will be given only brief treatment here. The presence of organically bound iodine in a biologically active constituent (termed *thyrojodin*) of the thyroid gland was established by Baumann in 1896, and Oswald, in the early 1900s, identified "thyreoglobulin" as the major iodinated protein of the thyroid gland. In 1914, Kendall isolated the first crystals of thyroxine and, based on an erroneous estimate of molecular weight and on other data suggesting the presence of an indole nucleus, assigned to it the structure of an iodinated indolepropionic acid (Fig. 6-2) and proposed the name thyroxyindole, later shortened to thyroxin. This

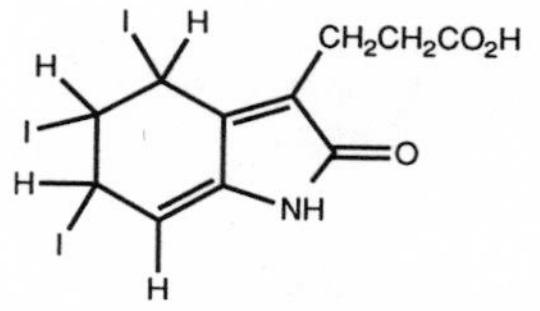

Figure 6-2. Thyroxyindole.

structure was accepted until 1926, when Harrington correctly identified the product of catalytic hydrogenation of thyroxine as the new amino acid thyronine. Harrington subsequently established the structure of thyroxine as 3,5,3',5'-tetraiodothyronine and confirmed the structure by total synthesis. In 1952, the isolation by Gross and Pitt-Rivers and by Rosche *et al.* of 3,5,3'-triiodothyronine from thyroid gland hydrolysates represented another major development, in that this iodoamino acid proved to be even more potent than thyroxine. Braverman *et al.* (1970) subsequently reported the conversion of T_4 to T_3 *in vivo*, and T_3 is now thought to be the physiologically active hormone. Virtually all of the deiodination products of T_4 have subsequently been identified in mammalian cells or in blood (Frieden, 1981; Jorgensen, 1981).

6.2.2 Biosynthesis of Thyroid Hormones

In the thyroid gland, thyroid peroxidase (TPO), a membrane-bound heme glycoprotein, catalyzes the oxidative iodination of Tyr residues present in thyroglobulin. In a second oxidative step, also catalyzed by TPO, these iodinated residues are coupled to give T_3 and T_4 residues. Thyroglobulin containing 1% iodine (52 atoms) on the average will contain 10–12 residues of monoiodotyrosine (MIT) and diiodotyrosine (DIT), 3–4 residues of T_4, and less than 1 T_3 residue. Proteolytic cleavage of the protein backbone releases the hormones. These biosynthetic steps, which will be discussed in greater detail below, are outlined in Fig. 6-3. It should be noted that several common enzymes readily iodinate tyrosine residues in proteins and that iodination *per se* does not necessarily indicate the presence of thyroid hormones (Frieden, 1981). Several reports discuss in detail the biosynthesis of thyroid hormones, including reviews by Jorgensen (1981), Neary *et al.* (1984), and Nunez (1984).

6.2.2.1 Uptake of Iodide into the Thyroid Gland

As discussed in Chapter 5, iodine is a trace element. To satisfy the requirements for thyroid hormone synthesis, the thyroid gland possesses an

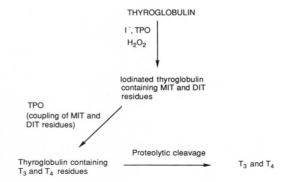

Figure 6-3. Schematic diagram of thyroid hormone biosynthesis.

active transport system that, in normal humans, maintains a concentration of iodide some 30-fold higher than in the circulation. The transport of iodide into the thyroid is regulated by the pituitary glycoprotein thyrotropin, also known as thyroid-stimulating hormone (TSH) (see Chapter 5, Section 5.3.3) (Jorgensen, 1981).

6.2.2.2 Iodination of Tyrosine Residues

The general properties of haloperoxidases will be discussed more completely in Chapter 7. Iodination requires hydrogen peroxide or a peroxide-generating source, I^-, a peroxidase, and an iodine acceptor. In T_4 biosynthesis, following oxidation of I^- to the iodine oxidation level, an enzyme-associated iodinating species $[TPO-I_{oxid}(I^+)]$ that is capable of iodinating free or protein-bound Tyr is formed (Fig. 6-4). In this scheme, the chemical nature of the oxidized iodine species has been a subject of considerable debate (Neary et al., 1984). Iodine produced by TPO-catalyzed oxidation of iodide could be capable of iodination of Tyr residues

$$TPO + H_2O_2 \longrightarrow TPO_{oxid}(Cpd\ 1) + H_2O$$

$$TPO_{oxid}(Cpd\ 1) + I^- \longrightarrow TPO\text{-}I_{oxid}(I^+)$$

$$TPO\text{-}I_{oxid}(I^+) + acceptor \longrightarrow I\text{-}acceptor + TPO$$

Figure 6-4. Sequence of reactions that lead to iodination of tyrosine residues (Neary et al., 1984).

without further participation of the enzyme. However, several lines of evidence have accumulated which suggest that I_2 is not the iodinating species (see, e.g., Deme et al., 1978, and references therein) and that I^+ or its equivalent is the iodinating species (Davidson et al., 1978). Direct evidence for a TPO-associated iodinating intermediate was obtained, and the proposal was made that a sulfenyl iodide ($-SI$), formed by cleavage of an active-site disulfide bond by I^+, is the reactive intermediate (reviewed by Neary et al., 1984; Jorgensen, 1981).

6.2.2.3 The Coupling Reaction

While the mechanism of the coupling reaction has been the subject of much research for nearly 50 years, some important aspects have been clarified only recently (Nunez, 1984). The reaction involves (1) either a one-electron oxidation of two proximal DIT residues or a two-electron oxidation of one DIT residue properly positioned near another DIT residue; (2) formation of a charge transfer complex between these two residues; and (3) collapse of this intermediate to T_4 and a side-chain residue (Fig. 6-5). The identification in 1980 of dehydroalanine in this "lost side chain" solved a problem that had persisted for four decades (Gavaret et al., 1980). According to this mechanism, regardless of whether a one- or two-electron oxidation is involved, the same charge transfer complex would be formed (Garavet et al., 1981).

6.2.2.4 The Role of Thyroglobulin in Hormone Synthesis

While free DIT is converted readily to T_4, the tertiary structure of thyroglobulin clearly plays an important part in the formation of T_4 in vivo. In the native state of thyroglobulin, the hormonogenic DIT residues not only must be in close proximity, but also must be held in the proper orientation for coupling. Thyroglobulin is a dimeric glycoprotein made up of two identical 300K subunits. Of the 120 Tyr residues in human thyroglobulin, only 25–30 are subject to iodination, and of these iodinated residues, only 8–10 couple to form T_3 and T_4 (Malthiery and Lissitzky, 1985, and references therein). Further, under conditions of iodine deficiency, hormone production efficiency is extremely high, with essentially all of the incorporated iodine found in iodothyronines. Other studies have shown that T_4 formation occurs in early stages of thyroglobulin iodination, an indication that the native state of the protein plays a role in preferential T_4 formation. Diiodothyronine stimulates the activity of iodoperoxidases,

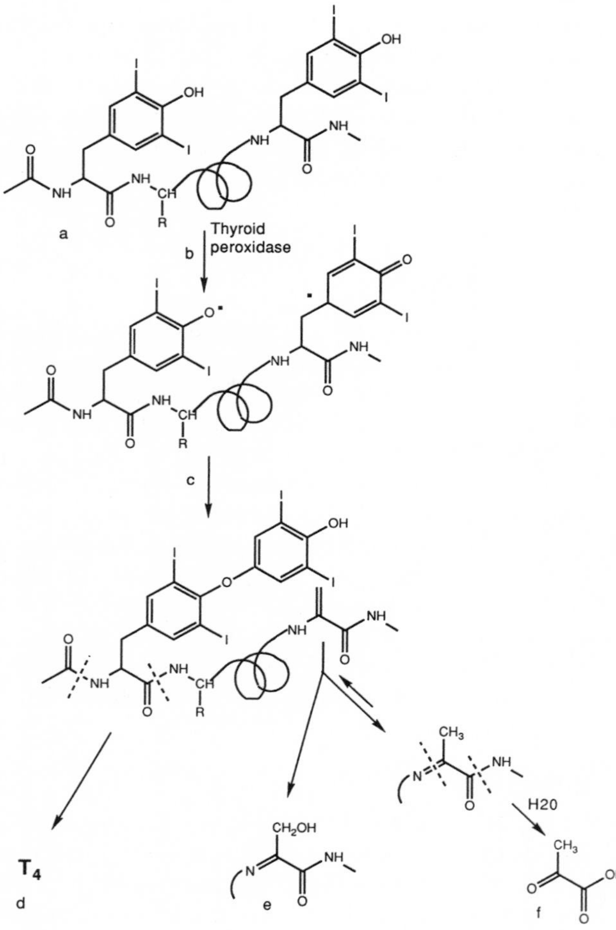

Figure 6-5. The coupling of 3,5-diodotyrosine (DIT) residues of thyroglobulin to form thyroxine (T$_4$). (a) DIT residues shown schematically in proximity on the folded thyroglobulin chain; (b) the formation of tyrosyl oxygen- and carbon-localized radicals by the action of thyroid peroxidase; (c) the radical coupling reaction to form thyroglobulin-bound T$_4$ and dehydroalanine; and proteolytic and hydrolytic reactions yielding (d) free T$_4$, (e) a serine residue, and (f) pyruvic acid. Adapted from E. C. Jorgensen, Thyromimetic and antithyroid drugs, in *Burger's Medicinal Chemistry*, Part III, 4th ed. (M. E. Wolff, ed.), pp. 103–145. Copyright 1981 John Wiley and Sons, Inc., New York.

including TPO, and this may contribute to preferential formation of the hormone (Nakamura *et al.*, 1984, and references therein).

A complete understanding of thyroid hormone formation clearly requires extensive knowledge of thyroglobulin structure. The large size and extensive post-translational processing of thyroglobulin have precluded the use of classical protein chemistry methodology. However, recent research using recombinant DNA technology has provided much information regarding the primary structure of thyroglobulin from various species, information that includes identification of the hormonogenic domains of the protein. For example, the entire primary structure of bovine thyroglobulin has been determined from the sequence of its 8,431-base complementary DNA. From this, four hormonogenic acceptor Tyr residues—corresponding to iodothyronines present in four hormone-containing tryptic peptides previously characterized—were identified and localized near the amino and carboxyl ends of the protein (Mercken *et al.*, 1985). Evidence for the location of "donor" hormonogenic Tyr residues in calf thyroglobulin was recently obtained by tritium labeling of the dehydroalanine product of the lost side chain, followed by CNBr cleavage (Palumbo, 1987).

Human thyroglobulin also has been studied extensively. For example, CNBr digest of iodine-poor human thyroglobulin had revealed an N-terminal 22K polypeptide containing the preferential hormonogenic Tyr "acceptor" of the protein. Iodination *in vitro* of this fragment led to very efficient synthesis of T_4, independent of the remaining protein (Marriq *et al.*, 1986). More recently, human thyroglobulin mRNA has been cloned and sequenced, and from this the amino acid sequence has been deduced (Malthiery and Lissitzky, 1987). Radioiodination of human thyroglobulin and fractionation of the tryptic peptides, coupled with knowledge of the primary structure, in turn has permitted an examination of the consensus sequences of amino acids that are involved in hormonogenesis and identification of both donor and acceptor tyrosyl residues (Lamas *et al.*, 1989).

Rat thyroglobulin mRNA also has been cloned and sequenced, and a 967-amino acid sequence at the carboxyl end has been determined (Di Lauro *et al.*, 1985). The secondary structure of this carboxyl-terminal third of rat thyroglobulin has been modeled, using the available primary sequence data. Two domains were predicted, including a highly structured domain containing the hormonogenic sites (Formisano *et al.*, 1985).

As is evident from the above examples, rapid progress is being made in the characterization of the structure of thyroglobulin. Taken with the classic mechanistic studies that have examined the iodination and coupling steps, continuing research efforts in this area may soon provide a complete and detailed picture of the biochemical process by which the thyroid hormones are formed.

6.2.3 Release and Biological Transport of Thyroid Hormones

6.2.3.1 Secretion of T_3 and T_4

The hormonogenic processes described above are extracellular, and the poorly diffusible precursor iodinated protein thereby formed functions as a storage depot for hormone (and also for iodine). Secretion of hormone results from endocytosis of iodinated thyroglobulin followed by complete hydrolysis in the lysosomes. Under normal conditions, less than 1 % of the stored thyroglobulin is required to satisfy demand for hormone. In a conservation step, released free MIT and DIT are deiodinated and the I^- formed is mixed with I^- entering from the circulation. Released Tyr also is recycled into thyroglobulin. T_3 and T_4 are resistant to these deiodinase enzymes and are released into the circulation (Jorgensen, 1981). In humans, results from several laboratories indicate the daily production rate to be between 83 and 93 μg of T_4. Of the daily production of 22.6–44.8 μg of T_3, 80% is derived from deiodination of T_4 and the remainder is secreted from the thyroid. Monodeiodination of T_4 produces approximately equal amounts of T_3 and rT_3 (Engler and Burger, 1984, and references therein).

6.2.3.2 Serum Transport of T_3 and T_4

Consistent with their hydrophobic structures, T_3 and T_4 are poorly soluble, and, following secretion into the circulation, they are quickly bound to specific plasma proteins that function as carriers. In humans, approximately 65% of the total 900 μg of circulating T_4 is bound to thyroxine-binding globulin (TBG), a protein present in low concentration, but which has a high binding affinity for T_4 ($K_d = \sim 10^{-10}$ M). T_3 is bound about one-tenth as efficiently as T_4. A second important T_4 carrier protein is thyroxine-binding prealbumin (TBPA), also termed transthyretin (TTR), a simple nonglycosylated protein made up of four identical subunits. TBPA binds T_4 with only 1 % of the affinity that TBG does, but because of its high serum concentration, it carries about 30% of the circulating T_4. Amino acid and nucleotide sequences of human TBPA have been reported, and its three-dimensional structure has been determined by X-ray crystallography.

The thorough characterization of TBPA has offered an excellent opportunity for investigation of the interactions of the hormones with the carrier protein. A binding site for T_4 has been located within an open channel passing through the protein along its long axis (Fig. 6-6). TBPA has been suggested as a model for the binding site for the nuclear receptor

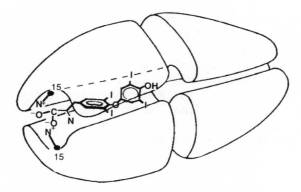

Figure 6-6. A schematic drawing of the four identical subunits of thyroxine-binding prealbumin (TBPA) associated to form a central channel. T_4 is shown in the channel with the carboxylate ion of its alanine side chain forming ion-paired associations with the ε-ammonium groups of paired lysine residues of TBPA. Adapted from E. C. Jorgensen, Thyromimetic and antithyroid drugs, in *Burger's Medicinal Chemistry*, Part III, 4th ed. (M. E. Wolff, ed.), pp. 103–145. Copyright 1981 John Wiley and Sons, Inc., New York.

protein (NRP) (see Section 6.2.4.2). Jorgensen (1981) has reviewed the comprehensive structure–activity relationships (SAR) that have been derived from studies of the relative binding efficiencies of T_3, T_4, and analogues to TBPA, as well as to TBG. TBPA is also involved with the transport and metabolic stabilization of retinol (vitamin A). Recently, TBPA has been related structurally to iodothyronine deiodinase type I. One of the three deiodinase isoenzymes responsible for the formation of T_3 and rT_3, type I enzyme accounts for the major part of the production of T_3. Analysis of SAR for iodothyronine analogues and metabolites revealed that the structural requirements for ligand binding to type I deiodinase had high analogy with the binding site of TBPA. Binding data obtained with TBPA have been used to design inhibitors of type I iodothyronine deiodinase (Köhrle *et al.*, 1987). The role of iodothyronine deiodinases in thyroid hormone metabolism and regulation of thyroid biosynthesis will be discussed in greater detail in Section 6.2.5.2.

6.2.3.3 Transcellular and Transnuclear Transport of Thyroid Hormones

Most of the hormonal activities of thyroid hormones are produced by the binding of T_3 to a chromatin-associated nuclear receptor (Section 6.2.4.2), and thus the hormones must gain entry into the cell. The mechanism of translocation of T_3 and T_4 from plasma to the intracellular compartment has been a matter of considerable debate and is the subject

of much current research. Passive diffusion was initially considered the
likely mechanism for transport, but more recently evidence has
accumulated for the existence of receptor-mediated limited-capacity active
transport systems in several cells (Blondeau *et al.*, 1988, and references
therein). Similarly, the assumption that free T_3 enters the nucleus from the
cytosol by simple diffusion has been shown not to hold for isolated
hepatocytes, but rather evidence was obtained for the intervention of
an energy-dependent stereospecific transport system (Oppenheimer and
Schwartz, 1985; Mooradian *et al.*, 1985).

6.2.4 Thyroid Hormone Receptors and Biological Response

6.2.4.1 The Nuclear Receptor

Recognition of the involvement of thyroid hormones with energy-
generating metabolism led early researchers to look for a direct interaction
of hormones with energy-producing reactions. The unique iodine-
containing diphenyl ether structural unit further led investigators to seek a
functional role for iodine or the ether oxygen or both. Several theories were
formulated and discarded over the years including a proposal by Niemann
in 1950 that a reversible oxidation of T_4 to a quinoid form coupled with
electron transfer to an energy-generating system. In 1957, Szent-Györgyi
invoked a unique role for iodine in his proposal that energy transfer
involved the excited triplet state of iodine [these and other proposals are
reviewed by Jorgensen (1981)]. However, accumulating evidence, such as
the presence of a lag period before the onset of response and the ability of
inhibitors of protein synthesis to block hormonal response, gradually made
it clear that the major effects of the hormones were being mediated through
stimulation of protein synthesis. Oppenheimer *et al.* (1972) reported
binding of T_3 to nuclei of rat liver and kidney cells and characterized the
high-affinity, low-capacity binding site as a 60,000–70,000-molecular weight
protein (1973). This and subsequent research which solidified the status of
the T_3 nuclear receptor complex as a basic unit of thyroid hormone action
has been reviewed by Oppenheimer (1979). Activation of this receptor by
T_3 binding is thought to stimulate the synthesis of specific mRNA sequences coding for T_3-inducible proteins, syntheses of which proteins in turn
lead to the observed biological effects. For example, a T_3-induced increase
in Na^+/K^+-ATPase activity has received considerable attention as a
possible cause of increased oxygen consumption associated with thyroid
hormone action. According to the nuclear receptor hypothesis, this could
result from a nuclear receptor response to T_3 that produces increased

mRNA that, in turn, directs the synthesis of more sodium pumps. However, a more direct action of T_3 on ion transport may also be involved (reviewed by Sterling, 1986).

Much recent research has concentrated on the molecular events that follow binding of T_3 to the nuclear receptor. A mechanism for the control of gene transcription involves the interaction of DNA-binding proteins with specific DNA sequences. The localization of T_3 binding sites on defined regions of chromatin suggests that the receptor may function as such a regulatory protein (Fig. 6-7) (De Nayer, 1987). Thus, the nuclear T_3 receptor has been located on promoter regions of human growth hormone genes and lactogen genes isolated from human lymphoblastoid IM-9 cells. In addition, the presence of T_3 was necessary for the binding of the T_3 receptor to the growth hormone gene promoter, suggesting that binding of T_3 to the receptor promotes recognition of specific gene regions by the receptor protein (Barlow *et al.*, 1986; Barlow and De Nayer, 1988). Adding significance to these results is the recent demonstration that the T_3 receptor belongs to a family of DNA-binding proteins that have sequence homology to a cellular oncogene (Sap *et al.*, 1986; Weinberger *et al.*, 1986). These

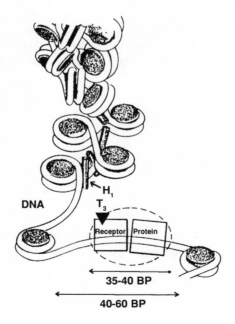

Figure 6-7. A schematic diagram showing association of the T_3 receptor with other protein(s) on a DNA region between two nucleosomes (linker DNA) (De Nayer, 1987; used with permission from S. Karger AG).

unifying concepts have done much to stimulate progress in the understanding of the vast and varied responses to thyroid hormones.

6.2.4.2 Structural Requirements for Binding to the Nuclear Receptor and for Biological Activity

The critical roles played by thyroid hormones in vertebrate physiology
have prompted extensive SAR studies. Jorgensen (1981) has compiled a
comprehensive summary of SAR data for iodothyronines and analogues, a
summary which includes nuclear binding efficiencies and data on relative *in
vivo* activities. The "mystique of iodine" has been shown to be illusionary,
first by the demonstrated activity of other halogenothyronines and, second,
by the potent activity of 3'-alkyl-3,5-diiodothyronines (Frieden, 1981). A
rank order of affinity for the rat liver nuclear receptor was shown to be
i-propyl $>$ I $>$ t-butyl $>$ methyl. The schematic representation of the receptor–ligand complex shown in Fig. 6-8 has been formulated from a vast
amount of SAR data. The requirements for a hormonally active molecule,
as summarized by Frieden (1981), include (1) a lipophilic core of two
mutually perpendicular rings bridged by O, S, or C; (2) an amino acid or
other ionic side chain of two or three carbons *para* to the bridging atom;
(3) a phenolic group *para* to the bridging atom; (4) nonpolar groups in the
3 and 5 positions; and (5) a bulky nonpolar group in the 3' position. As
indicated by Fig. 6-8, this latter group has been shown to adopt a distal,

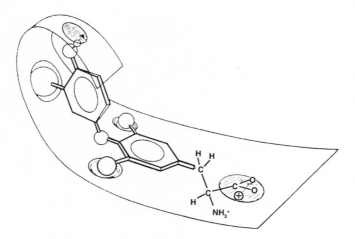

Figure 6-8. Schematic representation of the receptor–T_3 complex (Frieden, 1981; used with
permission of the publisher and author).

rather than proximal, orientation. For more thorough discussions of these SAR data, the reviews by Alexander (1984), Frieden (1981), and Jorgensen (1981) are recommended.

Recently, Chae and McKinney (1988) have explored the possible contribution of a "stacking complexation" mechanism to receptor binding. Thyroid hormone analogues were shown from NMR data to function as electron acceptors in molecular complexes with aromatic donors, and binding free energy of the complexes correlated well with binding efficiencies for the T_3 nuclear receptor. The stacking complexation mechanism had been proposed previously for the binding of halogenated aromatic hydrocarbons to the Ah (dioxin) receptor, and a possible role of thyroid hormone-binding proteins in mediating carcinogenicity and toxicity of certain aromatic compounds has been suggested (see Chapter 8).

6.2.4.3 Non-nuclear Receptors

In addition to cell membrane receptors that appear to mediate transmembrane movement of thyroid hormones (Section 6.2.3), membrane binding sites linked to other biological responses have been described. For example, Segal and Ingbar (1984) reported that T_3 induced an immediate increase in calcium accumulation in rat thymocytes, a response that was insensitive to inhibitors of protein synthesis. Direct effects have also been observed due to T_3 binding to saturable receptors in the mitochondrial membranes from several tissue types, where a direct effect on oxidative phosphorylation has been proposed (reviewed by Sterling, 1986), and binding of T_3 and T_4 to cytosolic proteins has been described [for a concise discussion of the possible role(s) of non-nuclear T_3 receptors in mediating thyroid hormone action, see Sterling, 1986]. A direct biological response has not been linked to binding to cytosolic receptors (De Nayer, 1987).

6.2.5 Regulation of Thyroid Hormone Synthesis and Release

6.2.5.1 TRH and TSH

The thyroid–pituitary–hypothalamus axis is responsible for the maintenance of homeostasis of thyroid hormones. TSH, released from the pituitary, stimulates the synthesis and release of thyroid hormones by mechanisms thought to be mediated primarily by cAMP as a second messenger (Section 6.2.2; Chapter 5, Section 5.3.3). The synthesis and release of TSH, in turn, is stimulated by thyrotropin-releasing hormone (TRH) released from the hypothalamus. In feedback control, thyroid hormones regulate TSH secretion by blunting the stimulating action of TRH on the pituitary. Short-term exposure blocks release of TSH, while longer exposure inhibits TSH synthesis in the pituitary through nuclear receptor-mediated TSH gene regulation. Low levels of circulating T_3 and T_4 cause an exaggerated response to TRH, increased TSH release, and stimulated hormone production. In addition to regulation at the pituitary level, there is recent evidence that T_3 and T_4 can regulate TSH secretion by inhibition of hypothalamic TRH secretion (Iriuchijima et al., 1985) (for a further discussion of TSH and TRH biochemistry, see Kannan, 1987).

6.2.5.2 Iodothyronine Deiodinases and Regulation

The importance of deiodinating enzymes in thyroid function is demonstrated unequivocally in the conversion of T_4 to T_3. The further deiodination of T_3 to inactive diiodothyronines represents one mechanism for attenuating hormonal activity. In addition, evidence has accumulated for the existence of a relationship between the regulation of deiodination of thyroid hormones in target cells and the intracellular effects of thyroid hormones on hypothalamic and pituitary functions (Larsen and Silva, 1983). Iodothyronine deiodinases have been grouped into three classes—types I, II, and III—distinguished by such parameters as substrate specificities, kinetic patterns, and deiodination site (for reviews, see Köhrle et al., 1987; Kaplan, 1984). In the rat, the ability of T_4 to inhibit TRH release and to stimulate growth hormone synthesis requires its conversion to T_3 in the pituitary by type II enzyme, and the same is probably true in humans. A regulatory role has been proposed that involves modulation of type II deiodination in the pituitary and hypothalamus (Kaplan, 1984). Deiodination of T_4 by type III enzyme occurs exclusively on the tyrosine ring to give rT_3. Since rT_3 is 50 times more effective as an inhibitor of type II deiodinase than is T_3, a physiological role for rT_3 has been proposed as part of a regulatory circuit. By this postulated mechanism, rT_3 produced by type III enzyme in surrounding CNS tissue enters the pituitary and inhibits type II deiodination of T_4, thus decreasing T_3 concentration and increasing TSH secretion (Köhrle et al., 1987).

The above are examples of the complex interplay of deiodination, hormonal activity, and regulation of hormone levels. It is clear that not only does iodine serve as an important structural feature to promote receptor binding, but also that the readily available routes for control of iodination levels present mechanisms for fine-tuning of hormonal levels.

6.2.6 Metabolism and Deactivation of Thyroid Hormones

Reductive monodeiodination reactions account for approximately 80% of the disposal of T_4 (reviewed by Engler and Burger, 1984). Further monodeiodination, producing a cascade of lesser iodothyronines, is responsible for the degradation of one-third to one-half of the T_3 and one-third of the rT_3. Approximately 20% of labeled T_4 is excreted as thyronine (T_0). Other degradative pathways combine side-chain modification and/or formation of O-sulfate and O-glucuronide conjugates with deiodination. Thus, tetraiodothyroacetic acid (tetrac) and triiodothyroacetic acid (triac) are formed in the liver by deamination and oxidative decarboxylation of T_4 and T_3, respectively (Fig. 6-9). These are further metabolized by deiodination or conjugation.

T_4 is also subject to oxidative deiodination, in a process that involves ether link cleavage (ELC), producing DIT, iodide, and iodinated protein (Fig. 6-10). This mechanism of T_4 deiodination accounts for 40–50% of all the T_4 degraded during human leukocyte phagocytosis (a process that generates hydrogen peroxide), and the suggestion has been made that ELC of T_4 may contribute to the bactericidal capacity of the leukocyte (Engler and Burger, 1984, and references therein). The ELC pathway has been demonstrated to be an alternative pathway of iodothyronine degradation *in vivo* in the rat (Balsam *et al.*, 1983).

T_3 and T_4 are derivatives of Tyr, and Tyr is the precursor of such aminergic neurotransmitters as dopamine and norepinephrine. In consideration of this, and of the central stimulating and peripheral sympathomimetic effects of thyroid hormones (see Gross, 1986, for a recent review), Dratman (1974) proposed that the enzymatic processes that are involved with the conversion of tyrosine to biogenic amines might also process T_3 or T_4 to neuroactive amines. Support for such a role came from the localization of T_3 in rat synaptoneurosomes following administration of either T_3 or T_4 (Dratman *et al.*, 1976). Further, the ethylene glycol (Fig. 6-11) identified as a urinary metabolite of T_3 must arise *in vivo* from

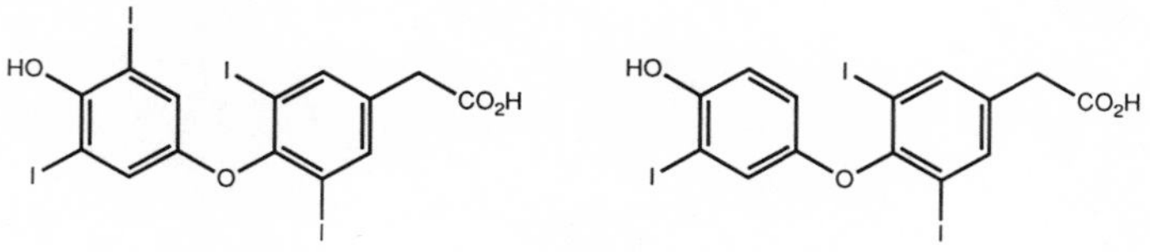

Tetrac   Triac

Figure 6-9. Tetraiodothyroacetic acid (tetrac) and triiodothyroacetic acid (triac), formed by deamination and oxidative decarboxylation of T_4 and T_3, respectively.

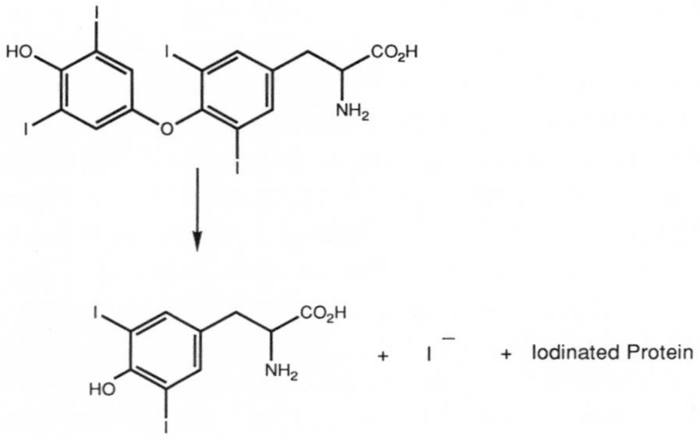

Figure 6-10. Ether link cleavage during oxidative deiodination of T_4.

decarboxylase and β-hydroxylase action on T_3 (Gordon *et al.*, 1985). In order to identify possible metabolites responsible for neuroactivity, a series of new side-chain-modified analogues of T_4 and T_3 were prepared (Han *et al.*, 1987). Included in this series was the T_3 analogue of norepinephrine (ethanolamine side chain) (Fig. 6-11). At $10^{-7} M$, this analogue amplified the effects of norepinephrine on the heart. However, at a higher concentration ($10^{-5} M$), this compound acted as an inhibitor of both rate and force of cardiac contraction, causing rapid cardiac arrest. No explanation of this anomalous behavior is readily apparent, and, to date, this analogue has not been detected *in vivo*. A related side-chain-modified T_3, T_3-amine (Fig. 6-11), has been shown to be a potent adrenergic and dopaminergic agonist (Cody *et al.*, 1984). This analogue also has not been detected *in vivo*. A precise adrenergic function(s) for thyroid hormones remains to be identified and confirmed.

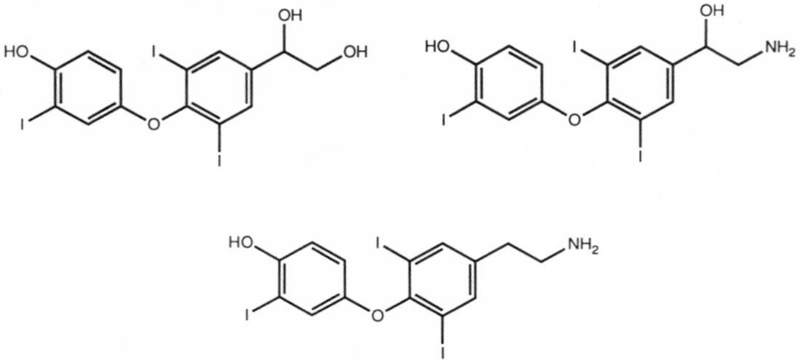

Figure 6-11. Side-chain modified analogues of T_3.

6.2.7 Conclusion

As vertebrates emerged from the marine environment, efficient biochemical mechanisms evolved that concentrate I^- in the thyroid gland, iodinate tyrosine residues in protein, and ultimately produce the thyroid hormones, T_3 and T_4. The essential involvement of these hormones in a host of biological processes, many mediated through T_3 receptor modulation of gene transcription, underlines the importance of this evolutionary event. I^- provides the bulky, nonpolar substituents required for receptor binding, and the available mechanisms for facile iodination (biosynthesis of hormone) and deiodination provide mechanisms for regulation of hormone levels that are interactive with TRH–TSH control mechanisms. Current research is providing detailed information on the mechanism of thyroid hormone actions, as well as providing leads to chemotherapeutic approaches to the management and cure of diseases related to thyroid dysfunction.

References

Alexander, N. M., 1984. Iodine, in *Biochemistry of the Essential Ultratrace Elements* (E. Frieden, ed.), Plenum Press, New York, pp. 33–53.

Balsam, A., Sexton, F., Borges, M., and Ingbar, S. H., 1983. Formation of diiodotyrosine from thyroxine. Ether-link cleavage, an alternate pathway of thyroxine metabolism, *J. Clin. Invest.* 72:1234–1245.

Barlow, J. W., and De Nayer, P., 1988. Characterization of thyroid hormone receptors in human IM-9 lymphocytes, *Acta Endocrinol.* 117:327–332.

Barlow, J. W., Voz, M. L. J., Eliard, P. H., Mathy-Hartert, M., De Nayer, P., Economidis, I. V., Belayew, A., Martial, J. A., and Rousseau, G. G., 1986. Thyroid hormone receptors bind to defined regions of the growth hormone and placental lactogen genes, *Proc. Natl. Acad. Sci. USA* 83:9021–9025.

Blondeau, J.-P., Osty, J., and Francon, J., 1988. Characterization of the thyroid hormone transport system of isolated hepatocytes, *J. Biol. Chem.* 263:2685–2692.

Braverman, L. E., Ingbar, ·S. H., and Sterling, K., 1970. Conversion of thyroxine (T_4) to triiodothyronine (T_3) in athyreotic human subjects, *J. Clin. Invest.* 49:855–864.

Chae, K., and McKinney, J. D., 1988. Molecular complexes of thyroid hormone tyrosyl rings with aromatic donors. Possible relationship to receptor protein interactions, *J. Med. Chem.* 31:357–362.

Cody, V., Meyer, T., Dohler, K. D., Hesch, R. D., Rokos, H., and Marko, M., 1984. Molecular structure and biochemical activity of 3,5,3'-triiodothyronamine, *Endocrine Res.* 10:91–99.

Davidson, B., Neary, J. T., Strout, H. V., Maloof, F., and Soodak, M., 1978. Evidence for a thyroid peroxidase associated "active iodine" species, *Biochim. Biophys. Acta* 522:318–326.

Deme, D., Pommier, J., and Nunez, J., 1978. Specificity of thyroid hormone synthesis. The role of thyroid peroxidase, *Biochim. Biophys. Acta* 540:73–82.

De Nayer, P., 1987. Thyroid hormone action at the cellular level, *Horm. Res.* 26:48–57.

Di Lauro, R., Obici, S., Condliffe, D., Ursini, V. M., Musti, A., Moscatelli, C., and Avvedimento, V. E., 1985. The sequence of 967 amino acids at the carboxyl end of rat thyroglobulin. Location and surroundings of two thyroxine-forming sites, *Eur. J. Biochem.* 148:7–11.

Dratman, M. B., 1974. On the mechanism of action of thyroxine, an amino acid analog of tyrosine, *J. Theor. Biol.* 46:255–270.

Dratman, M. B., Crutchfield, F. L., Axelrod, J., Colburn, R. W., and Thoa, N., 1976. Localization of triiodothyronine in nerve ending fractions of rat brain, *Proc. Natl. Acad. Sci. USA* 73:941–944.

Engler, D., and Burger, A. G., 1984. The deiodination of the iodothyronines and of their derivatives in man, *Endocrine Rev.* 5:151–184.

Formisano, S., Moscatelli, C., Zarrilli, R., Di Ieso, B., Acquaviva, R., Obici, S., Palumbo, G., and Di Lauro, R., 1985. Prediction of the secondary structure of the carboxy-terminal third of rat thyroglobulin, *Biochem. Biophys. Res. Commun.* 133:766–772.

Frieden, E., 1981. Iodine and the thyroid hormones, *Trends Biochem. Sci.* 6:50–53.

Garavet, J.-M., Nunez, J., and Cahnmann, H. J., 1980. Formation of dehydroalanine residues during thyroid hormone synthesis in thyroglobulin, *J. Biol. Chem.* 255:5281–5285.

Garavet, J.-M., Cahnmann, H. J., and Nunez, J., 1981. Thyroid hormone synthesis in thyroglobulin. The mechanism of the coupling reaction, *J. Biol. Chem.* 256:9167–9173.

Gordon, J. T., Dratman, M. B., and Wassell, M. S., 1985. Isolation, characterization and structure elucidation of a new urinary thyroid hormone metabolite whose in vivo production requires side chain decarboxylation and β-hydroxylation, Annual Meeting of the Endocrine Society, Baltimore, p. 21 (Abstract 847).

Gross, G., 1986. Effect of thyroid hormones on myocardial adrenoceptors and adrenoceptor-mediated cardiovascular responses in the rat, in *New Aspects of the Role of Adrenoceptors in the Cardiovascular System* (H. Grobecker, A. Philips, and K. Starke, eds.), Springer-Verlag, Berlin, pp. 121–128.

Han, S.-Y., Gordon, J. T., Bhat, K., Dratman, M. T., and Joullie, M. M., 1987. Synthesis of side chain-modified iodothyronines, *Int. J. Pept. Protein Res.* 30:652–661.

Iriuchijima, T., Rogers, D., and Wilber, J. F., 1985. TSH secretory regulation: New evidence that triiodothyronine (T_3) and thyroxine (T_4) can inhibit TRH secretion both in vivo and in vitro, in *Frontiers in Thyroidology*, Vol. 1 (G. Medeiros-Neto and E. Gaiton, eds.), Plenum Medical Book Co., New York, pp. 233–236.

Jorgensen, E. C., 1981. Thyromimetic and antithyroid drugs, in *Burger's Medicinal Chemistry*, Part III, 4th ed. (M. E. Wolff, ed.), John Wiley and Sons, New York, pp. 103–145.

Kannan, C. R., 1987. *The Pituitary Gland*, Plenum Medical Book Co., New York, pp. 145–169.

Kaplan, M. M., 1984. The role of thyroid hrmone deiodination in the regulation of hypothalamo-pituitary function, *Neuroendocrinology* 38:254–260.

Köhrle, J., Brabant, G., and Hesch, R.-D., 1987. Metabolism of the thyroid hormones, *Horm. Res.* 26:58–78.

Lamas, L., Anderson, P. C., Fox, J. W., and Dunn, J. T., 1989. Consensus sequences for early iodination and hormonogenesis in human thyroglobulin, *J. Biol. Chem.* 264:13541–13545.

Larsen, P. R., and Silva, J. E., 1983. Intrapituitary mechanisms in the control of TSH secretion, in *Molecular Basis of Thyroid Hormone Action* (J. H. Oppenheimer and H. H. Samuels, eds.), Academic Press, New York, pp. 352–385.

Malthiery, Y., and Lissitzky, S., 1985. Sequence of the 5′-end quarter of the human-thyroglobulin messenger ribonucleic acid and of its deduced amino-acid sequence, *Eur. J. Biochem.* 147:53–58.

Malthiery, Y., and Lissitzky, S., 1987. Primary structure of human thyroglobulin deduced from the sequence of its 8448-base complementary DNA, *Eur. J. Biochem.* 165:491–498.

Marriq, C., Lejeune, P.-J., Venot, N., Rolland, M., and Lissitzky, S., 1986. Characterization of a hormonogenic domain from human thyroglobulin, *FEBS Lett.* 207:302–306.

Mercken, L., Simons, M.-J., Swillins, S., Massaer, M., and Vassart, G., 1985. Primary structure of bovine thyroglobulin deduced from the sequence of its 8,431-base complementary DNA, *Nature* 316:647–651.

Mooradian, A. D., Schwartz, H. L., Mariash, C. N., and Oppenheimer, J. H., 1985. Transcellular and transnuclear transport of 3,5,3'-triiodothyronine in isolated hepatocytes, *Endocrinology* 117:2449–2456.

Nakamura, M., Yamazaki, I., Nakagawa, H., Ohtaki, S., and Ui, N., 1984. Iodination and oxidation of thyroglobulin catalyzed by thyroid peroxidase, *J. Biol. Chem.* 259:359–364.

Neary, J. T., Soodak, M., and Maloof, F., 1984. Iodination by thyroid peroxidase, *Methods Enzymol.* 107:445–475.

Nunez, J., 1984. Thyroid hormones: Mechanism of phenoxy ether formation, *Methods Enzymol.* 107:476–488.

Oppenheimer, J. H., 1979. Thyroid hormone action at the cellular level, *Science* 203:971–979.

Oppenheimer, J. H., and Schwartz, H. L., 1985. Stereospecific transport of triiodothyronine from plasma to cytosol and from cytosol to nucleus in rat liver, kidney, brain, and heart, *J. Clin. Invest.* 75:147–154.

Oppenheimer, J. H., Koerner, D., Schwartz, H. L., and Surks, M. I., 1972. Specific nuclear triiodothyronine binding sites in rat liver and kidney, *J. Clin. Endocrin. Metab.* 35:330–333.

Palumbo, G., 1987. Thyroid hormonogenesis. Identification of a sequence containing iodophenyl donor site(s) in calf thyroglobulin, *J. Biol. Chem.* 262:17182–17188.

Sap, J., Muñoz, A., Damm, K., Goldberg, Y., Ghysdael, J., Leutz, A., Beug, H., and Vennström, B., 1986. The c-erb-A protein is a high-affinity receptor for thyroid hormone, *Nature* 324:635–640.

Segal, J., and Ingbar, S. H., 1984. An immediate increase in calcium accumulation by rat thymocytes induced by triiodothyronine: Its role in the subsequent metabolic response, *Endocrinology* 115:160–166.

Sterling, K., 1986. The molecular mechanism of thyroid hormone action at the cellular level, in *The Thyroid Gland* (L. van Middlesworth, ed.), Year Book Medical Publishers, Chicago, pp. 203–229.

Weinberger, C., Thompson, C. C., Ong, E. S., Lebo, R., Gruol, D. J., and Evans, R. M., 1986. The c-erb-A gene encodes a thyroid hormone receptor, *Nature* 324:641–646.

Biohalogenation

7

7.1 Introduction

A vast number of halogenated compounds are present in the biosphere. While most of these have resulted from the work of synthetic chemists, a large number are naturally occurring—products of nature's halogenation processes. The majority of these naturally occurring halogenated compounds are produced by haloperoxidases, a group of enzymes widely distributed in nature that are capable of halogenating a broad spectrum of organic substrates. A comprehensive review of recent research involved with isolation, identification, and biological evaluation of naturally occurring halogenated compounds, now an active area of natural products chemistry, is beyond the scope of this chapter. Recent reviews are available and will be cited (e.g., Neidleman and Geigert, 1986, 1987). While examples of halometabolites produced by various species will be given —particularly compounds that have useful medical applications—the emphasis in this chapter will be placed on the biochemical processes that produce the halometabolites. Thus, the mechanism of halogenation by haloperoxidases will be reviewed, and specific examples of haloperoxidases will be given, including recently identified vanadium-containing nonheme haloperoxidases.

7.2 Naturally Occurring Halometabolites

Compounds synthesized by plants and animals are referred to as primary or secondary metabolites, depending on whether or not they are involved in essential metabolic functions of the producing species. While secondary metabolites previously had been considered by-products of detoxification, biosynthesis of these compounds is now thought to be an important evolutionary process that increases the survival chances of the

155

species (Fenical, 1982). Also, until recently, the belief was common that the occurrence of covalently bound halogen in living organisms was an infrequent event (e.g., Fowden, 1968). Now, however, there is increasing recognition that naturally occurring halogen-containing compounds are widespread in nature, particularly as secondary products of metabolism of bacteria, fungi, and marine organisms, and that species capable of producing these metabolites are also far less rare than originally thought (Siuda and DeBernardis, 1973; Neidleman, 1975; Neidleman and Geigert, 1986). The study of these metabolites and the search for additional compounds are spurred by the demonstration that many of these compounds —as well as nonhalogenated secondary metabolites—are potent antibacterial, antifungal, antitumor, and antiviral agents (e.g., Munro *et al.*, 1987). Selected examples of halometabolites are given in this section as an introductory illustration of the range of halogenated products that have been identified in nature. Additional examples will be presented in later sections dealing with biochemical mechanisms of halogenation and biosynthetic schemes that produce halometabolites. Much more extensive reviews of halometabolites are given in the literature cited in this discussion.

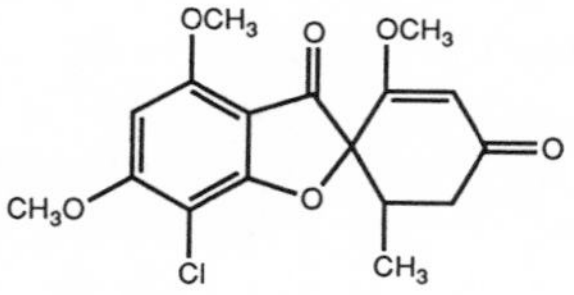

Griseofulvin

1

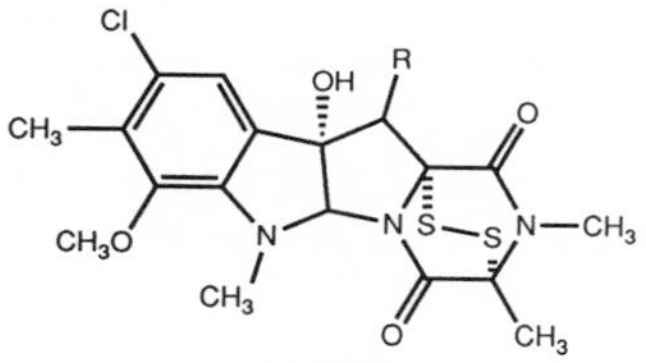

Sporidesmin: R = OH
Sporidesmin b: R = H

2

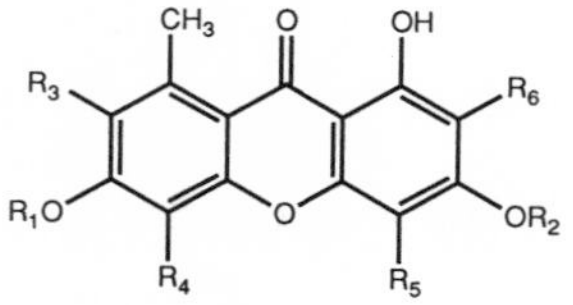

R$_1$ and R$_2$ = CH$_3$ or H
R$_3$, R$_4$, R$_5$ and R$_6$ = H or Cl

3

Figure 7-1. Examples of chlorometabolites of fungi and lichens. From Neidleman and Geigert (1986); see this source for references to original work and for additional examples.

7.2.1 Chlorometabolites Produced by Fungi and Lichens

An early survey of 139 species of fungi was carried out by Cutterbuck *et al.* (1940). From an assumption that at least a 25% conversion of Cl⁻ to organic chlorine must occur for utilization to be considered significant, the conclusion was reached that <4% of fungi surveyed produced chlorinated metabolites. However, a later reinterpretation of these data, using the concentrations of Cl⁻ in the final product as a measure of organochlorine production, led to the conclusion that >50% of the fungi produced chlorometabolites, with the possibility that almost all did (Neidleman, 1975). Examples of chlorometabolites produced by fungi are the antifungal agent griseofulvin (1), the toxic sporidesmins (e.g., 2) (responsible for facial eczema of grazing sheep), and several chlorinated xanthones 3 isolated from lichens (symbiotic associations of fungi with unicellular algae) (Fig. 7-1). There are no reported examples of bromo-metabolites produced by fungi, although the bromo analogue of griseo-fulvin can be produced by replacing Cl⁻ with Br⁻ in the fermentation medium (for a tabulation of additional chlorometabolites and their fungal sources, see Neidleman and Geigert, 1986).

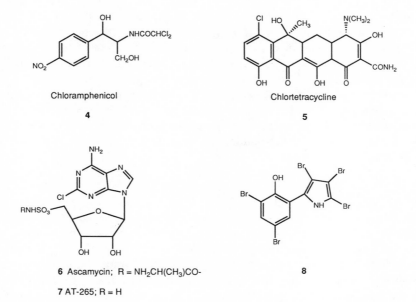

Figure 7-2. Examples of halometabolites of bacteria. From Neidleman and Geigert (1986); see this source for references to original work and for additional examples.

7.2.2 Halometabolites Produced by Bacteria

Although the prototypical antibiotic, penicillin, is produced by a fungus, most antibiotics are derived from bacteria and bacteria-like actinomycetes. Chloramphenicol (**4**), elaborated by *Streptomyces venezuelae*, chlortetracycline (**5**) from *Streptomyces aureofaciens*, and the nucleoside antibiotics ascamycin (**6**) and AT-265 (**7**) (Fig. 7-2) (Osada and Isono, 1985) from *Streptomyces* spp. are examples of potent chlorinated antibiotics that are produced from bacteria in fermentation processes. A diverse array of other chloro- and bromometabolites, for example, the extremely bromine-rich pyrrole **8** (Fig. 7-2) produced by *Pseudomonas bromoutilis*, are elaborated by marine and terrestrial bacteria (for further examples and for references to original work, see Neidleman and Geigert, 1986, 1987).

7.2.3 Halometabolites Produced by Marine Organisms

Marine plants and animals are particularly rich sources of halometabolites, a result of the abundance of halides in seawater (Cl^-, 19,000 mg/liter; Br^-, 65 mg/liter; I^-/IO_3^-, 5×10^{-4} mg/liter) (Fenical, 1982). In 1974, during an expedition of the research ship *Alpha Helix* to Baja California, over 100 marine organisms were screened for halometabolites. The discovery that approximately 25% of the organisms contained $> 10\,\mu g$ of organic halogen per gram of tissue clearly demonstrated the widespread occurrence of halogenated compounds in the marine environment (Theiler *et al.*, 1978, and references therein). Terpenes, acetogenins, tyrosine-derived compounds, and alkaloids are among important classes of marine natural products containing halogen. Bromine is the most commonly occurring halide in marine natural products, a point that will be addressed below.

7.2.3.1 Terpenes

Halides are incorporated into marine products derived from isoprene, and Br^- also appears to play an essential role in terpene biosynthesis in marine organisms (see Section 7.4). The participation of Br^- in marine terpene biosynthesis is reflected in the impressive number (several hundred) of halogenated terpenes that have been identified, many of which have potentially useful biological activities. For example, the polyhalogenated monoterpenes **9** and **10** from the red alga *Plocamium cartilagineum* and polyhalogenated sesquiterpenes from the red alga *Laurencia obtussa*, including isoobtusol (**11**) and isoobtusol acetate (**12**) (Fig. 7-3), showed potent activity against HeLa 229 human carcinoma cells (Gonzalez *et al.*, 1978, 1982; reviewed by Munro *et al.*, 1987). (For additional examples, see Neidleman and Geigert, 1986; Fusetani, 1987; Munro *et al.*, 1987).

7.2.3.2 Acetogenins

Another major class of marine natural products are polyketides, including acetogenins—linear, nonbranched compounds produced by condensation of acetyl CoA units. While this primary metabolic pathway frequently is modified in both terrestrial and marine organisms to produce secondary metabolites with diverse and unusual structures (see below), the biosynthesis of simple halogenated ketones and hydrocarbons also has evoked much interest. For example, 1-iodo-3,3-dibromo-2-heptanone (**13**), 1,1,3,3-tetrabromo-2-heptanone (**14**), and similar compounds (Fig. 7-4) were isolated during the *Alpha Helix* expedition from the red alga *Bonnemaisonia hamifera*, an organism having very high halogen content that also displayed high antimicrobial activity. These represented the first simple brominated aliphatic ketones found in nature, and the iodinated ketone was notable as an early example of an iodinated natural product not related to thyroid hormones (Siuda *et al.*, 1975). Halogenated ketones of the general structure CX_3COCX_2R, where R may be halogen or hydrogen, also have been isolated from several other marine organisms. Antibacterial and antifungal properties displayed by many of these compounds may

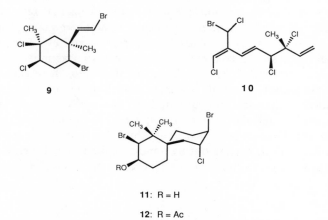

Figure 7-3. Examples of monoterpenes from marine organisms. From Munro *et al.* (1987); see this source for original references and additional examples; see also Fusetani (1987) and Neidleman and Geigert (1986) for additional examples.

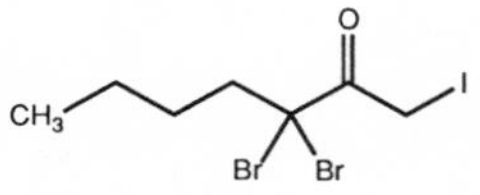

13 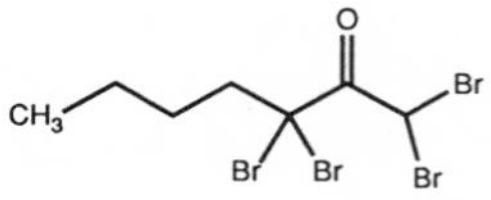   **14**

BrCH=CHCBr=CH(CH₂)₄CH=CHC≡C(CH₂)₃CO₂H

15

Figure 7-4. Halogenated acetogenins isolated from marine organisms (Siuda *et al.*, 1975; Schmitz and Gopichand, 1978).

reflect their ability to function as alkylating agents, similar to iodoacetate (see Chapter 1 in Vol. 9B of this series) (Beissner *et al.*, 1981, and references therein).

Brominated examples of more complex acetogenins include the dibromoacetylenic acid **15** (Fig. 7-4), produced by the sponge *Xestospongia muta*. This compound showed little cytotoxicity or antitumor activity (Schmitz and Gopichand, 1978).

Volatile halogenated organic compounds, for example, the halomethane derivatives **16–20** (Fig. 7-5), were isolated initially from tropical marine plants, exemplified by the red seaweed *Asparagopsis taxiformis*. Bromoform is the major volatile metabolite in this seaweed (McConnell and Fenical, 1977b). More recently, a number of marine macroalgae—including brown algae, red algae, and green algae—from temperate zones have been identified that produce volatile halocarbons and release these to the seawater at rates of nanograms to micrograms per gram of dry algae per day. Based on the rates of halocarbon release and estimates of global macroalgal biomass, the conclusion was reached that as much as 10^{10} g of volatile organobromine may be released to the atmosphere each year, an amount that would constitute a major contribution to the atmospheric accumulation of this class of compound (by comparison, 5×10^{12} g/year of methyl chloride is released to the atmosphere). The function of these compounds in the algae is not clear. They may be simply side products of biosynthetic reactions carried out by nondiscriminating haloperoxidases, or they may provide protection from herbivores or microbes (Gschwend *et al.*, 1985).

CHBr₃	CHBr₂I	CHBr₂Cl	CHCl₃	CCl₄
16	**17**	**18**	**19**	**20**

Figure 7-5. Volatile halogenated hydrocarbons from the red seaweed *Asparagopsis taxiformis* (McConnell and Fenical, 1977b).

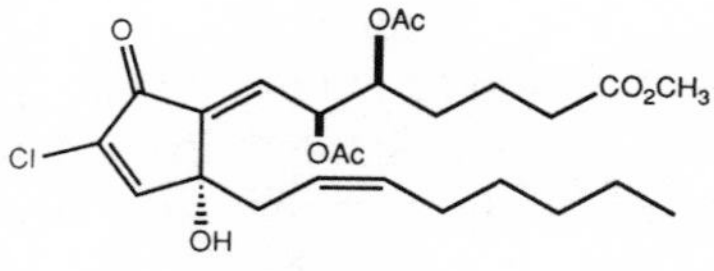

(7-E)-PUNAGLANDIN-4

21

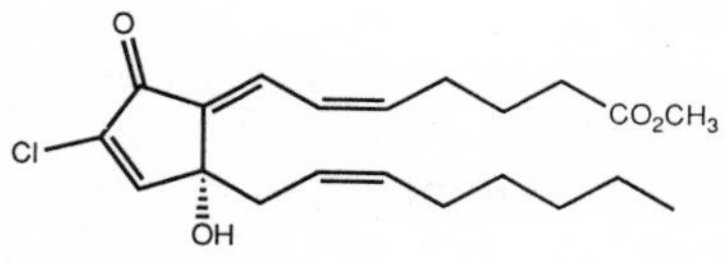

CHLOROVULONE

22

Figure 7-6. Examples of halogenated marine prostaglandins (Suzuki *et al.*, 1988, and references therein; Iguchi *et al.*, 1985).

7.2.3.3 Marine Prostaglandins

A series of halogenated marine prostaglandins [punaglandins and chlorovulones, for example, **21** and **22** (Fig. 7-6)] have recently been characterized from coral (Suzuki *et al.*, 1988; Iguchi *et al.*, 1985). The biochemistry of these compounds, some of which have impressive antitumor activity, is discussed in Chapter 4 of Vol. 9B of this series.

7.2.3.4 Halogen-Containing Marine Alkaloids

Several highly brominated indole alkaloids have been isolated from marine organisms. For example, among a series of bromoindoles obtained from the red alga *Laurencia brongniartii*, 2,3,5,6-tetrabromoindole (**23**) (Fig. 7-7) was found to be cytotoxic to L1210 leukemia cells and to have

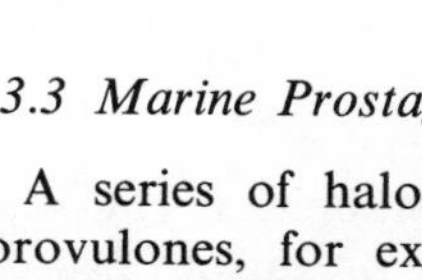

23

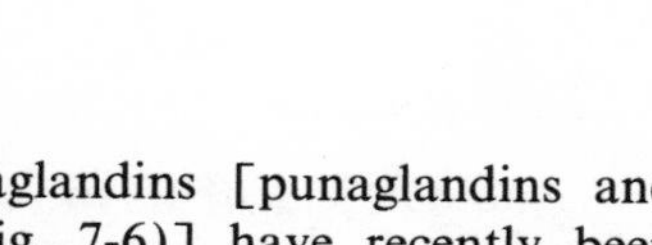

24 R = H

25 R = Br

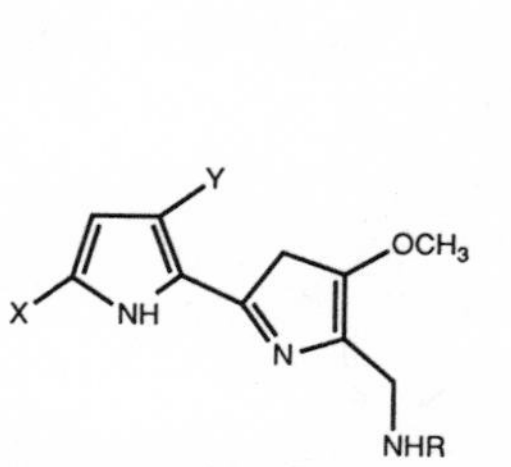

TAMBJAMINES A - D

A: X = Y = R = H
B: X = Br, Y = R = H
C: X = Y = H, R = Bu
D: X = H, Y = Br, R = Bu

26

Figure 7-7. Examples of halogenated marine alkaloids (Carter *et al.*, 1978; Tymiak *et al.*, 1985; Carte and Faulkner, 1983).

antimicrobial activity (Carter *et al.*, 1978). Brominated tryptamine derivatives, for example, *N,N*-dimethyl-5-bromotryptamine (**24**) and *N,N*-dimethyl-5,6-dibromotryptamine (**25**), have been isolated from sponges (Fig. 7-7). The dibromo analogue had significantly greater antimicrobial activity than did the monobromotryptamine (reviewed by Tymiak *et al.*, 1985).

Tambjamines A–D (**26**) (Fig. 7-7) are present in bryozoan, an organism that is a food source for nudibranchs. The ingested tambjamines then function to protect the nudibranchs from predators. These bipyrroles exhibit antimicrobial activity (Carte and Faulkner, 1983).

7.2.3.5 Halometabolites Derived from Tyrosine

Another class of nitrogen-containing halometabolites consists of compounds derived from tyrosine, a class particularly prevalent in sponges. Examples include the cytotoxic aeroplysinin-1 (**27**) and the tyrosine-derived amide dibromoverongiaquinol (**28**) (Fig. 7-8), a compound having potent antimicrobial activity (reviewed by Munro *et al.*, 1987).

7.2.4 Halometabolites Produced by Higher Plants

The chlorinated 1,4-benzoxazin-3-one (**29**) (Fig. 7-9), isolated from young corn roots, is one of the rare examples of halometabolites produced by higher terrestrial plants. Careful exclusion of all organochlorine compounds from the growth media during growth-chamber experiments confirmed that **29** is produced naturally (Le-Van and Wratten, 1984). A series of chlorinated sesquiterpenes, **30** (Fig. 7-9), were reported from Compositae (Bohlmann *et al.*, 1981) (for additional examples, see Neidleman and Geigert, 1986).

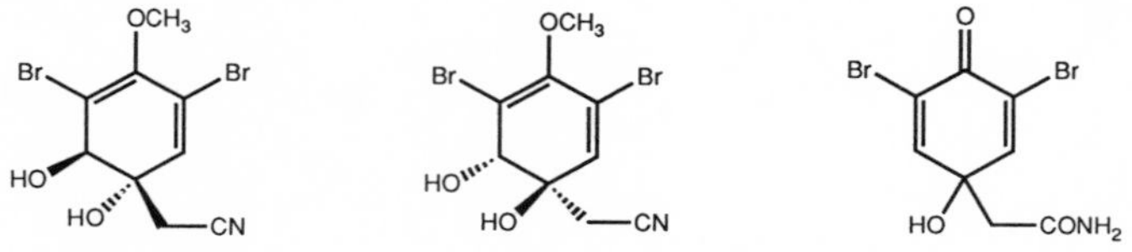

AEROPLYSININ-1   DIBROMOVERONGIAQUINOL

27   28

Figure 7-8. Marine halometabolites derived from tyrosine (reviewed by Munro *et al.*, 1987).

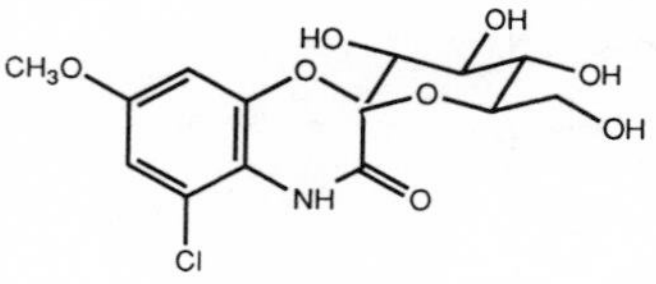

29

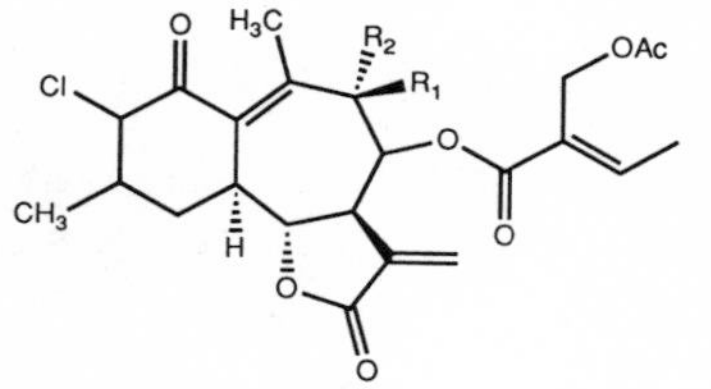

30

$R_1 = OAc, R_2 = H$
$R_1 = R_2 = H$
$R_1 = OH, R_2 = H$
$R_1, R_2 = =O$

Figure 7-9. Example of halometabolites of higher plants (Le-Van and Wratten, 1984; Bohlmann *et al.*, 1981; for additional examples, see Neidleman and Geigert, 1986).

7.2.5 Iodometabolites

The thyroid hormones (thyroxine, triiodothyronine) and derived metabolites have been discussed in detail in Chapter 6. In addition to these important iodometabolites, many iodinated secondary metabolites have been isolated from marine sources. Examples include acetogenins, for example, **13** (Fig. 7-4) and the halogenated enone **31** (Fig. 7-10) isolated from red alga, sesquiterpenes, and several iodotyrosine-derived products (Neidleman and Geigert, 1986, and references therein).

7.2.6 Fluorometabolites

Fluoroacetic acid, identified in the 1940s as a toxic component of several plants found in the Southern Hemisphere, was the first fluorometabolite found in nature. The biochemistry of fluoroacetate, and the metabolically derived fluorocitrate, is discussed in Chapter 1 of Vol. 9B

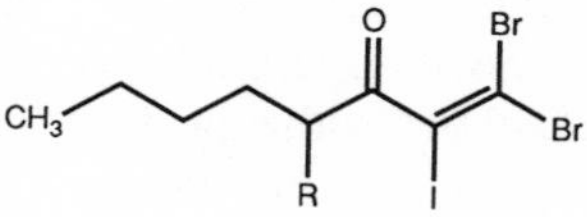

R = H, Cl

31

Figure 7-10. Additional examples of iodinated marine metabolites (reviewed by Neidleman and Geigert, 1986; see this source for original references and additional examples).

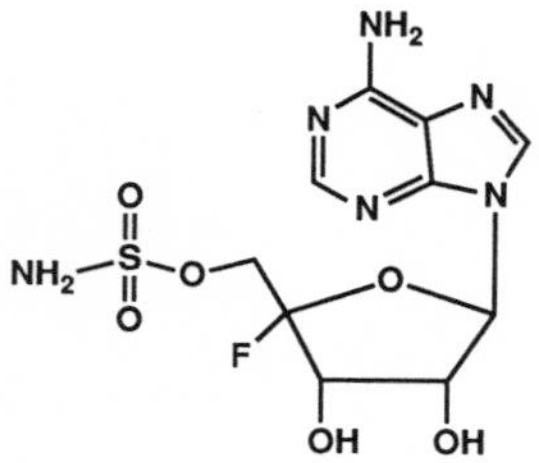

32

Figure 7-11. Nucleocidin.

of the series. Several other fluorometabolites have been found in fluoroacetate-producing plants, including fluorinated fatty acids, fluoroacetone, and fluorocitrate. The antibiotic nucleocidin, **32** (Fig. 7-11), containing a novel fluorocarbohydrate moiety, was isolated from *Streptomyces calvus* in 1957 (Neidleman and Geigert, 1986, and references therein).

7.3 Haloperoxidases

The representative sampling of halometabolites given in the previous section illustrates the diversity and flexibility of halogenating systems in nature. The vast majority of these halogenations are carried out by haloperoxidases, enzymes once thought to be rare but, similarly to their products, now recognized as being quite common. Haloperoxidases catalyze the peroxidative oxidation of I^-, Br^-, and/or Cl^- to electrophilic halogenating species, which, in the presence of nucleophilic halogen acceptors, produce halogenated products (Fig. 7-12). For reaction to occur, there must be present an organic substrate, halide, and hydroperoxide. Haloperoxidases are divided into three groups based on their halide selectivity—chloroperoxidases can oxidize Cl^-, Br^-, and I^-; bromoperoxidases can oxidize Br^- and I^-; and iodoperoxidases can oxidize only I^-. There are no peroxidases capable of oxidizing F^-. This pattern of reactivity reflects the ease of oxidation of the halides, which follows the order $I^- > Br^- > Cl^- \gg F^-$. Thus, I^- and Br^- can be oxidized by an enzyme

$$H_2O_2 + AH + H^+ + X^- \xrightarrow{\text{Haloperoxidase}} AX + 2H_2O$$

Figure 7-12. Halogenation of halogen acceptor (AH).

having an oxidation potential sufficient to oxidize Cl⁻, but an enzyme (an iodoperoxidase) could possess an oxidation potential sufficient to oxidize I⁻ but not Br⁻ or Cl⁻ (reviewed by Neidleman and Geigert, 1986).

7.3.1 Occurrence

Recognition of the broad occurrence of haloperoxidases has been a recent event. Thus, only three haloperoxidases were known in the early 1950s—horseradish peroxidase (an iodoperoxidase), lactoperoxidase from milk, saliva, and tears (a bromoperoxidase), and myeloperoxidase from white blood cells (a chloroperoxidase). A decade later, three additional haloperoxidases had been identified: thyroid peroxidase (TPO), chloroperoxidase (CPO) from the fungus *Caldariomyces fumago*, and eosinophil peroxidase (EPO) from white blood cells. During the past 15 years, however, research by several groups has resulted in identification of a large number of additional peroxidases from a variety of sources. In addition to horseradish root and the thyroid gland of vertebrates, sources of iodoperoxidases now include other plant sources and several species of brown algae. Marine organisms are a particularly rich source of bromoperoxidases, while a large proportion of fungi contain chloroperoxidases (Neidleman and Geigert, 1986, and references therein).

Despite the abundance of chlorometabolites, many of them commercially important, that are produced by bacteria, haloperoxidases have only recently been isolated from bacteria. Thus, bromoperoxidases recently isolated from *Pseudomonas aureofaciens* and *Streptomyces phaeochromogenes* were the first haloperoxidases detected in halometabolite-producing procaryotes (van Pee and Lingens, 1985, 1984)]. The first procaryotic chloroperoxidase has recently been isolated from the bacterium *Pseudomonas pyrrocinia*, an organism that produces the antifungal antibiotic pyrrolnitrin (33) (Fig. 7-13). This nonheme chloroperoxidase is capable of chlorinating indole to 7-chloroindole. It also catalyzes the chlorination of 4-(2-amino-3-chlorophenyl)pyrrole (34) to give aminopyrrolnitrin, 35, the immediate precursor to the antibiotic (Fig. 7-13) (Wiesner *et al.*, 1986, 1988).

Three mammalian haloperoxidases—myeloperoxidase (MPO), eosinophil peroxidase (EPO), and thyroid peroxidase (TPO)—were discussed in previous chapters. A fourth mammalian enzyme, lactoperoxidase (LPO), has a role in host defense similar to that of MPO and EPO. LPO is found in human tears, saliva, and milk and thus has extracellular action. While MPO utilizes Cl⁻ to produce HOCl as a toxic agent and EPO oxidizes Br⁻ preferentially, with formation of HOBr and singlet oxygen, LPO

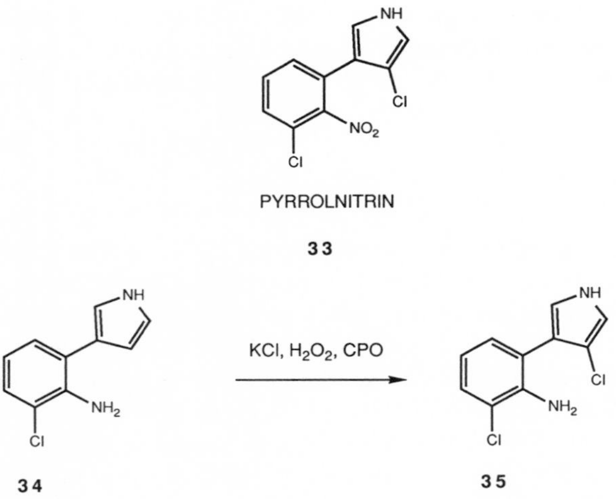

PYRROLNITRIN

3 3

KCl, H₂O₂, CPO

3 4   **3 5**

Figure 7-13. Pyrrolnitrin; CPO-catalyzed formation of pyrrolnitrin precursor (Weissner *et al.*, 1986, 1988).

oxidizes the pseudohalogen thiocyanate (SCN$^-$) to hypothiocyanous acid (HOSCN), a compound that may target protein sulfhydryl groups to mediate its toxic effects (Thomas *et al.*, 1983).

7.3.2 Structure

7.3.2.1 Prosthetic Group and Metal

Most haloperoxidases consist of a glycoprotein and ferriproto-porphyrin IX (Fig. 7-14) as a heme component, although many exceptions exist with respect to the metal and heme components. Important examples

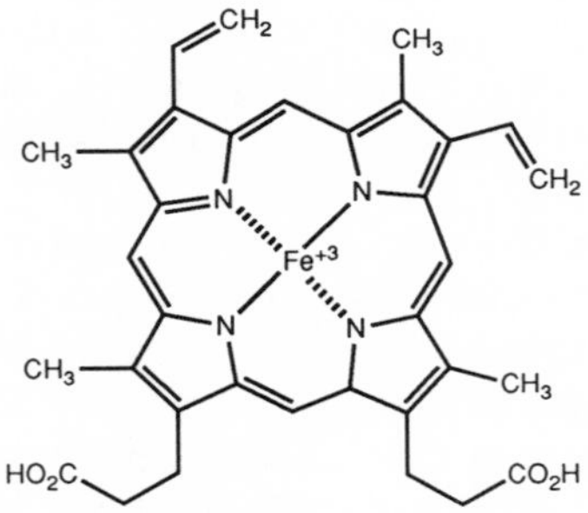

Figure 7-14. Ferriprotoporphyrin IX.

of ferriprotoporphyrin IX-containing haloperoxidases include EPO, chloroperoxidase from the fungus *C. fumago*, LPO, ovoperoxidase, TPO, horseradish peroxidase (HPO), the bacterial bromoperoxidase (*P. aureofaciens*), and the bromoperoxidase from the green marine alga *Penicillus capitatus*. In these enzymes, the metal (Fe^{3+}) of heme activates the heterolytic cleavage of H_2O_2 and stores one oxidizing equivalent, the porphyrin regulates the oxidation potential and stores an oxidizing equivalent, and the protein environment of the heme plays an important role in controlling the reactivity of the metal and in stabilizing reactive intermediates (see below) (Neidleman and Geigert, 1986).

Several examples of haloperoxidases that contain metals other than iron are known, and many of these are nonheme enzymes. For example, a series of bromoperoxidases from brown seaweed and several red seaweeds do not possess a heme prosthetic group, and, for many of these, reconstitution experiments have shown that vanadium is required for halogenating activity (de Boer *et al.*, 1988; Wever *et al.*, 1987, 1988; Neidleman and Geigert, 1986; and references therein). The bromoperoxidase isolated from *Xanthoria parietina*, a lichen symbiotically linked to a green alga and an ascomycete, also requires vanadium for activity, demonstrating the presence of vanadium-containing bromoperoxidases in the terrestrial environment (Plat *et al.*, 1987). The nonheme bromoperoxidase isolated from the actinomycete *S. aureofaciens* (in an unsuccessful attempt to isolate a chloroperoxidase from this chlortetracycline-producing organism) was inactivated by dialysis against EDTA, but activity was not restored by incubation with vanadium (Krenn *et al.*, 1988).

A nonheme zinc-containing chloroperoxidase isolated from the hyphomycete *Curvularia inaequalis* is the first reported example of a nonheme chloroperoxidase (Liu *et al.*, 1987). The absence of heme imparted chemical stability to this enzyme, protecting it from oxidative degration by HOCl and H_2O_2.

7.3.2.2 Glycoprotein Structure

There have been extensive structural studies on the glycoprotein components of haloperoxidases. Neidleman and Geigert (1986) have compiled a summary of molecular weights and subunit characteristics of several haloperoxidase glycoproteins. No apparent correlation could be drawn between molecular weight and halide ion specificity.

The complete amino acid sequence of horseradish peroxidase has been determined, using peptide mapping following tryptic and thermolytic digestion (Welinder, 1976). More recently, complementary CPO (*C. fumago*) (Fang *et al.*, 1986) and human TPO DNA (Kimura *et al.*, 1987) have been cloned and sequenced, revealing the complete amino acid sequences of these enzymes. The results of this research should be far-reaching. For example, enzymes having altered active-site amino acids, produced by site-directed mutagenesis, can now be used to study structure–activity relationships with respect to the catalase, peroxidase, and halogenating activity. This work also confirmed the identity of the fifth axial heme ligand of CPO as one of the three cysteine residues of the glycoprotein (Blanke and Hager, 1988).

7.3.3 Formation of the Halogenating Intermediate

Reaction of a heme-containing haloperoxidase with peroxides (e.g., hydrogen peroxide) leads to an unstable intermediate (half-life of one-half minute or less), termed Compound I, that is two oxidation equivalents above the native enzyme (Fig. 7-15). This intermediate is a ferryl porphyrin π-cation radical, in which one oxidation equivalent is stored in the ferryl ion (Fe^{IV}) and the other as a porphyrin-centered π-cation radical. In the presence of one-electron donor substrates, two one-electron reductions occur to give native enzyme and the products of the peroxidase cycle, the process proceeding through another intermediate, Compound II (Fig. 7-15). In contrast to Compound I, Compound II is quite stable, with a half-life of several hours. Two-electron donors, such as halide ions, produce native enzyme directly, without formation of detectable levels of Compound II and the halogenating intermediate (Fig. 7-15). In a catalase cycle, found in a limited number of haloperoxidases, including CPO (*C. fumago*), reaction of Compound I with H_2O_2 leads to the net conversion of two molecules of H_2O_2 to O_2 and H_2O (Fig. 7-15) (Neidleman and Geigert, 1986).

7.3.4 The Halogenating Intermediate

Much research has been carried out to determine the precise nature of the molecular species that is responsible for the halogenation step. Several haloperoxidases have been investigated, but a consensus on the mechanism of halogenation has been achieved for only a few. In this section, a summary will be given of the research done with the heme-containing CPO (*C. fumago*) as well as results of preliminary research done with nonheme haloperoxidases, including vanadium-containing peroxidases. The mechanisms that are largely accepted for myeloperoxidase and thyroid peroxidase will be summarized.

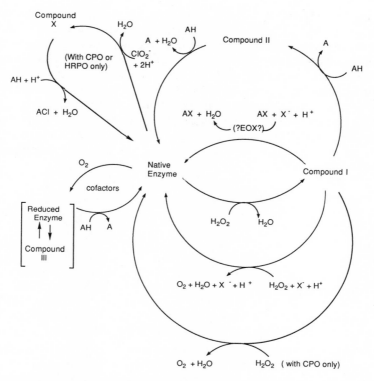

Figure 7-15. Reaction cycles of haloperoxidases. Adapted, with permission, from Neidleman and Geigert (1986; p. 105).

7.3.4.1 Chloroperoxidase (C. fumago)

CPO isolated from *C. fumago* by Hager in 1966 (Morris and Hager, 1966) has been the most thoroughly studied of the haloperoxidases, due to its ease of isolation and purification (cultures grown on fructose as the sole carbon source produce as much as 500 mg of CPO per liter) (Fang *et al.*, 1986, and references therein). In addition, the many similarities of this CPO to cytochrome P-450 (see Chapter 9) and its catalase activity have added to the interest in this enzyme. Despite this keen interest, there remains much disagreement on the nature of the actual halogenating species. There is agreement that formation of molecular I_2 is mandatory for CPO-catalyzed iodinations. However, free halogens are not involved in CPO-catalyzed bromination or chlorination reactions (Libby *et al.*, 1982, and references therein).

In the absence of a halogen-acceptor substrate, CPO catalyzes the oxidation of Br^- or Cl^- to the corresponding hypohalous acid. There has been much debate on whether it is the free hypohalous acid (HOX) or enzyme-bound hypohalite (EOX) that is responsible for chlorination or bromination in the presence of substrates. These two mechanisms (A and B) are summarized in Fig. 7-16. In mechanism A, reaction of Compound I with halide produces free hypohalous acid whereas in mechanism B, EOX is formed. The observed differences in substrate preferences between HOBr and HOCl halogenations, on the one hand, and peroxidase-catalyzed halogenations, on the other, and the fact that chlorination of

Mechanism A: Hypohalous acid (HOX) Intermediate

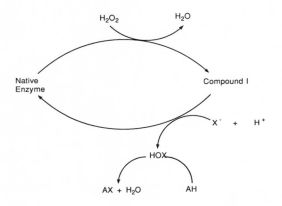

Mechanism B: Enzyme-bound (Compound EOX) Intermediate

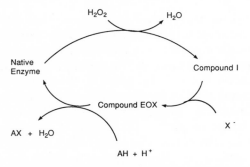

Figure 7-16. Two alternate mechanisms for haloperoxidase-catalyzed halogenation. Adapted, with permission, from Neidleman and Geigert (1986, p. 106).

2-chlorodimedone was much faster than the rate of enzymatic oxidation of Cl^- to HOCl were cited by Hager and co-workers as strong evidence supporting mechanism B (Libby et al., 1982). However, Neidleman and Geigert (1986) noted that HOCl produced in the latter experiment can attack CPO, causing both consumption of the product and inhibition of enzyme activity—both effects leading to a lower observed rate of HOCl production. In addition, caution was advised in making comparisons between the results of enzymatic and chemical reactions when reaction conditions can markedly affect product formation, as is the case with halogenation reactions. Differences observed between the chemical and TPO-catalyzed iodination of thyroglobulin provide a good example of this phenomenon (Turner et al., 1983).

In another kinetic study, the effects of halide, peracid, and 2-chlorodimedone concentrations on the initial rates of halogenation under steady-state conditions were measured (Lambeir and Dunford, 1983; Dunford et al., 1987). Kinetic parameters were derived from observed initial rate increases caused by increasing halide and 2-chlorodimedone concentrations that were consistent with a model in which EOX is the halogenating species. However, Neidleman and Geigert (1986) have criticized this interpretation, mainly due to a failure to account for a shift in the optimal pH for the enzyme reaction that would result from a change in the $[Cl^-]/ROOH$ ratio (see Section 7.3.6). Thus, addition of Cl^- at constant pH would shift the pH optimum toward the pH of the reaction, leading to an apparent increase in initial rates.

The stereochemical and regiochemical outcome of CPO halogenations appears to favor mechanism A, the formation of free HOX. The absence of stereoselectivity in CPO halogenations was demonstrated with substrates expected to produce optically active products under "normal" enzymatic reactions. These substrates—two bicycloalkenes, methionine, and an alkylated cyclopentan-1,3-dione related to the natural substrate dimedone—produced no optically active products when subjected to CPO-catalyzed halogenation (Fig. 7-17) (Ramakrishnan et al., 1983). Rearrangements that accompany the reaction of allyl halides with hypohalous acids to produce halohydrins were used in another series of experiments to compare the behavior of chemical and CPO-catalyzed halogenations. For example, migration of the allylic halogen to the central carbon occurs essentially to the same extent during chemical and enzymatic chlorination of allyl bromide (Lee et al., 1983). Similarly, in early experiments that revealed the electrophilic character of CPO-catalyzed halogenations, chlorination of anisole catalyzed by CPO produced the same ortho:para ratio of monochloroanisoles as did chlorination with HOCl (Brown and Hager, 1967).

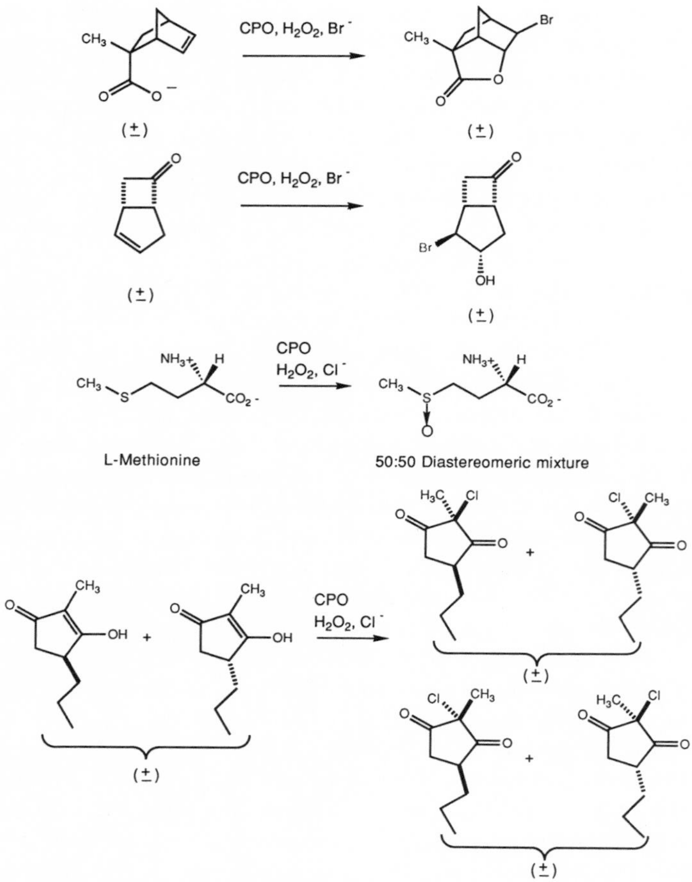

Figure 7-17. Evidence for absence of stereoselectivity in CPO-catalyzed halogenations (Ramakrishnan *et al.*, 1983).

In reviewing the debate regarding the halogenating intermediate in CPO-catalyzed halogenations, an evaluation given by Neidleman and Geigert (1986) is relevant. Those who have studied the kinetics of the reaction tend to favor EOX as the reactive intermediate while those who analyze products favor HOX.

7.3.4.2 Bromoperoxidase (*P. capitus*)

Many of the spectral and catalytic properties of the marine algal bromoperoxidase (*P. capitatus*) are similar to those of CPO. Thus, in addition to catalyzing the bromination of organic substrates, this enzyme has catalase activity (0.1 % that of catalase and 10 % that of CPO) and, at low pH, is capable of catalyzing the chlorination of 2-chlorodimedone. This represents the first peroxidase isolated from marine algae that can catalyze both bromination and chlorination reactions. The rate of Br_3^- formation is faster than the rate of bromination of 2-chlorodimedone, and Br_3^- thus was proposed as the halogenating species (Manthey and Hager, 1989).

7.3.4.3 Nonheme Bromoperoxidase (Corallina pilulifera)

The bromoperoxidase from the marine alga C. pilulifera does not have heme as a prosthetic group but does contain 2.3 ± 0.2 atoms of iron and 1.6 ± 0.1 atoms of magnesium per molecule of enzyme. Bromination of aromatic substrates catalyzed by this enzyme is consistent with the presence of electrophilic bromine (Br^+). The proposal was made that non-heme iron bound to cysteine or tyrosine residues of the enzyme mediates the oxidation–reduction of halide ions and hydrogen peroxide (Itoh et al., 1986). In a study of the halogenation of a series of aromatic heterocycles with this bromoperoxidase, no evidence was found for the presence of free Br_2, in contrast to the behavior of CPO (C. fumago) with the same substrates. These results suggested that the bromoperoxidase halogenation occurs at the active site of the enzyme (Itoh et al., 1987). However, Wever et al. (1988) have noted that the inability to detect free Br_2 from bromoperoxidase may be a result of the relatively high pH (6.5) at which the bromoperoxidase reactions were performed, a pH at which the reaction between HOBr and H_2O_2 is very rapid. In contrast, the CPO halogenations were carried out at pH 3.0, conditions under which HOBr and H_2O_2 react quite slowly.

7.3.4.4 Nonheme Vanadium-Containing Bromoperoxidases

In the nonheme vanadium-containing bromoperoxidases isolated from seaweeds, V^{5+} apparently serves as a binding site for both Br^- and H_2O_2. Consistent with this proposal, several peroxovanadium(V) complexes have been reported that bind I^- at the metal center with subsequent oxidation of I^- to hypoiodite. A lack of specificity toward substrate structure shown by the bromoperoxidase from Ascophyllum nodosum indicates that this enzyme probably produces free HOBr. The ligands from the protein to vanadium have yet to be identified (de Boer et al., 1986; Wever et al., 1988).

7.3.4.5 Thyroid Peroxidase

As was discussed in Chapter 6, there has also been debate regarding the nature of the iodinating species formed during TPO-catalyzed iodination of tyrosine. A sulfenyl iodide $(-SI)$ formed by reaction of TPO-produced I^+ with an active-site disulfide bond has been proposed as the iodinating species (reviewed by Neary *et al.*, 1984).

7.3.4.6 Myeloperoxidase

There is general agreement that the reaction of Cl^- with MPO in the presence of H_2O_2 produces HOCl as a reactive intermediate. The role of HOCl in leukocyte defense mechanisms was discussed in Chapter 3.

7.3.5 Glycoprotein Structure and Reactivity

The value of elucidating the structure of the glycoprotein component of haloperoxidases can be appreciated by considering a recent analysis of structural factors resident in the protein framework that influence the reactivities of four heme enzymes, especially with respect to the fate of Compound I (Dawson, 1988). Of the four enzymes treated—cytochrome P-450 (P-450), CPO, HPO, and secondary amine monooxygenase—two, CPO and HPO, are peroxidases and two, cytochrome P-450 and secondary amine monooxygenase, are oxygenases. Despite the presence of heme as a common prosthetic group and many spectral and other structural similarities, these enzymes possess quite diverse catalytic activity. For example, both peroxidases and P-450 form Compound I as an intermediate, but only CPO is capable of chlorinating organic substrates in the presence of Cl^-. All will *N*-dealkylate alkylamines, but this reaction is mediated by an O_2-dependent path with the monooxygenases and by a H_2O_2-dependent path with the peroxidases. P-450 will oxygenate unactivated alkanes but has no peroxidase activity.

A knowledge of the proximal iron ligands and of the protein environments at both the proximal and distal sides of the heme is essential in understanding this diversity of reactivity originating from a single prosthetic group. Critical structural features are the nature of the proximal axial ligand, the polarity of the distal and proximal heme environments, and the accessibility of the heme edge. For example, for P-450, the thiolate ligand (cysteine) serves as a strong internal electron donor to facilitate cleavage of the $O-O$ bond of the iron-bound peroxide moiety. Both the proximal and distal sides of the heme are nonpolar, facilitating the acceptance of O_2 and nonpolar hydrocarbons as substrates to form the iron-oxo intermediate. HPO, which contains a proximal axial histidine ligand, accomplishes $O-O$ bond scission through proton donation of a distal histidine and the "pull" of a distal charged group.

Compound I, once formed, is both a good oxidant and a good oxo donor, with one-electron transfer reactions (peroxidase activity) presumed to occur at the heme edge, as shown by Ortiz de Montellano *et al.* (1987) for HPO. In P-450, the heme is inaccessible and the iron-oxo group is exposed to substrate, so oxo transfer, rather than peroxidase activity, is the dominant reaction [the mechanism of P-450 oxo transfer is discussed further in Chapter 9 (Section 9.2)]. With HPO, the converse is true—the heme edge is exposed, favoring peroxidase activity, while steric hindrance prevents substrates from approaching the iron-bound oxo group. CPO, which has a thiolate as the proximal axial ligand, has a P-450 metal coordination structure in a peroxidase heme environment and thus can carry out both one- and two-electron chemistry (Sono *et al.*, 1986). While the stereospecificity of epoxidation reactions (oxo transfer) of CPO implies that the substrate has access to the iron-oxo group, Dawson (1988) suggests that the lack of stereospecificity of halogenations may indicate that this chemistry occurs at the exposed heme edge.

The above is a brief summary of the analysis given by Dawson, a summary intended to illustrate the critical influences that the environment of the heme prosthetic group has on the catalytic behavior of these enzymes. As additional structures of protein components of haloperoxidases are elucidated, further understanding of such factors as halide specificity undoubtedly will emerge.

7.3.6 pH Dependence of Rates of Halogenation

While most enzymatic transformations proceed most efficiently at one given pH (that is, the pH that gives the maximum catalytic turnover), a characteristic of haloperoxidase-catalyzed halogenations is the dependence of the optimal pH of the reaction on the ratio of the concentrations of the two requisite substrates, peroxide and halide. This dependence on substrate ratios has been demonstrated for several peroxidases, including CPO, MPO, EPO, and LPO, with each enzyme having a preferred pH range (Neidleman and Geigert, 1986). The dependence is given by the relationship: optimal $pH = \beta + \log$ [halide ion/H_2O_2], where β is the optimal pH when the concentrations of the halide and H_2O_2 are equal. An explanation for the dependence is found in the model advanced by Andrews and Krinsky (1982). Two binding sites for halide ions are proposed—one that is a site for substrate binding that leads to hypohalous acid formation, and a second that inhibits the binding of peroxide as a substrate when occupied by halide. The model of Andrews and Krinsky requires that binding of halide at the inhibitory binding site must be preceded by protonation of that site. When the halide concentration is increased, making more halide available for inhibition, the rate of reaction will be increased by increasing the pH, an increase which will inhibit binding at the inhibitory site. Competitive experiments with CN^- have indicated that the inhibitory site is the sixth coordination position of Fe^{3+} in the native enzyme (Lambeir and Dunford, 1983).

CPO (*C. fumago*) has a particularly low pH optimum (pH 3–4). The suggestion has been made that a carboxylate situated on the protein can function as a ligand when unprotonated, thus blocking the binding of halide at the active site. Protonation of the carboxylate causes its exit from the active site and restores activity of the enzyme. Most haloperoxidases prefer a more neutral pH (pH 4.0–7.5) (Neidleman and Geigert, 1986, and references therein).

7.3.7 Substrate Specificity and Scope of Reaction

In the previous sections, the structures of enzymes that carry out biological halogenation reactions have been discussed, along with the molecular events that produce reactive electrophilic halogenating intermediates. In this section, a summary of structural units that are reactive toward biological halogenation will be given. Examples of how these halogenation reactions are incorporated into biosynthetic schemes to produce halometabolites will be given in Section 7.4.

Much of the chemistry of haloperoxidase halogenations is understood readily in terms of the normal organic chemistry of electrophilic halogenations. Thus, reaction of alkene substrates produces α, β-halohydrins, as shown for isolated, conjugated, and cumulative unsaturated systems according to the normal intermediate halonium ion mechanism (Fig. 7-18). Representative examples are given in Fig. 7-19, while a more thorough compendium of haloperoxidase reactions can be found in the review of Neidleman and Geigert (1986). Halolactonization reactions occur readily when a carboxylate function is situated such that it can attack the halonium intermediate. Carbon loss can accompany halogenation when oxygen is attached to the end of the double bond (Fig. 7-19). The participation of a neighboring bromide during the CPO-catalyzed chlorination of allyl bromide was discussed above. If halide concentration is sufficiently high, halide can compete with water for the halonium intermediate to give

Figure 7-18. Halonium ion mechanism for Markownikoff addition of HOX to a double bond.

a dihalide (or a mixed dihalide; see Section 7.4.3). Haloperoxidase-catalyzed halogenations of alkyne substrates give α-haloketones, and cyclopropanes give α, γ-halohydrins.

Haloperoxidase-catalyzed halogenation of activated aromatic compounds occurs readily, as can be seen by the large number of metabolites that contain halogenated aromatic rings. Notable examples include the thyroid hormones, griseofulvin, chlortetracycline, and many bromophenols from the marine environment (see Section 7.2). Several heteroaromatic systems serve as substrates, including thiazoles, as do several nitrogen-

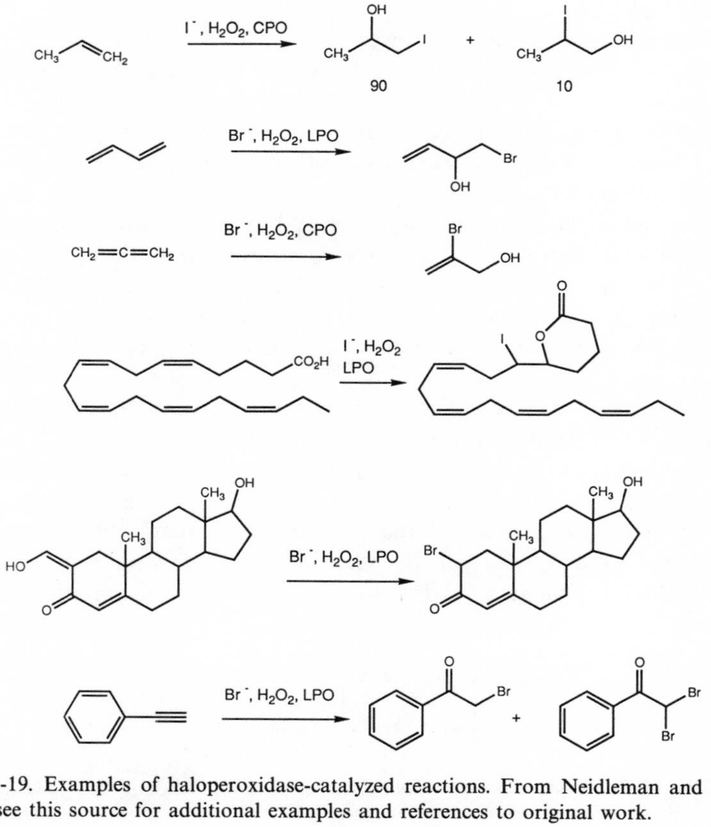

Figure 7-19. Examples of haloperoxidase-catalyzed reactions. From Neidleman and Geigert (1986); see this source for additional examples and references to original work.

containing heterocycles, including bases present in nucleic acids (Itoh *et al.*, 1987). The regioselectivity of aromatic halogenation conforms to that expected for normal aromatic electrophilic substitution reactions.

As with chemical halogenations, the ease of enzymatic halogenations of carbonyl compounds increases with increasing enol content of the substrate. Thus, 1,3-dicarbonyl substrates, such as β-diketones and β-ketoacids, react readily whereas simple aliphatic ketones are unreactive. 2-Chlorodimedone has been used extensively as a reagent in assays for haloperoxidase activity because of its high reactivity and because of a dramatic drop in ultraviolet absorption upon formation of dichlorodimedone that facilitates spectroscopic monitoring of reaction rates.

In a process important to mammalian defense mechanisms against microorganisms, the amino group of amino acids reacts with elelctrophilic halogen to produce unstable haloamines (see discussions in Chapters 3 and 4). Decarboxylative elimination of HX produces aldehydes as the final product (Fig. 7-20).

CPO has been studied most extensively with respect to substrate selectivity. As has been noted, and as is illustrated in many of the examples of this section, there is a very wide substrate specificity for this enzyme, indicative of reactions removed from the enzyme active site. Differences in substrate specificity may be expected as other enzyme are investigated more thoroughly. For example, Itoh *et al.* (1987) found that the nonheme bromoperoxidase (*C. pilulifera*) apparently has a narrower substrate range (is less reactive) than does CPO (*C. fumago*) in brominations of nucleoside bases. This was used as one argument to support the proposal that brominations by the bromoperoxidase occur at the active site, since HOBr or Br_2 in solution should have the same reactivity regardless of enzyme source.

The halide substrate selection is determined both by inherent redox potentials of the enzyme in question, as discussed in Section 7.3, and also by the availability of halide. Thus, replacement of Cl^- with Br^- in fermentation media has been used to produce bromometabolites (e.g., bromotetracycline) in place of the normal chlorometabolites (Neidleman and Geigert, 1986). As will be seen in the following section, halide availability also influences product formation by a nonenzymatic, "passive" mechanism.

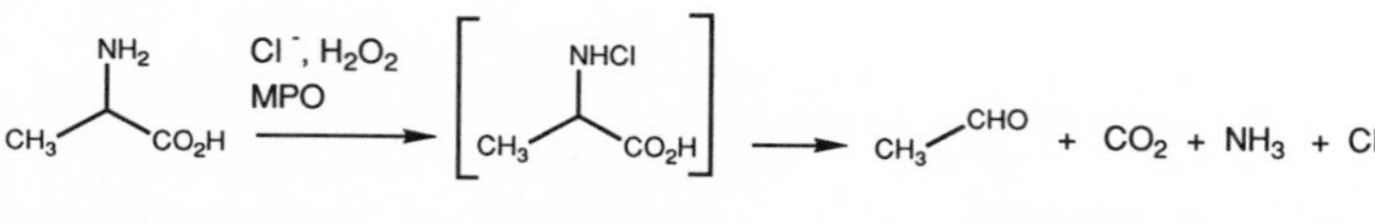

Figure 7-20. Haloperoxidase-catalyzed halogenation of amino acids. From Neidleman and Geigert (1986); see this source for additional examples and references to original work.

7.4 Biosynthesis of Halometabolites in Marine Organisms

The marine environment provides the greatest number and greatest variety of halometabolites. To illustrate the operation of biohalogenation in natural product biosynthesis, selected examples of marine halometabolite biosynthesis will be given. Neidleman (1975) has provided an example of bacterial halometabolite biosynthesis in his discussion of the biosynthesis of chloramphenicol, while examples of halogenated fungal metabolite biosynthetic schemes can be found in the monograph by Turner (1971).

7.4.1 Acetogenin Biosynthesis

In marine organisms, initial enzyme-catalyzed brominations of typical fatty acid precursors (e.g., β-ketoacids) ultimately produce a wide variety of acetate-derived secondary metabolites, from simple volatile halocarbons to quite complex branched structures. For example, Beissner et al. (1981) demonstrated the ability of bromoperoxidase to catalyze the formation of halogenated ketones, and also bromoform, from 3-oxooctanoic acid. Based on product distribution, the mechanisms shown in Fig. 7-21 were

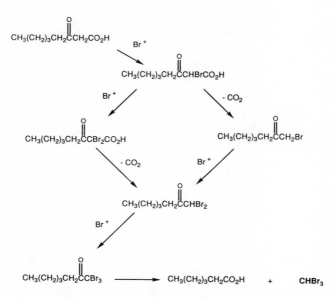

Figure 7-21. Biosynthesis of halogenated ketones and of bromoform from 3-oxooctanoic acid (Beissner et al., 1981).

proposed. Mono- or dibromination of the active methylene group followed by decarboxylation and bromination of the mono- or dibromomethyl-ketone intermediate to the tribromomethylketone is followed by non-enzymatic hydrolysis to bromoform. Iodomethanes that occur in seawater presumably arise from bromoperoxidase-catalyzed incorporation of I^-. Reaction of iodomethanes and bromomethanes with Cl^- also could produce chloromethanes (Theiler *et al.*, 1978).

The introduction of bromine into biosynthetic intermediates can trigger additional reactions that lead to metabolites of considerable complexity. Thus, in addition to halogenated ketones and other simple acetogenins, red seaweeds produce more complex halometabolites, for example, the unsaturated acids **36** and **37** (Fig. 7-22). Biomimetic synthetic studies have provided convincing evidence that these are formed by Favorskii rearrangement of 1,1,3,3-tetrabromo-2-heptanone (**38**) and 1,1,3-tribromo-2-heptanone (**39**), respectively, according to the scheme shown in Fig. 7-22. By analogy, the proposal was made that the unsaturated haloketone **40** and the halogenated lactone **41**, also produced by red seaweeds, are formed by Favorskii rearrangements of 1,1,3,3,5-penta-bromo-2,4-nonanedione (**42**) (Fig. 7-23) (McConnell and Fenical, 1997a; reviewed by Fenical, 1979).

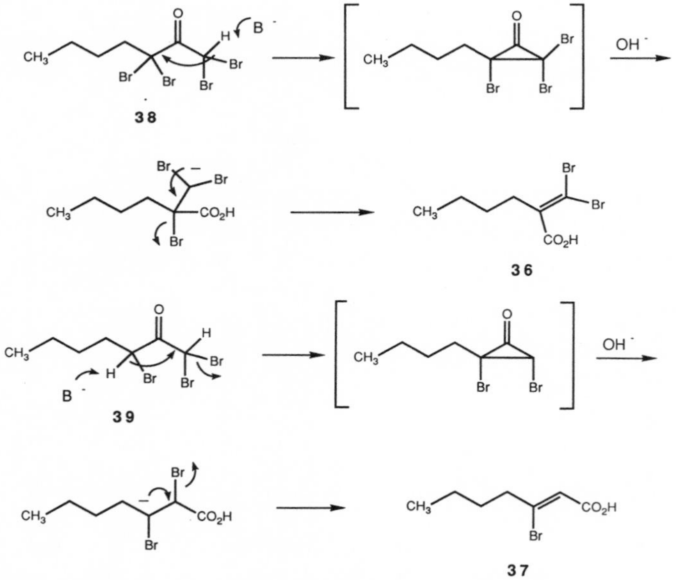

Figure 7-22. Favorskii rearrangements in the biosynthesis of complex acetogenins (McConnell and Fenical, 1977a; reviewed by Fenical, 1979).

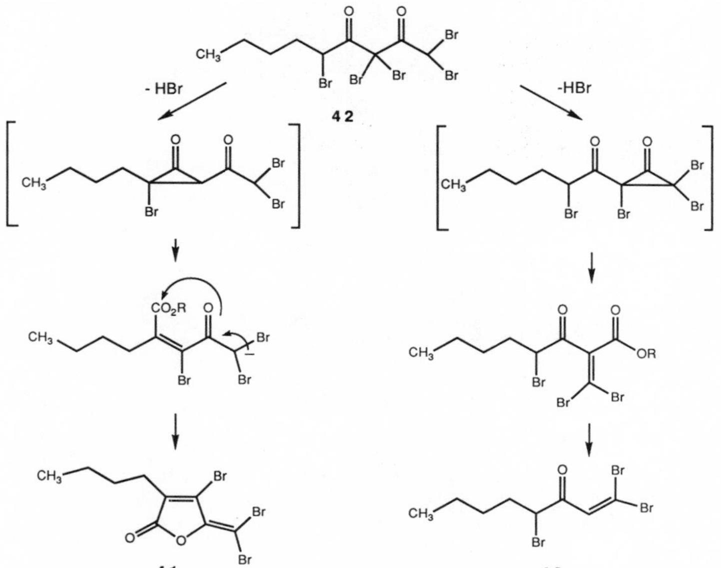

Figure 7-23. Additional examples of the Favorskii rearrangement in natural product biosynthesis.

7.4.2 Terpene Biosynthesis

A key step in terpene biosynthesis in terrestrial organisms is the proton-catalyzed cyclization of an appropriate monoterpene precursor. The scheme shown in Fig. 7-24 for the biosynthesis of the terpene aldehyde **43** illustrates this process. Comparable cyclizations play key roles in terpene biosynthesis in the marine environment. However, here the positive center that induces cyclization is generated by a bromonium ion, as illustrated by the biosynthesis of the halogenated monoterpene **44** (Fig. 7-24) (reviewed by Fenical, 1979, 1982).

As with acetogenin biosynthesis, the increased reactivity introduced with the presence of halogen can lead to more complex metabolites. Thus, the frequent occurrence of nonhalogenated marine terpenes that deviate from the "normal" head-to-tail arrangement of isoprene units has been explained by rearrangement of unstable bromoterpenoid intermediates, as illustrated by formation of the rearranged sesquiterpene **45** (Fig. 7-25) (reviewed by Fenical, 1982). Similarly, formation of unusual azulene derivatives, such as guaiazulene (**46**) and the perhydroguaiazulene **47**, can

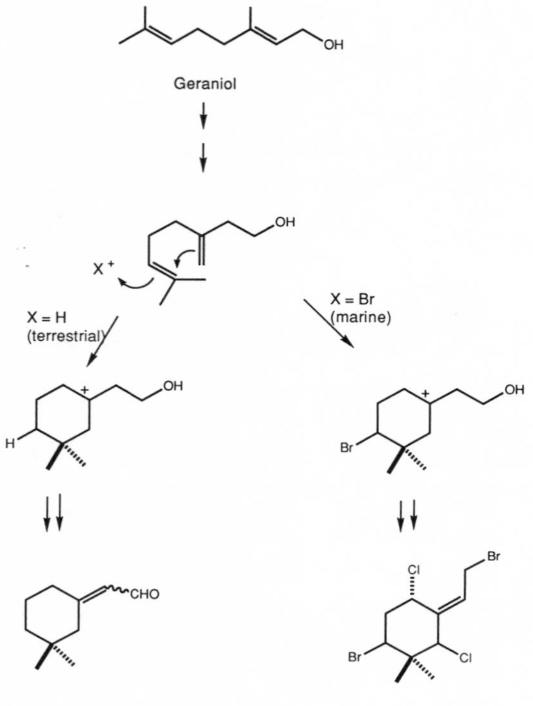

Figure 7-24. Bromonium ion-induced cyclization during biosynthesis of marine terpenes (reviewed by Fenical, 1979, 1982).

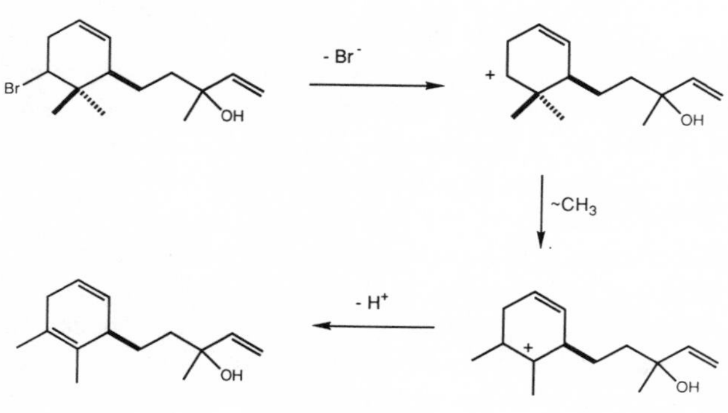

Figure 7-25. Methyl migration during biosynthesis of "irregular" terpenes (reviewed by Fenical, 1982).

be rationalized on the basis of a bromine-induced rearrangement of the bromoselinane derivative **48** (Fig. 7-26) (reviewed by Fenical, 1979).

The above reactions are illustrative of the fascinating chemistry that can occur subsequent to biological halogenation. Fenical (1979, 1982) provides many additional examples of mechanisms by which complex secondary metabolites—halogenated and nonhalogenated—can result from rearrangements made possible by the presence of reactive halogens in biosynthetic intermediates.

7.4.3 Formation of Mixed Halides

While many chlorinated products are found in the marine environment, most, if not all, marine halogenations are catalyzed by bromoperoxidases. To account for this, the proposal has been made that chlorinated products arise through a nonenzymatic mechanism involving nucleophilic attack by Cl^- on an intermediate enzyme-produced bromonium ion. Similar attack of Cl^- on Br^- on iodonium ion intermediates generated during haloperoxidase iodinations would produce bromoiodo and chloroiodo compounds. Subsequent nucleophilic displacement with halide or water could result in a variety of substitution patterns (Fig. 7-27; Neidleman and Geigert, 1983). The greater reactivity of bromometabolites may be advantageous to the organism, both in defense

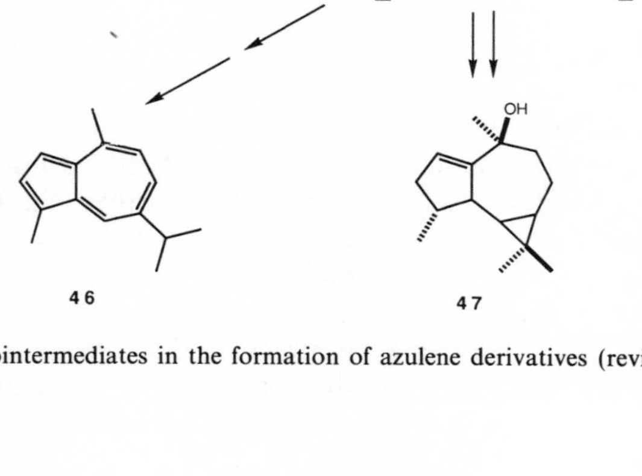

Figure 7-26. Halointermediates in the formation of azulene derivatives (reviewed by Fenical, 1979).

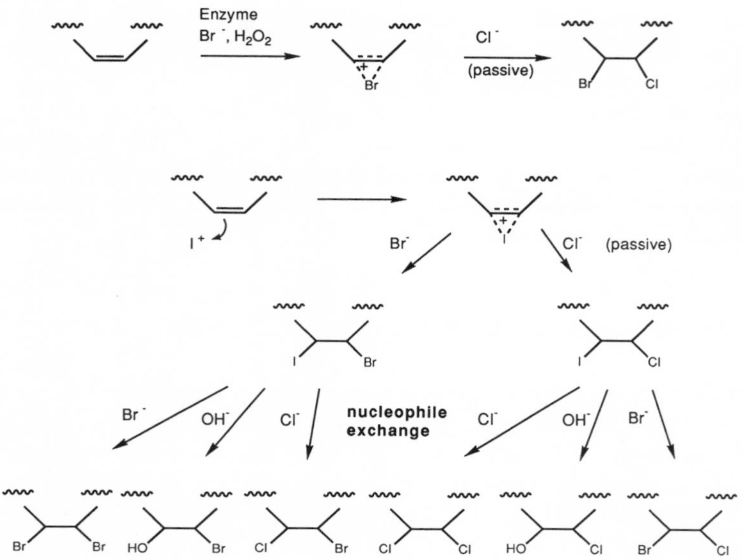

Figure 7-27. Proposed mechanism for the formation of mixed halides in the marine environment. After Neidleman and Geigert (1986; p. 117), with permission.

mechanisms and in biosynthetic schemes (see above), and this may explain in part why these organisms use Br^- rather than the more abundant Cl^- in the initial biohalogenation step. In this regard, a bromoperoxidase from the marine alga *Penicillus capitatus* has been characterized recently that has the ability to catalyze both Cl^- and Br^- oxidation reactions, and could explain the occurrence of products having chlorine and bromine substituted on nonadjacent carbons (Manthey and Hager, 1989).

7.5 Biological Fluorination

The biosynthesis of fluoroacetate has intrigued chemists and biochemists since the discovery in the 1940s that this highly toxic compound occurs naturally. Fluoroacetate, and other fluorine-containing metabolites, are found in plants grown where the concentration of F^- in the soil is high. Neidleman and Geigert (1983) have recently produced fluoroiodo derivatives using horseradish peroxidase-catalyzed addition of I^- to a double bond, followed by attack of F^- on the iodonium intermediate. Based on this result, they have proposed a speculative scheme for the biosynthesis of fluoroacetate consisting of a similar reaction on an ω-olefinic fatty acid. Deiodination *in vivo* of the intermediate fluoroiodo compound would produce the ω-fluoro fatty acid as a precursor of fluoroacetate and related naturally occurring fluorometabolites (Fig. 7-28) (Neidleman and Geigert, 1976).

Sanada *et al.* (1986) have reported that the thienamycin producer *Streptomyces cattleya* synthesizes fluoroacetate and fluorothreonine if a source of F^- is present in the fermentation mixture. The production of nucleocidin by *Streptomyces calvus* was reported first in 1957, but fluorine was not detected in its structure until 1969. The mechanism of biological fluorination by these two microorganisms apparently is unknown.

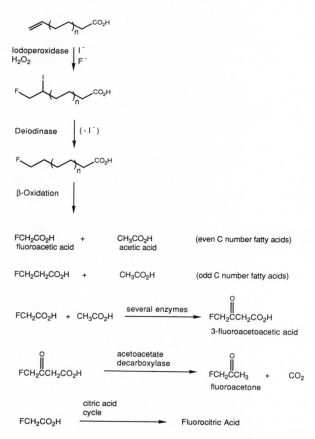

Figure 7-28. Proposed mechanism for formation of fluorinated plant metabolites. After Neidleman and Geigert (1986), with permission.

7.6 Biohalogenation—Summary

Biohalogenation is an important process, vital to the existence of both terrestrial and marine life. The critical and unique role played by iodine in the biosynthesis, biological activity, regulation, and metabolism of thyroid hormones, essential to growth and development of vertebrates, was discussed in Chapter 6. Mammalian defense mechanisms are dependent on haloperoxidases to generate toxic halogenating and oxidizing reactive species to mediate killing of invading microorganisms. In marine organisms, the diversity and prevalence of biohalogenations are most readily apparent, and biohalogenation appears to serve a dual purpose. Secondary metabolites containing halogens often are toxic to microorganisms or other marine life, thus protecting the producing organisms from disease or predators. In this role, halogenation often renders an organic substrate more toxic (Neidleman and Geigert, 1986), a phenomenon appreciated by man, as seen, for example, in the development of chemotherapeutic agents (see Vol. 9B of this series) and insecticides (Chapter 8). A second role of biohalogenation in marine organisms is in providing biochemical pathways, through halogen-induced cyclizations or rearrangements, to additional complex secondary metabolites, some of which are halogen-free.

Recognition of the importance of biological halogenation is a fairly recent phenomenon. The search for halogenated metabolites and for halogenating enzymes now is intense, and the number of halometabolites and halogenating enzymes identified is increasing at a rapid rate. Rapid progress is also being made in the elucidation of halogenation mechanisms. Medical and industrial benefits undoubtedly will also increase with this increasing research activity.

References

Andrews, P. C., and Krinsky, N. I., 1982. A kinetic analysis of the interaction of human myeloperoxidase with hydrogen peroxide, chloride ions, and protons, *J. Biol. Chem.* 257:13240–13245.

Beissner, R. S., Guilford, W. J., Coates, R. M., and Hager, L. P., 1981. Synthesis of brominated heptanones and bromoform by a bromoperoxidase of marine origin, *Biochemistry* 20:3724–3731.

Blanke, S. R., and Hager, L. P., 1988. Identification of the fifth axial heme ligand of chloroperoxidase, *J. Biol. Chem.* 263:18739–18743.

Bohlmann, F., Jakupovic, J., King, R. M., and Robinson, H., 1981. New germacranolides, guaianolides and rearranged guaianolides from *Lasiolaena santosii*, *Phytochemistry* 20:1613–1622.

Brown, F. S., and Hager, L. P., 1967. Chloroperoxidase. IV. Evidence for an ionic electrophilic substitution mechanism, *J. Am. Chem. Soc.* 89:719–720.

Carte, B., and Faulkner, D. J., 1983. Defensive metabolites from three nembrothid nudibranchs, *J. Org. Chem.* 48:2314–2318.

Carter, G. T., Rinehart, K. L., Jr., Li, H. L., Kuentzel, S. L., and Conner, J. L., 1978. Brominated indoles from *Laurencia brongniartii*, *Tetrahedron Lett.* 1978:4479–4482.

Cutterbuck, P. W., Mukhopadhyay, S. L., Oxford, A. E., and Raistrick, H., 1940. Studies in the biochemistry of micro-organisms. 65 (A) A survey of chlorine metabolism by moulds. (B) Caldariomycin, $C_5H_8O_2Cl_2$, a metabolic product of *Caldariomyces fumago* Woronichin, *Biochem. J.* 34:664–677.

Dawson, J. H., 1988. Probing structure–function relations in heme-containing oxygenases and peroxidases, *Science* 240:433–439.

de Boer, E., Tromp, M. G. M., Plat, H., Krenn, G. E., and Wever, R., 1986. Vanadium(V) as an essential element for haloperoxidase activity in marine brown algae: Purification and characterization of a vanadium(V)-containing bromoperoxidase from *Laminaria saccharina*, *Biochim. Biophys. Acta* 872:104–115.

de Boer, E., Boon, K., and Wever, R., 1988. Electron paramagnetic resonance studies on conformational states and metal ion exchange properties of vanadium bromoperoxidase, *Biochemistry* 27:1629–1635.

Dunford, H. B., Lambeir, A.-M., Kashem, M. A., and Pickard, M., 1987. On the mechanism of chlorination by chloroperoxidase, *Arch. Biochem. Biophys.* 252:292–302.

Fang, G.-H., Kenigsberg, P., Axley, M. J., Nuell, M., and Hager, L. P., 1986. Cloning and sequencing of chloroperoxidase cDNA, *Nucleic Acids Res.* 14:8061–8071.

Fenical, W., 1979. Molecular aspects of halogen-based biosynthesis of marine natural products, *Recent Adv. Phytochem.* 13:219–239.

Fenical, W., 1982. Natural products chemistry in the marine environment, *Science* 215:923–928.

Fowden, L., 1968. The occurrence and metabolism of carbon–halogen compounds, *Proc. Roy. Soc. B* 171:5–18.

Fusetani, N., 1987. Marine metabolites which inhibit development of echinoderm embryos, in *Bioorganic Marine Chemistry*, Vol. 1 (P. J. Scheuer, ed.), Springer-Verlag, Heidelberg, pp. 61–92.

Gonzalez, A. G., Arteaga, J. M., Martin, J. D., Rodriguez, M. L., Fayos, J., and Martinez-Ripolls, M., 1978. Two new polyhalogenated monoterpenes from the red alga *Plocamium cartilagineum*, *Phytochemistry* 17:947–948.

Gonzalez, A. G., Darias, V., and Estevez, E., 1982. Chemotherapeutic activity of polyhalogenated terpenes from Spanish algae, *Planta Med.* 44:44–46.

Gschwend, P. M., MacFarland, J. K., and Newman, K. A., 1985. Various halogenated compounds released to seawater from temperate marine macroalgae, *Science* 227:1033–1035.

Iguchi, K., Kaneta, S., Mori, K., Yamada, Y., Honda, A., and Mori, Y., 1985. Chlorovulones, new halogenated marine prostanoids with an antitumor activity from the stolonifer *Clavularia virdis* Quoy and Gaimard, *Tetrahedron Lett.* 26:5787–5790.

Itoh, N., Izumi, Y., and Yamada, H., 1986. Characterization of nonheme type bromoperoxidase in *Corallina pilulifera*, *J. Biol. Chem.* 261:5194–5200.

Itoh, N., Izumi, Y., and Yamada, H., 1987. Haloperoxidase-catalyzed halogenation of nitrogen-containing aromatic heterocycles represented by nucleic bases, *Biochemistry* 26:282–289.

Kimura, S., Kotani, T., McBride, O. W., Umeki, K., Hirai, K., Nakayama, T., and Ohtaki, S., 1987. Human thyroid peroxidase: Complete cDNA and protein sequence, chromosome mapping, and identification of two alternately spliced mRNAs, *Proc. Natl. Acad. Sci. USA* 84:5555–5559.

Krenn, B. E., Plat, H., and Wever, R., 1988. Purification and some characteristics of a non-haem bromoperoxidase from *Streptomyces aureofaciens*, *Biochim. Biophys. Acta* 952:255–260.

Lambeir, A.-M., and Dunford, H. B., 1983. A steady state kinetic analysis of the reaction of chloroperoxidase with peracetic acid, chloride, and 2-chlorodimedone, *J. Biol. Chem.* 258:13558–13563.

Lee, T. D., Geigert, J., Dalietos, D. J., and Hirano, D. S., 1983. Neighboring group migration in enzyme-mediated halohydrin formation, *Biochem. Biophys. Res. Commun.* 110:880–883.

Le-Van, N., and Wratten, S. J., 1984. Compound 30.4, an unusual chlorinated 1,4-benzoxazin-3-one derivative from corn (*Zea mays*), *Tetrahedron Lett.* 25:145–148.

Libby, R. D., Thomas, J. A., Kaiser, L. W., and Hager, L. P., 1982. Chloroperoxidase halogenating reactions. Chemical *versus* enzymic halogenation intermediates, *J. Biol. Chem.* 257:5030–5037.

Liu, T.-N. E., M'Timkulu, T., Geigert, J., Wolf, B., Neidleman, S. L., Silva, D., and Hunter-Cevera, J. C., 1987. Isolation and characterization of a novel nonheme chloroperoxidase, *Biochem. Biophys. Res. Commun.* 142:329–333.

Manthey, J. A., and Hager, L. P., 1989. Characterization of the catalytic properties of bromoperoxidase, *Biochemistry* 28:3052–3057.

McConnell, O. J., and Fenical, W., 1977a. Halogenated metabolites—including Favorsky rearrangement products—from the red seaweed *Bonnemaisonia nootkana*, *Tetrahedron Lett.* 1977:4159–4162.

McConnell, O. J., and Fenical, W., 1977b. Halogen chemistry of the red alga *Asparagopsis*, *Phytochemistry* 16:367–374.

Morris, D. R., and Hager, L. P., 1966. Chloroperoxidase I. Isolation and properties of the crystalline glycoprotein, *J. Biol. Chem.* 241:1763–1768.

Munro, M. H. G., Luibrand, R. T., and Blunt, J. W., 1987. The search for antiviral and anticancer compounds from marine organisms, in *Bioorganic Marine Chemistry*, Vol. 1 (P. J. Scheuer, ed.), Springer-Verlag, Heidelberg, pp. 93–176.

Neary, J. T., Soodak, M., and Maloof, F., 1984. Iodination by thyroid peroxidase, *Methods Enzymol.* 107:445–476.

Neidleman, S. L., 1975. Microbial halogenation, *CRC Crit. Rev. Mircrobiol.* 5:333–358.

Neidleman, S. L., and Geigert, J., 1983. The enzymatic synthesis of heterogenous dihalide derivatives: A unique biocatalytic discovery, *Trends Biotechnol.* 1:21–25.

Neidleman, S. L., and Geigert, J., 1986. *Biohalogenation: Principles, Basic Roles, and Applications*, Ellis Horwood, Chichester, England.

Neidleman, S. L., and Geigert, J., 1987. Biological halogenation: Roles in nature, potential in industry, *Endeavor* 11:5–15.

Ortiz de Montellano, P. K., Choe, Y. S., DePillis, G., and Catalano, C. E., 1987. Structure–mechanism relationships in hemoproteins. Oxygenations catalyzed by chloro-peroxidase and horseradish peroxidase, *J. Biol. Chem.* 262:11641–11646.

Osada, H., and Isono, K., 1985. Mechanism of action and selective toxicity of ascamycin, a nucleoside antibiotic, *Antimicrob. Agents Chemother.* 27:230–233.

Plat, H., Krenn, B. E., and Wever, R., 1987. The bromoperoxidase from the lichen *Xanthoria parietina* is a novel vanadium enzyme, *Biochem. J.* 248:277–279.

Ramakrishnan, K., Oppenhuizen, M. E., Saunders, S., and Fisher, J., 1983. Stereoselectivity of chloroperoxidase-dependent halogenation, *Biochemistry* 22:3271–3277.

Sanada, M., Miyano, T., Iwadara, S., Williamson, J. M., Arison, B. H., Smith, J. L., Douglas, A. W., Liesch, J. M., and Inamine, E., 1986. Biosynthesis of fluorothreonine and fluoroacetic acid by the thienamycin producer, *Streptomyces cattleya*, *J. Antibiot.* 39:259–265.

Schmitz, F. J., and Gopichand, Y., 1978. (7*E*, 13ξ, 15*Z*)-14,16-Dibromo-7,13,15-hexadecatriene-5-ynoic acid. A novel dibromo acetylenic acid from the marine sponge *Xestospongia muta*, *Tetrahedron Lett*. 1978:3637–3640.

Siuda, J. F., and DeBernardis, J. F., 1973. Naturally occurring halogenated organic compounds, *Lloydia* 36:107–143.

Siuda, J. F., VanBlaricom, G. R., Shaw, P. D., Johnson, R. D., White, R. H., Hager, L. P., and Rinehart, K. L., Jr., 1975. 1-Iodo-3,3-bromo-2-heptanone, 1,1,3,3-tetrabromo-2-heptanone, and related compounds from the red alga *Bonnemaisonia hamifera*, *J. Am. Chem. Soc.* 97:937–938.

Sono, M., Dawson, J. H., Hall, K., and Hager, L. P., 1986. Ligand and halide binding properties of chloroperoxidase: Peroxidase-type active site heme environment with cytochrome P-450 type endogenous axial ligand and spectral properties, *Biochemistry* 25:347–356.

Suzuki, M., Morita, Y., Yanagisawa, A., Baker, B. J., Scheuer, P. J., and Noyori, R., 1988. Synthesis and structural revision of (7*E*)- and (7*Z*)-punaglandin 4, *J. Org. Chem.* 53:286–295.

Theiler, R., Cook, J. C., Hager, J. F., and Siuda, J. F., 1978. Halohydrocarbon synthesis by bromoperoxidase, *Science* 202:1094–1096.

Thomas, E. L., Pera, K. A., Smith, K. W., and Chwang, A. K., 1983. Inhibition of *Streptococcus mutans* by the lactoperoxidase antimicrobial system, *Infect. Immun.* 39:767–778.

Turner, C. D., Chernoff, S. B., Taurog, A., and Rawitch, A. B., 1983. Differences in iodinated peptides and thyroid hormone formation after chemical and thyroid peroxidase-catalyzed iodination of human thyroglobulin, *Arch. Biochem. Biophys.* 222:245–258.

Turner, W. B., 1971. *Fungal Metabolites*, Academic Press, New York.

Tymiak, A. A., Rinehart, K. L., Jr., and Bakus, G. J., 1985. Constituents of morphologically similar sponges. *Aplysina* and *Smenospongia* species, *Tetrahedron* 41:1039–1047.

van Pee, K.-H., and Lingens, F., 1984. Detection of a bromoperoxidase in *Streptomyces phaeochromogenes*, *FEBS Lett.* 173:5–8.

van Pee, K.-H., and Lingens, F., 1985. Purification of bromoperoxidase from *Pseudomonas aureofaciens*, *J. Bacteriol.* 161:1171–1175.

Welinder, K. G., 1976. Covalent structure of the glycoprotein horseradish peroxidase (EC 1.11.1.7), *FEBS Lett.* 72:19–23.

Wever, R., de Boer, E., Plat, H., and Krenn, B. E., 1987. Vanadium—an element involved in the biosynthesis of halogenated compounds and in nitrogen fixation, *FEBS Lett.* 216:1–3.

Wever, R., Krenn, B. E., de Boer, E., Offenberg, H., and Plat, H., 1988. Structure and function of vanadium-containing bromoperoxidases, *Prog. Clin. Biol. Res.* 274:477–493.

Wiesner, W., van Pee, K.-H., and Lingens, F., 1986. Detection of a new chloroperoxidase in *Pseudomonas pyrrocinia*, *FEBS Lett.* 209:321–324.

Wiesner, W., van Pee, K.-H., and Lingens, F., 1988. Purification and characterization of a novel bacterial non-heme chloroperoxidase from *Pseudomonas pyrrocinia*, *J. Biol. Chem.* 263:13725–13732.

Persistent Polyhalogenated Compounds: Biochemistry, Toxicology, Medical Applications, and Associated Environmental Issues

8.1 Introduction

This chapter will focus on several classes of halogenated compounds that have had widespread practical applications in industry and commerce. Because many of these compounds are chemically and thermally very stable, their extensive use has evoked serious concerns over potential long-term harmful effects to the environment. Indeed, many of these compounds have proven to be quite toxic, and environmental concerns are increased by their bioaccumulation in the food chain. Research designed to elucidate mechanisms of toxicity has helped define environmental issues and has produced significant biochemical data not only on the toxicology of the compounds of concern, but also on cellular mechanisms of the organisms involved, including humans. The environmental impact of certain halogenated compounds makes this topic a matter of extreme relevance, and inclusion of this subject in this volume clearly is appropriate. Biochemistry, toxicity, and potential environmental impact of several classes of halogenated insecticides will be considered first. This will be followed by a review of the problems associated with the toxicity of polyhalogenated hydrocarbons and related phenols, dioxins, and dibenzofurans. The development of halogenated volatile anesthetics and the use of perfluorinated aliphatic hydrocarbons as artificial oxygen carriers will be reviewed briefly. The biochemistry of simple halogenated aliphatic compounds, such as chloroform and carbon tetrachloride, is related in large part to metabolic activation to toxic species and will be discussed in Chapter 9.

8.2 Biochemistry and Toxicology of Chlorinated Insecticides

Chemical control of insects has been known for centuries (for a concise historical review, see Costa, 1987a). For example, pyrethrum, composed of naturally occurring pyrethroids found in the genus *Chrysanthemum*, was used by the Persians in 400 B.C. for louse control. With increasing global population, the struggle to protect human health and food and fiber resources from the onslaught of a variety of pests has assumed increasing urgency. Over the past several decades, synthetic chemical pesticides, including several classes of halocarbon insecticides, have played a critical role in this struggle, and a degree of complacency accompanied some early spectacular successes. However, development of pesticide-resistant strains of insects, along with recent increased awareness of the risks associated with adding synthetic chemicals to the environment —especially chemicals designed to kill some form of life—have spurred interest in alternatives to chemical pesticides.

In this section, the development, biochemistry, and toxicology of chlorinated insecticides will be reviewed. These will be considered in three categories: (1) DDT and related compounds, (2) hexachlorocyclohexanes (e.g., lindane), and (3) cyclodienes (e.g., aldrin, chlordane, chlordecone, Mirex) (Fig. 8-1). These compounds in general act as neurotoxins, and often evidence can be found for more than one mechanism for disruption of proper neuronal function. Many of these agents also affect the endocrine system. In the discussion that follows, the focus is on primary mechanisms of toxicity.

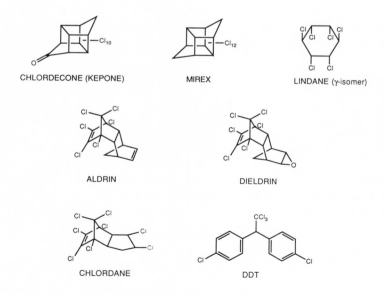

Figure 8-1. Examples of halogenated insecticides.

8.2.1 2,2-Bis(*p*-chlorophenyl)-1,1,1-trichloroethane (DDT) and Analogues

8.2.1.1 Historical Background

The insecticidal properties of DDT, first synthesized by Othmar Zeidler in 1873, were discovered in 1939 by Paul Müller of the J. R. Geigy Company in Switzerland. DDT played an important role in disease control during and immediately after World War II—the conquest of typhus in Naples is a notable example. The initial effectiveness of DDT in malaria control led to impressive improvements in public health of protected populations, as well as in their food production. The activity shown by DDT against a wide spectrum of insects gave early indications that this compound would be a potent weapon suitable for a sustained and ultimately successful battle against a host of pests. Indeed, the estimate has been made that, from the 1940s to the 1960s, public health and agricultural use of DDT amounted to about 4 billion pounds and that by 1972 there were about 1 billion pounds of DDT in the biosphere (Quraishi, 1977, and references therein). This early enthusiasm was tempered in the late 1940s by several instances of insect resistance to DDT, giving impetus to research directed toward determining the mechanism of toxicity and to the development of analogues capable of overcoming this resistance. At the same time, concern over potential problems associated with the widespread use of chemical insecticides was growing. A combination of lipophilicity, persistence in the environment, and the fact that DDT is stored primarily in fat had resulted in global ecodisposition of DDT by the 1960s, along with severe bioconcentration. In 1962, publication of Rachel Carson's *Silent Spring*, a condemnation of chemical contamination of the environment—with special emphasis given to bioaccumulation of DDT and its effects on birds—spearheaded an environmental movement calling for governmental action against the use of persistent pesticides. The manufacture and sale of DDT was banned by the U.S. Government in 1973. Nonetheless, DDT is thought to have saved more lives than penicillin—discovered at about the same time—and Müller was awarded the Nobel Prize in medicine in 1948 because of the impact of his discovery on public health. DDT is still used widely in developing countries, particularly those in which malaria remains a serious health problem. In addition, DDT has become a significant research tool in neuropharmacology (reviewed by Costa, 1987b; Woolley, 1982; Bickel and Muehlebach, 1980; Quraishi, 1977).

8.2.1.2 Biochemistry of DDT—Effect of DDT on Neural Electrochemistry

DDT is a broad-spectrum insecticide that is toxic to invertebrates and vertebrates (for a discussion of early theories of DDT toxicity, see

Quraishi, 1977). While DDT has favorable toxicological properties—topical applications and prolonged exposure to DDT dust have been shown to have minimal harmful effects on humans—it is acutely toxic when injected into mammals. Symptoms of DDT poisoning include hyperresponsiveness to external stimuli, hyperthermia, tremor, loss of coordination, convulsions, and ultimately death—a sequence of events indicative of neurotoxicity. Direct evidence for neurotoxicity came from the demonstration by Yeager and Munson (1945) that DDT-treated nerve cells exhibited continued firing following a single depolarizing impulse. Subsequent research has confirmed and extended these results.

A brief review of the processes involved in the "firing" of a neuron will benefit the discussion of DDT-induced alterations in these events. At resting potential, the interior of the neuron, as of other cells, has a much higher concentration of potassium than of sodium, concentration gradients being maintained by the Na^+/K^+-ATPase pump (see also the discussion in Chapter 3). In the resting state, the nerve axon membrane is much more permeable to potassium than to sodium, and the resting potential (-60 mV) is close to the equilibrium potential of potassium (-75 mV). A nerve impulse, or action potential, is generated by depolarization of the axon membrane—initiated by a number of factors, such as neurotransmitter stimulation, receptor response, or injury—to a critical threshold value (-40 mV), at which time the potential rapidly (in ~ 1 ms) becomes positive ($+30$ mV). This action potential is produced when the initial depolarization of the membrane potential to a threshold value leads to the opening of sodium channels. Sodium ions flow into the nerve cell as a response to the electrochemical gradient, further depolarizing the membrane, with opening of more sodium channels. The process ceases when the sodium equilibrium potential of $+30$ mV is reached. The sodium channels then begin to close spontaneously, blocking further entry of sodium into the neuron, and potassium gates begin to open, albeit more slowly. Potassium ions then are forced outward by the electrochemical gradient for about 2 ms until the amount of sodium which entered during the action potential is matched by exiting potassium. The resulting potential approaches the equilibrium potential of potassium (-75 mV). The resting potential is restored shortly thereafter by sodium/potassium pumps. The time course of the change in membrane potential and ionic conductances is shown in Fig. 8-2 (Stryer, 1988).

In a typical neuron, the depolarization accompanying the action potential also increases the permeability of the nerve terminal to calcium ions. Calcium which enters the terminal in response to a large inward-directed electrochemical gradient triggers the release of a chemical neurotransmitter.

For the past three decades, extensive electrophysiological studies of DDT action have shown that DDT has no effect on the resting potential of neurons. However, Narahashi and Yamasaki (1960) demonstrated that DDT prolonged and increased the magnitude of the negative afterpotential in the giant axon of the cockroach, and Narahashi and Haas (1967, 1968) showed that this is associated with a slowing down of the normal closing of sodium channels. Dubois and Bergman (1977) made direct measurements of "gating" currents—charges that are believed to control the opening and closing of the sodium channel—and confirmed that DDT has an effect only on charges related to closing of a channel previously opened. In addition to providing an explanation for the DDT-induced prolongation of the negative afterpotential, this is also consistent with the insensitivity of the resting potential to DDT. The prolonged depolarized state of the neuron often causes repetitive discharges in the nerve fibers, resulting in a disrupted synaptic and neuromuscular transmission. The effects of DDT on the action potential of a single nerve fiber (the sciatic nerve of a toad) are shown in Fig. 8-3 to illustrate these phenomena (van den Bercken, 1970). A more thorough discussion of the research that has helped solidify this as a mechanism of DDT toxicity can be found in several excellent reviews (e.g., Woolley, 1982; Joy, 1982).

DDT was also shown to slow the opening of potassium channels in the lobster axon (Narahashi and Haas, 1968). However, similar effects in

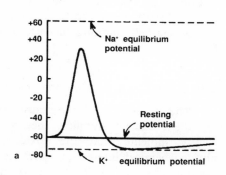

Figure 8-2. Time course of the change in membrane potential (mV) (a) and the change in Na$^+$ and K$^+$ conductances (b) during depolarization of an axon membrane to produce an action potential. From *Biochemistry*, by Lubert Stryer. Copyright 1975, 1981, 1988 by Lubert Stryer. Reprinted by permission of W. H. Freeman and Co.

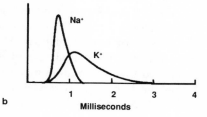

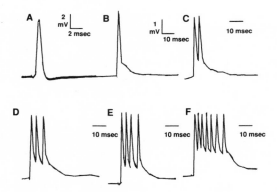

Figure 8-3. Repetitive activity in DDT-treated nerve fibers. (A) A normal action potential. (B) Action potential with negative afterpotential, 65 min after treatment with $5 \times 10^{-4} M$ DDT. (D–F) Repetitive action potentials superimposed on negative afterpotential 85, 90, 105, and 135 min after treatment, respectively. Reprinted, with permission, from van den Bercken (1970).

other systems were not observed, or were of modest magnitude (reviewed by Woolley, 1982).

There is evidence from several studies that DDT may also inhibit cell membrane Ca^{2+}-ATPase, and thus cause lowered external calcium concentrations. This could contribute to, or cause, repetitive firing. Inhibition of Ca^{2+}-ATPase appears to be also related to the mechanism by which DDT causes eggshell thinning in certain avian species (reviewed by Woolley, 1982).

8.2.1.3 Physicochemical Interactions of DDT with Membrane Channels

The reversibility of the neurotoxic effects of nonlethal doses of DDT and the absence of significant neurostructural damage in surviving animals clearly indicate that the toxic effects are produced by reversible physicochemical interactions with neural membranes. Several theories for the interaction of DDT with nerve membranes have been proposed based on the three-dimensional structure of DDT and on structure–activity relationships of analogues. One example of these models is that proposed by Holan (1969), a model that was modified by Coats et al. (1977) to account for activity of nonsymmetrical DDT analogues (Fig. 8-4; Fukuta, 1976). In this model, the overall size of the model is important, and SAR studies confirmed that a minimum molecular volume is required to induce toxicity. Using the hypothetical model for the sodium channel proposed by Hille (1977), Woolley (1982) has proposed that DDT can penetrate the lipid

Figure 8-4. The Holan model for the membrane receptor site for DDT. The Cl_3 apex of the molecules fits into a lipid pore having a diameter approximately equal to that of a hydrated sodium ion. The phenyl groups interact with the overlying protein layer, and the dashed lines indicate the limit in size of substituents on the phenyl rings. Reprinted, with permission, from Fukuta (1976).

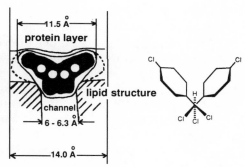

phase of the membrane to reach the gating mechanism near the inner surface of the membrane. Interaction with the open position of the gate would then slow its return to the closed position (Fig. 8-5).

8.2.1.4 Neurochemical Effects of DDT

The research cited above is a portion of a large volume of work that has established the nerve membrane as a primary target for DDT action. Included in studies in several disciplines to determine how the effects of DDT on the nerve membrane are related to symptomatic phenomena have been extensive investigations of the effects of DDT on neurochemical balances in the brain. For example, Hrdina et al. (1973) found that DDT decreased norepinephrine (NE) levels in brain stem and acetylcholine levels (ACh) in the cortex and striatum, but had no apparent effect on brain stem

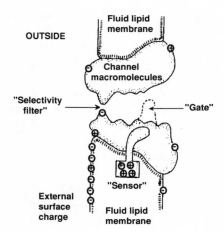

Figure 8-5. A hypothetical model of the sodium channel. Because of its high lipid solubility, the proposal is made that DDT could penetrate the lipid phase of the membrane to reach the gating mechanism situated near the inner surface of the membrane, interacting with the open position of the gate and slowing its return to the closed position. Reprinted, with permission, from Hille (1977).

serotonin (5-hydroxytryptamine; 5HT) or striatal dopamine. However, brain stem levels of the 5HT metabolite 5-hydroxyindoleacetic acid (5HIAA) were increased. In order to define more clearly the relationship between DDT toxicity and alterations in brain neurotransmitters, Hong *et al.* (1986) have studied the dose- and time-dependent relationships between tremorigenic doses of DDT and brain levels of neurotransmitters and their metabolites in the rat. Whereas there was no detectable increase in NE levels, a good correlation was found between DDT-induced tremor and increases in levels of metabolites of NE [3-methoxy-4-hydroxyphenyl-glycol (MHPG); dihydroxyphenylacetic acid (DOPAC)] in several brain areas. The data suggest, but do not prove, that the adrenergic nervous system may be involved in mediating DDT-induced tremor.

A similar correlation was found between metabolites of 5HT and, to a lesser extent, of dopamine. DDT also increased levels of the excitatory amino acids glutamate and aspartate in the brain stem, an increase that correlated with hyperthermia and tremor in a dose- and time-dependent manner. That a similar increase in excitatory amino acids was not found in the hypothalamus or striatum may indicate that DDT-induced tremor originates in the brain stem (Hong *et al.*, 1986; Hudson *et al.*, 1985).

The increased levels of neurotransmitter turnover, as indicated by the higher levels of their metabolites, are consistent with the proposal that DDT toxicity is mediated by a blockade of the closing of the activation gate of the sodium channel. Adding support to this is the demonstration that pretreatment of rats with the anticonvulsant 5,5-diphenylhydantoin significantly reduces DDT-induced tremors (Tilson *et al.*, 1986b). This agent is thought to block repetitive neuronal discharge by binding to the inactivation gates of sodium. In DDT-treated animals, this effect, an interference with recovery from inactivation, would counter the effect of DDT—an interference with recovery from activation. Blockade of the excitatory influence of α_1-adrenergic receptors with α_1-adrenergic antagonists attenuated the tremorigenic effects of DDT, whereas blockade of α_2-adrenergic receptors exacerbated DDT-induced tremors. These results implicate excitatory and inhibitory roles for α_1- and α_2-adrenergic receptors, respectively, in modulation of the tremorigenic activity of DDT (Herr *et al.*, 1987, and references therein). In contrast, blockade of serotonergic, dopaminergic, or cholinergic muscarinic receptors increased the toxicity of DDT (Hong *et al.*, 1986).

From the neurochemical studies cited here, it is evident that DDT affects the rate of turnover and balance of several chemical neurotransmitters. Because the proper functioning of ionic mechanisms controlling membrane potential is critical to neural activity, interference with the gating of the sodium channel would be expected to produce a spectrum of alterations in levels of neurotransmitters and their metabolites. The complexity of these neurochemical consequences has made difficult the determination of the ultimate biochemical mechanism(s) leading to the observed symptoms. Interpretation of the results of neuropharmacological experiments is also complicated by subtle factors that can give seemingly contradictory results—evaluation of the role of serotonergic neurons has been particularly difficult (Tilson *et al.*, 1986b, and references therein). For further discussions of these complex issues, the reviews by Woolley (1982), Joy (1982), and Van Woert *et al.* (1982) are recommended.

8.2.1.5 Analogues of DDT

The simplicity of the DDT molecule, the impressive insecticidal properties of DDT, and the problems associated with toxicity and environmental persistence of DDT are among the factors that have stimulated an immense amount of research designed to develop analogues with increased environmental degradability and with retained insecticidal properties. Results from the biological evaluation of such analogues has, in turn, aided in the development of models for the interaction of DDT with axonal membranes (see Section 8.2.1.3). Impressive examples of analogue syntheses and evaluation are found in the reports by Coats *et al.* (1977) and by Holan (1971). For reviews, see Quraishi (1977) and Brooks (1986).

An E2 elimination of HCl from DDT to form 1,1-bis(*p*-chlorophenyl)-2,2-dichloroethylene (DDE), catalyzed by the enzyme DDT dehydrochlorinase (DDTase) with glutathione (GSH) functioning as a base, is a principal mechanism by which insects become resistant to DDT. A similar OH^--catalyzed elimination also produces the environmentally stable DDE. Replacement of *p*-chloro substituents of DDT with electron-donating ethoxy groups results in an analogue (ethoxychlor, Fig. 8-6) which not only is biodegradable, but which also is active against insects having DDTase-based resistance. Several analogues based on this concept have been developed, including analogues having various modifications in the CCl_3 moiety (Brooks, 1986, and references therein). For example, as

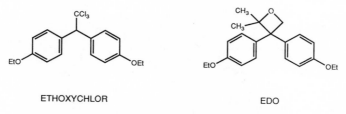

ETHOXYCHLOR   EDO

Figure 8-6. Examples of structural analogues of DDT.

part of a program of rational design of DDT analogues, Holan (1971) synthesized 2,2-bis(*p*-ethoxyphenyl)-3,3-dimethyloxetane (EDO) (Fig. 8-6), an analogue that readily undergoes degradation and retains the insecticidal potency of DDT but has only half the mammalian toxicity of DDT. In addition, since it is halogen-free, it cannot be metabolized by DDTase, and it was found to be very effective against a strain of housefly resistant to DDT. Electrophysiological studies have indicated that EDO greatly prolongs the depolarizing afterpotential of the axonal membrane. Thus, similar to DDT, EDO appears to block inactivation of axonal membrane sodium channels (Wu *et al.*, 1980, and references therein).

Halogen position is a critical parameter for production of neurotoxic effects of DDT, since the isomeric *o, p'*-DDT has much less activity. On the other hand, *o, p'*-DDT is much more effective than *p, p'*-DDT as an estrogenic agent, producing a variety of estrogenic responses (McBlain, 1987, and references therein), responses that are mediated through direct action on the estrogen receptor (Nelson, 1974). Recently, estrogenic activity of *o, p'*-DDT has been shown to reside with the levo enantiomer (McBlain, 1987). Conformational restrictions imposed by the *o*-chlorophenyl group have been proposed as important determinants of estrogenic activity (Korach *et al.*, 1988).

8.2.1.6 *o, p'-DDD in Chemotherapy*

In toxicological studies of DDT in dogs, the observation was made in 1948 that "technical grade" DDT caused severe adrenal cortical atrophy. This subsequently was attributed to the presence of 1-(*o*-chlorophenyl)-1-(*p*-chlorophenyl)-2,2-dichloroethane (*o, p'*-DDD) (Fig. 8-7) in the samples. Subsequent research revealed that *o, p'*-DDD blocks adrenocorticotropin (ACTH)-induced secretion of glucocorticoids from atrophied glands (Hart *et al.*, 1973). Based on these observations, *o, p'*-DDD (mitotane) was introduced for clinical use in 1960, and it is now an FDA-approved adrenocorticolytic used in the treatment of hyperadrenocorticism (Cushing's syndrome) resulting from adrenal tumor or hyperplasia.

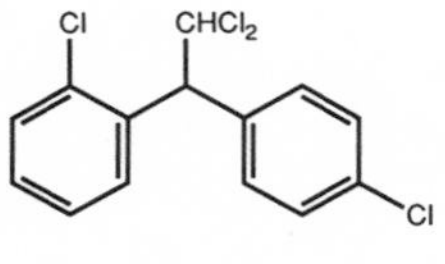

o,p-DDD (Mitotane) 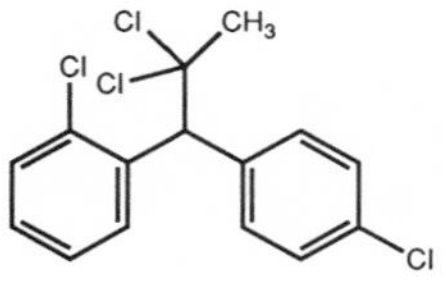   MITOMETH

Figure 8-7. Mitotane and Mitometh, active against adrenal cortical carcinomas.

The chemotherapeutic efficacy of o, p'-DDD is compromised by serious side effects, undoubtedly caused in part by the high doses (2–10 g/day) needed. Rapid metabolism is also a complicating feature of this drug, and there is evidence that a principal metabolite, o, p'-diphenylacetic acid, may be responsible, at least in part, for the observed toxic reactions. In an attempt to improve on the therapeutic properties of o, p'-DDD, the structurally related compound 1-(o-chlorophenyl)-1-(p-chlorophenyl)-2,2-dichloropropane (Mitometh) (Fig. 8-7) was synthesized. Ultrastructural effects of Mitometh on the adrenal cortex of guinea pigs appeared identical to those elicited by o, p'-DDD, but Mitometh-treated animals displayed far fewer symptoms of toxicity. Metabolism to o, p'-diphenylacetic acid is blocked, and a large proportion of drug is excreted unchanged from treated animals. Together, these data suggest that Mitometh may have improved chemotherapeutic properties over o, p'-DDD for the treatment of adrenal cortical carcinomas (Jensen et al., 1987).

8.2.2 Chlorinated Cyclodiene, Hexachlorocyclohexane, and Polychloronorbornane Insecticides

8.2.2.1 Chlorinated Cyclodienes

A Diels–Alder cyclization of hexachlorocyclopentadiene with cyclopentadiene is the starting point for the synthesis of a group of highly chlorinated polycyclic insecticides (Fig. 8-8). These compounds date from the 1940s and enjoyed early extensive use for control of a variety of pests. Examples include chlordane (LD_{50} for the rat between 457 and 590 mg/kg), used against house pests, termites, pests of man and domestic

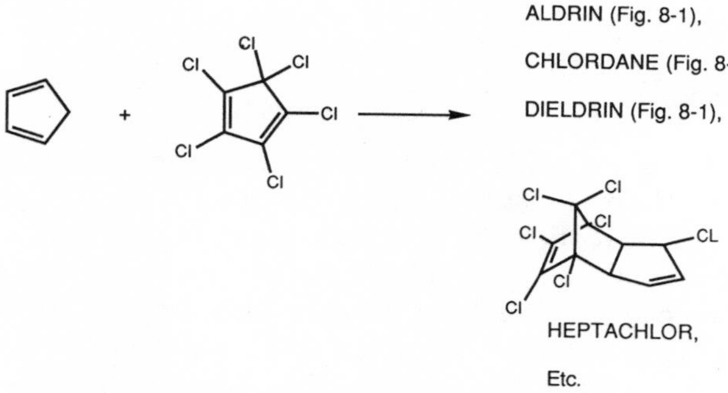

ALDRIN (Fig. 8-1),

CHLORDANE (Fig. 8-1),

DIELDRIN (Fig. 8-1),

HEPTACHLOR,

Etc.

Figure 8-8. Examples of cyclodiene insecticides.

animals, and lawn and agricultural pests; aldrin (acute oral LD_{50} for the rat of 55 mg/kg) and its epoxide, dieldrin (acute oral LD_{50} for the rat of 46 mg/kg), used to control soil insects; and chlordecone (Kepone) (acute oral LD_{50} for the rat of 132 mg/kg) (Fig. 8-1). Thus, these compounds are toxic both to invertebrates and vertebrates, as is DDT, but, in many cases, the acute toxicity is greater than that of DDT. There also are significant differences in the toxic reactions to DDT and the cyclodiene insecticides. For example, whereas DDT induces initial tremoring in vertebrates, cyclodiene toxicity is characterized by a sudden and violent onset of convulsions. This difference reflects, in part, differences in proposed modes of actions of the insecticides—whereas DDT has distinctive activity on sensory nerves and on peripheral receptors, the primary site of action of cyclodienes is in the central nervous system (CNS) (reviewed by Joy, 1982). In Section 8.2.2.4, recent research on the mechanism of action of selected cyclodiene insecticides will be reviewed to help define environmental issues, as well as to illustrate the impact of this class of compound on neuropharmacology.

8.2.2.2 The γ-Isomer of 1,2,3,4,5,6-Hexachlorocyclohexane (Lindane)

Hexachlorocyclohexane (HCH) was prepared by Michael Faraday in 1825, and its insecticidal properties were discovered in 1942. The crude mixture of HCH isomers is prepared by the chlorination of benzene in the presence of ultraviolet light. The presence of the γ-isomer (Fig. 8-1), discovered by Van der Linden, for whom lindane is named, is an absolute requirement for toxicity. Lindane is used for controlling pests of cotton, cattle, and hogs. Toxic symptoms in insects include tremors, ataxia, convulsions, and prostration, indicative of neurotoxicity. The LD_{50} for rats is 88 mg/kg (Quraishi, 1977).

8.2.2.3 Toxaphene

Toxaphene is defined as a complex mixture of chlorinated camphenes (chlorine content of 67–69%) containing more than 170 different C_{10} hydrocarbons. One particularly toxic component is 2,2,5-endo,6-exo,8,9,10-heptachlorobornane (Fig. 8-9) (Hayes, 1982). Chronic dietary exposure of laboratory animals to toxaphene produces degenerative changes in liver and kidneys. Toxaphene has an acute oral toxicity in the rat (LD_{50}) of 69 mg/kg, and it is particularly toxic to fish. Toxaphene has been used for control of a wide range of major pests of cotton and livestock.

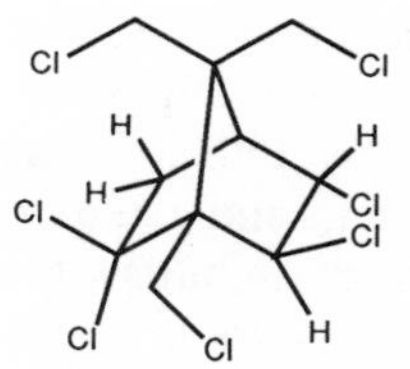

Figure 8-9. Structure of the toxaphene component, 2,2,5–endo, 6–exo, 8,9,10-heptachlorobor-nane.

8.2.2.4 Cyclodiene and Related Insecticides and the γ-Aminobutyric Acid (GABA) Receptor–Ionophore Complex

Cyclodiene-type insecticides promote neurotransmitter release in the CNS, an action related to their ability to increase intracellular calcium concentrations in the presynaptic terminal. This may result from an inhibition of ATPase (Costa, 1987a, and references therein). However, the primary site of action appears to be on GABAergic systems.

As discussed in Chapter 3, GABA is one of the major inhibitory neurotransmitters in the CNS of both vertebrates and invertebrates. To review briefly, interaction of GABA with a postsynaptic membrane increases the permeability of that membrane to Cl^- and thus moves the membrane potential farther from the action potential. The $GABA_A$ receptor complex, located postsynaptically, consists of four subunits containing the GABA binding site, a benzodiazepine binding site, and a Cl^- channel, gating of which is controlled by interaction of GABA with its binding site (see Chapter 3).

The observation that cyclodiene-resistant German cockroaches were also resistant to the action of picrotoxinin (PTX), a naturally occurring neuroexcitant, provided the initial lead pointing to the involvement of $GABA_A$ receptors in cyclodiene toxicity (Matsumura and Ghiasuddin, 1983). Thus, PTX is known to suppress the inhibitory activity of GABA by blocking the chloride channel of the $GABA_A$–ionophore complex. Cyclodiene and PTX resistances were related to fewer [3H]dihydroPTX ([3H]-DHPTX) binding sites in resistant insects, further implicating the PTX binding site in cyclodiene toxicity (Kaddus et al., 1983). In addition, cyclodiene insecticides and lindane, but not DDT, DDE, or hexachlorobenzene, inhibited [3H]-DHPTX binding to the coaxal muscle of the American cockroach in vitro and in vivo and blocked the GABA-induced increased permeability of Cl^- ion in the coaxal muscle of the American cockroach (reviewed by Costa, 1987b).

There is much evidence that cyclodiene insecticide-induced convulsions in mammals are also mediated by disruption of the GABA-regulated chloride channel. For example, binding studies using [36]S-labeled *t*-butyl-bicyclophosphorothionate ([36]S]-TBS), a recently developed ligand for the GABA-gated chloride channel, indicated overlap between cyclodiene insecticides, PTX, and TBS in mammalian brain preparations, and *in vivo* toxicity has been related directly to insecticide binding to PTX sites in mouse brain (Cole and Casida, 1986, and references therein). In addition, several chlorinated insecticides were found to block the GABA-induced [36]Cl]chloride influx into mouse brain vesicles. The cyclodienes endrin and dieldrin were the most effective (Bloomquist and Soderlund, 1985; see also Costa, 1987b, and references therein for additional studies).

In addition to cyclodiene-type insecticides, the hexachlorocyclohexane insecticide lindane and the polychlorobornane toxaphene components also bind to the PTX-binding site of the $GABA_A$ receptor. This provides a basis for the convulsant activities of these insecticides as well. Furthermore, the well-defined three-dimensional structures of these PTX binding site ligands should provide useful information concerning the topography of the GABA-gated chloride channel. Adding further significance to these results is the fact that from the 1940s to 1984 an estimated 3 billion pounds of the three classes of insecticides had been applied to soil and crops (Lawrence and Casida, 1984).

8.2.2.5 Biochemistry and Toxicology of Chlordecone (Kepone)

Chlordecone (Fig. 8-1) is normally classified as a cyclodiene-type insecticide. However, the complexity of the biological sequelae produced by toxic doses of chlordecone and the clear evidence that the primary mechanism(s) of toxicity differs from those of DDT or other cyclodiene insecticides warrant a separate discussion of this agent. Introduced in 1958 for use against such pests as leafeating insects, ants, and cockroaches, it was manufactured mainly for export and received little use in the United States. Consequently, relatively little early research has been done on chlordecone toxicology. However, in 1975, contamination of a production plant in Hopewell, Virginia, resulted in serious environmental contamination, in particular to the James River, and severe ailments to numerous workers due to chlordecone exposure. Included in the symptoms experienced by exposed workers were disabling neurological disorders such as tremors, ataxia, loss of coordination, and slurred speech, as well as testicular damage, sterility, and production of abnormal sperm. This industrial incident, together with the discovery that the related insecticide Mirex (Fig. 8-1)—used extensively in the southwest United States to control fire ants—is converted to a significant degree by a photochemical process to chlordecone, stimulated a vast amount of research in an effort to determine the biological consequences of chlordecone ingestion and the mechanisms of toxicity.

Chlordecone toxicity in experimental animals is characterized by severe DDT-like tremors. As with DDT, toxic effects elicited by nonlethal doses are essentially reversible. Nonetheless, several lines of evidence indicate that the principal toxic mechanisms of chlordecone and DDT are different. For example, whereas pretreatment of experimental animals with 5,5-diphenylhydantoin (see Section 8.2) significantly attenuates the tremorigenic activity of DDT, this same pretreatment exacerbates that of chlordecone (Tilson et al., 1986). Similarly, pretreatment of rats by intracerebroventricular injection of calcium decreased the peak tremor power due to DDT but increased the tremor response due to chlordecone (Herr et al., 1987). Chlordecone, unlike other pesticides in the cyclodiene class, is a poor inhibitor of GABA-dependent Cl^- uptake (Bloomquist et al., 1986).

These comparisons suggest that chlordecone does not act by blocking the inactivation of the sodium channel or by binding to the GABA-gated Cl^- channel. However, accumulating evidence indicates that chlordecone has a spectrum of other effects, functioning as a neurotoxin, an estrogen, a hepatoxin, and a potentiator of halothane toxicity (see Chapter 9). One of the primary targets is the CNS, wherein chlordecone has been implicated in several actions, including inhibition of membrane-bound ATPases, alteration of neurotransmitter and neuropeptide levels, and disruption of calcium regulation (reviewed by Desaiah, 1982). However, unlike DDT, whose neurotoxic effects clearly are more relevant to toxicity than are endocrine effects, chlordecone has pronounced estrogenic effects in vitro and in vitro (Hammond et al., 1979).

Several biochemical studies have demonstrated that chlordecone inhibits Na^+/K^+-ATPase activity in a dose-dependent manner, and a linear relationship between this decrease and tremor in rats has been demonstrated (Jordon et al., 1981). Na^+/K^+-ATPase, which regulates the active transport of sodium and potassium across cell membranes, is present in most tissue cell membranes that carry out active transport. Amino acids, including neurotransmitter amino acids, are among the compounds whose active transport is linked to this enzyme. In addition, a Mg^{2+}-dependent Na^+/K^+-ATPase has been linked to the active transport of catecholamine neurotransmitters. Thus, the ability of micromolar concentrations of chlordecone to inhibit these enzymes could block transport of neurotransmitters. In support of this, chlordecone has been shown to cause a decrease in catecholamine binding in rats (Desaiah, 1982, and references therein).

There is evidence that chlordecone–mediated alterations in turnover of neurotransmitters may be related to its toxicity. For example, chlordecone increased the turnover of 5HT, and a dose- and time-related correlation was found between increases of 5HIAA striatal levels and tremor (Hong *et al.*, 1984). A reduction in striatal GABA accumulation may be caused by this increased serotonergic activity (Gandolfi *et al.*, 1984). Involvement of noradrenergic systems was indicated by the increases of concentrations of the norepinephrine metabolite MHPG in several brain areas (Chen *et al.*, 1985).

While blockade of neurotransmitter transport by inhibition of ATPase clearly could alter neurotransmitter turnover, Komulainen and Bondy (1987) suggest that a chlordecone-induced increase in intraneuronal calcium may contribute to the higher activity of serotonergic and noradrenergic neurons. In synaptosomes, such an increase was shown to result from an increased influx of extraneuronal calcium by a nonspecific leakage through the plasma membrane. This potential to increase intraneuronal calcium appears unique to chlordecone among the halogenated insecticides and could have relevance to its special toxic properties, although Komulainen and Bondy point out that extrapolation of these data to the living organism should be done with caution.

A link has been found between calcium mobilization and induction of polyamine biosynthesis in several tissues, including the brain. For example, irreversible inhibition of ornithine decarboxylase, the rate-limiting enzyme in polyamine biosynthesis, with difluoromethylornithine (DMFO) (see Chapter 7 in Vol. 9B of this series) inhibits calcium-stimulated release of D-aspartate from nerve endings. The fact that DMFO produces a putrescine-sensitive inhibition of chlordecone-induced tremors, but not DDT-induced tremors, indicates a potential link between chlordecone neurotoxicity, calcium mobilization, and polyamine biosynthesis (Tilson *et al.*, 1986a).

Long-term dietary exposure of rats to subchronic doses of chlordecone causes a decrease in hypothalamic β-endorphin levels, with no apparent alteration in dopamine or 5HT turnover (Ali *et al.*, 1982). This is an example of evidence that the hypothalamus may be an important site of action for chlordecone (Bondy and Hong, 1987, and references therein).

Reproductive disorders were among the symptoms of chlordecone toxicity manifested by exposed workers in the Hopewell plant, indicative of steroid hormone malfunction. Previous observations that chlorinated pesticides can affect adversely the reproductive functions of several animal species further illustrate the potential for these agents to interact with steroid hormones. Several studies had indicated that chlordecone might possess estrogenic activity, even though, unlike *o, p'*-DDT, it has little structural resemblance to steroidal or nonsteroidal estrogens. As part of a detoxification mechanism, chlorinated insecticides cause potent induction of hepatic monooxygenase enzymes, an induction that also increases the rate of metabolism of several steroids (reviewed by Kupfer, 1975; see also Chapter 9). However, Hammond *et al.* (1979) demonstrated that estrogenic activity of chlordecone results from direct action on the estrogen receptor. Thus, chlordecone binds to the cytoplasmic estrogen receptor with a higher affinity than does *o, p'*-DDT (Section 8.2.1.5), translocates these receptor sites to the nucleus *in vitro* and *in vivo*, increases uterine weight, and induces progesterone receptor synthesis. Although the affinity of chlordecone for the estrogen receptor is only 0.01–0.04% that of estrogen, bioaccumulation and long half-life combine to produce considerable long-lasting estrogenic activity, such that chronic exposure to this insecticide could lead to reproductive disorders.

As with other nonspecific neurotoxic agents in this series of organochlorine compounds, more than one site of action of chlordecone appears likely. Together with the interrelated cascade of neurochemical events each may initiate, this makes sorting out specific mechanisms of action an exceedingly complex task.

8.3 Biochemistry and Toxicology of Halogenated Biphenyls, Terphenyls, Naphthalenes, Dibenzodioxins, and Related Compounds

8.3.1 Industrial Applications of Halogenated Aromatic Compounds and the Environmental Consequences of Their Use

Polychlorinated biphenyls, terphenyls, polybrominated biphenyls, polychlorinated phenols, and polychlorinated naphthalenes are important members of a series of halogenated aromatic compounds that have, or have had, numerous industrial applications. The thermal and chemical stability of these compounds contributes greatly to their potential industrial importance, but it is this same stability which causes them to persist in the environment. This resistance to degradation, the toxicity associated with many of these compounds, and the potential for biomagnification through food chains has led to curtailment of production and use of these materials. Adding to environmental concerns is the fact that extremely toxic polyhalogenated dioxins and dibenzofurans are present as impurities in certain of the industrial sources of these compounds (Firestone, 1984). Many toxic manifestations of these compounds appear to be mediated by a common mechanism, and they can be discussed conveniently as a group. A brief description of industrial origins of certain of these compounds follows.

8.3.1.1 Polychlorinated Biphenyls

Polychlorinated biphenyls (PCBs) (Fig. 8-10) are a class of compounds having the empirical formula $C_{12}H_{10-n}Cl_n$. The extreme thermal and chemical stability of PCB mixtures suggested many industrial applications, and commercial production of PCBs was begun in the United States in the 1920s. The domestic uses of PCBs are classified as closed systems (e.g., capacitors and transformers), nominally closed systems (e.g., hydraulic fluids), and open-ended systems (e.g., plasticizers and adhesives). PCBs were discovered in environmental samples in 1966, and the mass food poisoning in Japan in 1968 caused by consumption of a commercial brand of rice oil contaminated with PCBs (the "Yusho" incident) brought the potential health problems of PCBs to full public attention. There has been a subsequent drastic reduction of the production and use of PCBs. Polychlorinated terphenyls ($C_{18}H_{14-n}Cl_n$) (PCTs) (Fig. 8-10) have similar, but more restricted, industrial applications. For a more complete discussion of these issues, the review by Brinkman and De Kok (1980) is recommended.

8.3.1.2 Polybrominated Biphenyls

The polybrominated biphenyls ($C_{12}H_{10-n}Br_n$) (PBBs) (Fig. 8-10) are very effective flame retardants, and this property led to their extensive use

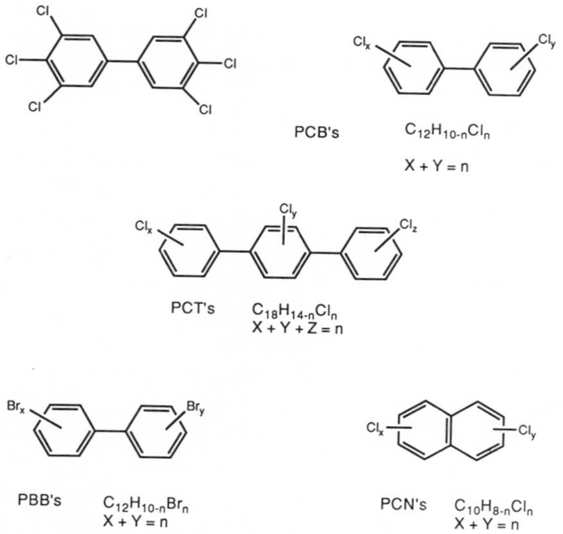

Figure 8-10. Structures of polyhalogenated aromatic compounds.

in such synthetic products as plastics, coatings and lacquers, and polyurethane foam. The use of PBBs has been restricted to materials that do not come in contact with food or feed and has been prohibited in fabrics to which humans could be exposed. The persistence of PBBs in the environment, the potential for bioaccumulation, and the structural similarity to the environment-polluting PCBs all contributed to serious environmental concerns over the use of PBBs. These concerns were increased drastically when several hundred pounds of PBBs were introduced accidentally into cattle feed in Michigan in 1973, with devastating effects on cattle ("the Michigan catastrophe"). Use of PBBs in the United States now has been halted (Brinkman and De Kok, 1980).

8.3.1.3 Polychlorinated Naphthalenes (PCNs)

The physical and chemical properties of PCNs (Fig. 8-10) and PCBs are similar. Before and during World War II, PCNs were used extensively in capacitor and cable manufacturing. Several incidences of illness caused by exposure to PCNs have been reported—including "pernapkrankheit," chloracne resulting from the use of PCNs as a coating for gas masks during World War I (Kimbrough, 1980).

8.3.1.4 Chlorophenols

As effective bactericides and fungicides, chlorophenols have had many industrial applications, including extensive use as wood preservatives. Chlorophenols are also starting materials for a large number of organic products—the use of 2,4,5-trichlorophenol for the production of 2,4,5-trichlorophenoxyacetic acid (2,4,5-T) is a notable example. As a mixture (Agent Orange) with 2,4-dichlorophenoxyacetic acid, this latter compound was used extensively as a defoliant by the U.S. military during the Vietnam War. Health and environmental considerations have halted production and use of 2,4,5-T (Firestone, 1984).

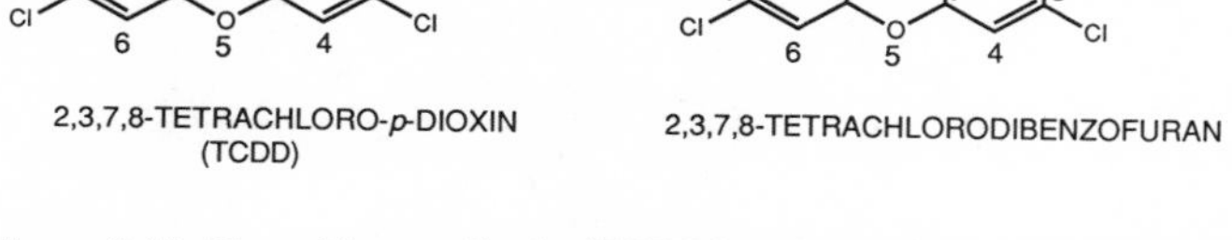

2,3,7,8-TETRACHLORO-*p*-DIOXIN
(TCDD)

2,3,7,8-TETRACHLORODIBENZOFURAN

Figure 8-11. Tetrachloro-*p*-dioxin (TCDD) and tetrachlorodibenzofuran.

8.3.1.5 Halogenated Dibenzodioxins and Furans

As noted above, dioxins and furans (Fig. 8-11) have been found as contaminants of industrial preparations of polyhalogenated biphenyls and phenols. Of these, the most toxic has been found to be 2,3,7,8-tetrachlordibenzo-*p*-dioxin (TCDD). The major source of TCDD is a side reaction that can occur during the production of trichlorophenol, as shown in Fig. 8-12. A notorious example of this is found in the "Seveso episode" in 1976. In this serious industrial accident, a reaction being used to produce 2,4,5-T became uncontrolled, overheated (> 180 °C), and released products containing TCDD over 700 acres in the town of Seveso, Italy. In another episode, the National Academy of Sciences has estimated that 220–360 pounds of TCDD were released between 1965 and 1971 during the spraying of Agent Orange in Vietnam. These and other examples of environmental contamination with TCDD have been reviewed by Reggiani (1980).

8.3.2 Toxic Manifestations of Halogenated Aromatic Hydrocarbons

Since the halogenated aromatic hydrocarbons (Ah) being considered in this section produce similar patterns of toxic responses and are thought to act through a common mechanism, much of the following discussion will focus on work done with TCDD, the most toxic and the most thoroughly studied, as the prototype of this class. There is wide variation in the sensitivity of animal species to the toxic effects of halogenated aromatic hydrocarbons, and this has complicated extrapolation of data from different studies. Nonetheless, a commonality of toxic responses can be discerned, and these effects are summarized here (reviewed by Poland and Knutson, 1982). Toxic and biological effects of TCDD include a wasting syndrome, hepatotoxicity and porphyria, dermal lesions (chloracne), loss of lymphoid tissue and suppression of thymus-dependent immunity (Vos, 1984), teratogenicity and reproductive effects, carcinogenicity, and the induction of several enzymes. For example, following a lethal dose of TCDD, wasting (weight loss and a depletion of adipose tissue) occurs after a latent period of a week or more. The lethality of TCDD, however, is not directly related to the wasting syndrome. Chloracne is the most frequently

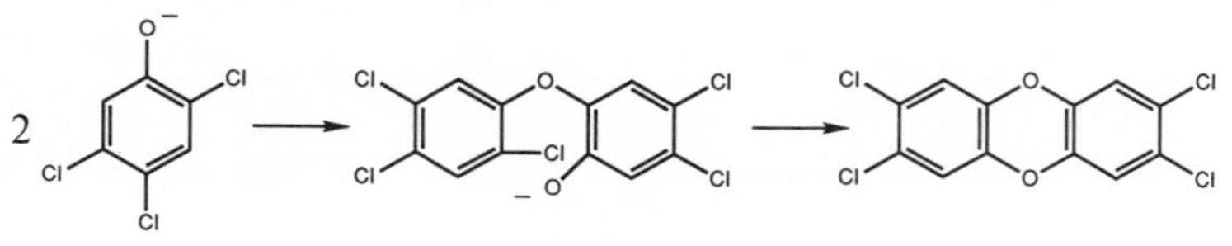

Figure 8-12. Mechanism for the formation of TCDD during trichlorophenol production.

observed lesion in humans exposed to TCDD (Kimbrough, 1984). The carcinogenic properties of TCDD appear to reflect its ability to serve as a tumor promoter, rather than tumorigenic properties (Poland and Knutson, 1982).

8.3.3 The TCDD (Ah) Receptor and Enzyme Induction

The brief overview of the biological effects of TCDD and related compounds presented above illustrates the wide range of profound biological responses that these compounds can elicit. The recognition that these responses must be related to fundamental cellular events has intensified interest in the mechanism of TCDD toxicity. Poland and Glover (1973) reported that TCDD and related dioxins are potent inducers of δ-aminolevulinic acid synthetase and aryl hydrocarbon hydroxylase (AHH), and subsequent research has revealed that the induction of several enzymes, including cytochrome P-448-mediated microsomal aromatic monooxygenases, most notably AHH, accompanies toxicity. [For a thorough discussion of induction of monooxygenase (P-450) activities, see Nebert et al., 1981.] The induction of AHH was shown to be initiated by reversible binding of TCDD to a cytosolic protein (or nuclear protein; see below)—termed the induction receptor—followed by translocation of the receptor–ligand complex to the nucleus (Poland et al., 1976). Two independent lines of evidence have suggested that the toxicity of TCDD and relatives is mediated through the induction receptor. First, for a large number of halogenated aromatic hydrocarbons, the toxicity can be correlated directly with receptor affinity or potency in induction of AHH activity (however, see Section 8.3.5.4 below). Second, toxic responses to TCDD in mice have been shown to segregate with the gene locus that determines the receptor (the Ah locus); for example, inbred mice having a low-affinity Ah receptor are relatively insensitive to TCDD (Knutson and Poland, 1982, and references therein).

8.3.4 The Aromatic Hydrocarbon (Ah) Receptor, AHH Induction, and TCDD Toxicity

8.3.4.1 Issues to Be Addressed by a Receptor-Mediated Model for Toxicity

While a relationship between the Ah receptor and TCDD toxicity seems clearly established, it is not obvious whether this is a causal relationship or if AHH induction and toxicity are coordinate independent responses to TCDD. Thus, this relationship *per se* does not explain TCDD toxicity and, in fact, raises several important questions. For example, many tissues and cell lines contain receptors that respond to TCDD with AHH induction but show no toxic response, suggesting that the binding to the receptor may be necessary but not sufficient for toxicity [for a more thorough discussion of these issues, see the review by Poland and Knutson (1982)]. Furthermore, the concentration of the Ah receptor and binding affinity for TCDD to the receptor are similar in several animal species, including the guinea pig and the hamster, yet there is a 5000-fold difference in LD_{50} for these two species. This latter phenomenon is not attributable to different metabolic rates of TCDD in the two species.

If AHH induction is necessary for TCDD toxicity, how does increased AHH activity produce the wide spectrum of toxic responses? The induction of AHH by polycyclic aromatic hydrocarbons (PAH) is initiated by the same receptor binding as is induction by TCDD. While AHH induction by PAH serves the purpose of increasing the rate of metabolism of the potentially toxic PAH, the extreme metabolic stability of TCDD virtually eliminates any possible role of TCDD metabolites in the toxic response. The possibility that toxicity might arise by a TCDD-induced AHH increase in the rate of metabolism of a vital endogenous compound would be difficult to reconcile with the wide spectrum of toxic responses (Poland and Knutson, 1982).

8.3.4.2 Models for Toxicity Based on Coordinate Gene Expression

Based on results from *in vitro* and *in vivo* experimental models of toxicity, Knutson and Poland (1982) have constructed a general model of toxicity based on multiple-gene activation. Breeding of hairless male mice (*hr/hr*) with heterozygous, haired females (*hr/+*) produces *hr/hr* and *hr/+* genetically identical litter mates except for one allele at the *hr* locus. Whereas both genotypes (and mice which are wild-type at the hairless locus) respond to topically applied TCDD with the induction of AHH activity, only *hr/hr* mice have an additional hyperplastic–metaplastic response resulting in chloracne. Since the mice differ in only one gene locus (the hairless trait), the *hr/+* (and wild-type) mice presumably have the genes necessary to produce the toxic response. Since all strains recognize TCDD and activate AHH, the conclusion was reached that in *hr/hr* mice an additional battery of genes are expressed in response to activation of the cytosolic Ah receptor.

From these and other data, a model of coordinate gene expression, controlled by the Ah locus—the structural gene for the receptor—was constructed. According to this model (Fig. 8-13), the cytosol receptor binds a

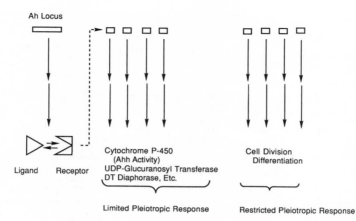

Figure 8-13. Model for pleiotropic responses to TCDD and other halogenated aromatic compounds. In the limited response (shown on the left of the diagram), genes are activated that express enzymes related to drug response (e.g., AHH). In the toxic response, an additional battery of genes are expressed that are involved in cell division and differentiation. After Knutson and Poland (1982), with permission; Copyright 1982 Cell Press.

ligand (e.g., TCDD), and the receptor–ligand complex, in a temperature-dependent process, is translocated to the nucleus. In the limited pleiotropic response, genes are activated that express enzymes related to drug response (e.g., AHH). In the toxic response, an additional battery of genes involved in cell division and differentiation are expressed. This model would account for species differences in susceptibility to halogenated aromatic compounds, since genes under control of the cytosol receptor would not be the same for each species, nor would they be the same in all tissues of the same species (Knutson and Poland, 1982; Poland and Knutson, 1982).

Safe *et al.* (1985) studied quantitative structure–activity relationships (QSAR) in a series of PCBs and found excellent correlation between binding affinities for the TCDD receptor and AHH induction potentials. The most active isomers were approximate isostereomers of TCDD. A receptor-mediated model, based on the mechanism of action of steroid hormones, was proposed, again involving binding of the inducer to a cytosolic receptor and translocation of the receptor–ligand complex to the nucleus, followed by binding of this complex to a nuclear receptor site, presumably a specific region of the nuclear DNA. This final step triggers the *de novo* synthesis of mRNA and protein, leading to the pleiotypic responses (Fig. 8-14).

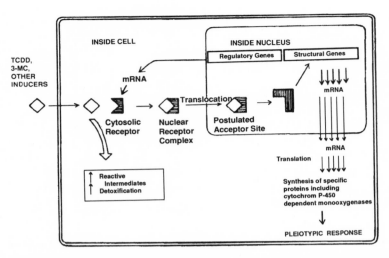

Figure 8-14. Proposed mechanism of action of toxic halogenated aromatic compounds. After binding of the inducer to a cytosolic receptor, translocation of the receptor–ligand complex to the nucleus occurs. Binding of this complex to a nuclear receptor triggers *de novo* synthesis of mRNA and protein, leading to pleiotypic responses (Safe *et al.*, 1985; used with permission).

8.3.4.3 Model Based on a Nuclear TCDD Receptor

Whitlock and Galeazzi (1984) reported evidence that experimental design may have compromised results placing the TCDD receptor in the cytosol and suggested that the "cytosolic" receptor in fact might be a nuclear receptor. Thus, in mouse hepatoma cells, at least 80% of the receptors were found associated with nuclei. TCDD–receptor complexes bound more tightly to the nuclei than did unoccupied receptors, and a temperature-dependent event increased the binding of the complex. Three types of receptors were studied. In addition to the normal receptor, a variant was defective in binding TCDD, but nonetheless became concentrated in the nucleus. A second variant bound TCDD effectively, but the TCDD–receptor complex did not accumulate in the nucleus. Based on these and other data, Whitlock and Galeazzi have proposed a mechanism for TCDD action which differs from those discussed above. In this model, initial binding of TCDD to the receptor occurs primarily inside the nucleus. The temperature-dependent second process thus could not be translocation of the complex from the cytosol to the nucleus. Instead, an event which somehow enhances binding of the TCDD–receptor complex to the nucleus is proposed, for example, a conformational change, an enzymatic processing of the complex, or subunit association or dissociation. The existence of separate ligand-binding and chromatin-binding domains on the receptor would explain the behavior of the variant receptors discussed above. Thus, the first would have a defect in the ligand-binding domain, while the second would have a defect in the chromatin-binding domain.

8.3.5 Halogenated Aromatic Hydrocarbons and Thyroid Function

8.3.5.1 Evidence for a Relationship between Halogenated Aromatic Hydrocarbon Toxicity and Thyroid Function

The potential for certain halogenated aromatic hydrocarbons to function as agonists or antagonists for thyroid hormones was recognized over two decades ago by Marshall and Tompkins (1968). Indeed, the spectrum of symptoms resulting from halogenated aromatic hydrocarbon toxicity are similar to signs of thyroid dysfunction. Early research revealed a direct effect of TCDD on thyroid hormone metabolism. Thus, UDP-glucuronyltransferase is one of the battery of enzymes induced by TCDD, and hepatic conjugation of T_4 to glucuronic acid, followed by secretion of the conjugate in bile, provides a major route for T_4 metabolism. Bastomsky (1977) found that, nine days after TCDD treatment of the rat, excretion of T_4 was increased fourfold. However, in an experiment to see if thyroid hormones could reverse the toxicity of TCDD, Neal et al. (1979) found that T_3 had no dramatic effect on survival of TCDD-treated rats, as might have been expected based on the results of Bastomsky. In fact, recent evidence that a relationship between TCDD toxicity and thyroid function exists includes the demonstration that thyroid hormones can potentiate TCDD toxicity, as measured with a mouse teratogenicity system (Birnbaum et al., 1985). Radiochemical thyroidectomy protects athyroid rats from the toxic effects of TCDD whereas thyroidectomized rats maintained on T_4 or euthyroid rats were not protected, an indication that any effect of TCDD on thyroid function is related to thyroid hormones and not to the thyroid gland (Rozman et al., 1984).

8.3.5.2 Structure–Activity Relationships (SAR)

Several SAR studies related to the biochemical behavior of halogenated aromatic hydrocarbons have produced a more refined picture of requirements for binding to the Ah receptor, and the relationship of this binding to toxicity. Certain of these results have been used to support a model for halogenated hydrocarbon toxicity that directly implicates interference with proper thyroid hormone function.

Goldstein (1980) has provided an extensive review of SAR data of dibenzodioxins for the induction of AHH and for toxicity. With respect to AHH induction, dibenzodioxins containing three or four halogens at positions 2, 3, 7, and 8 have the highest activity, with the 2,3,6,8-tetrachloro isomer (TCDD) being the most active. Further increase in halogen substitution decreases activity, and octachlorodibenzodioxin is inactive. Similar trends were seen with dibenzofurans. Halogenation at the 2, 3, 7, and 8 positions is also important for induction of the toxic response, as seen by the significantly greater toxicity of the 1,2,3,7,8- as compared to the 1,2,4,7,8-isomer. In addition, TCDD is much more toxic than 2,3,7-trichlorodibenzodioxin.

Two separate SARs exist for the induction of cytochrome P-450 and P-448 monooxygenases by halogenated biphenyls. Several isomers, in particular those bearing *ortho* halogens, for example, 2,4,5,2',4',5'-hexachlorobiphenyl, are "phenobarbital-type" inducers (see Nebert *et al.*, 1981; also Chapter 9), increasing cytochrome P-450 and aminopyridine *N*-methylase activity, but having no effect on AHH levels. In contrast 3,4,5,3',4',5'-hexachlorobiphenyl interacts with the Ah receptor and induces cytochrome P-448-linked AHH (Poland and Glover, 1977; reviewed by Goldstein, 1980).

From data such as these, Poland and Glover (1977) proposed that the TCDD receptor is a planar rectangle (3×10 Å) that requires halogens in three or preferably four corners for efficient binding. Thus, the structural requirements for binding in the biphenyl series include at least two adjacent halogens in the lateral positions of each benzene ring (positions 3,4,3',4' or 3,4,5,3',4',5'). The presence of halogens in the positions *ortho* to the biphenyl bridge would restrict coplanarity of the aromatic rings, and thus such substitution is detrimental to binding. The active biphenyls are approximate isosteres of TCDD, the prototypic rigid, planar, and more potent (relative to biphenyls) inducer of AHH.

8.3.5.3 Charge-Transfer-Mediated Binding to the Ah Receptor

McKinney *et al.* (1984) noted that the rectangle binding model described above does not have universal applicability. In particular, the fact that several nonhalogenated polycyclic aromatic hydrocarbons and benzoflavones also bind to the Ah receptor, some with affinity similar to that of halogenated aromatics, indicates that the role of halogen may not be interpreted so readily. An alternative model, derived from molecular parameters and molecular mechanics of Ah receptor ligands, is based on molecular polarizability and predicts a stacking charge transfer mode of interaction (McKinney *et al.*, 1985a, 1984). In this model, the availability of an accessible planar face capable of undergoing a dispersive interaction with a protein is of primary importance, rather than size and shape as in the rectangular box model. Polarization in a lateral direction is dominant in these essentially linear molecules, and this would be enhanced by the presence of lateral halogens.

8.3.5.4 Structure–Induction versus Structure–Toxicity Relationships— Evidence for Two Receptors

While a strong correlation between induction of Ah receptor binding affinity and toxicity of halogenated aromatic compounds has been cited in numerous studies, there are notable exceptions. For example, both toxic halogenated and nontoxic nonhalogenated aromatic hydrocarbons bind to the Ah receptor. Thus, 3-methylcholanthene is relatively nontoxic, yet binds to the Ah receptor with affinity comparable to that of TCDD. This lack of toxicity is not explained on the basis of differences in metabolism and has been cited as evidence that toxicity and induction responses are not causally related. From examination of these and other SAR data and consideration of the stacking binding model described above, McKinney et al. (1985b) have proposed the existence of a second receptor that mediates toxicity. Whereas binding to the Ah receptor is dependent on the polarizability of the molecule as a whole, the toxic response appears to depend more on specific polarizability components involving the lateral halogens. Possible configurations of the two receptor sites are depicted in Fig. 8-15.

8.3.5.5 A Cooperative Protein Receptor Mechanism Model for TCDD Toxicity

The possible existence of a TCDD-binding protein more toxicologically relevant than the Ah receptor, together with the T_4 depletion associated with TCDD toxicity (Bastomsky, 1977), prompted McKinney et al. (1985c) to investigate the interactions of TCDD and related compounds with thyroxine-binding prealbumin. As discussed in Chapter 6, this protein is an important plasma carrier of T_4 and is considered a good model for the thyroid hormone nuclear receptor. Computer modeling predictions and experimentally determined binding affinities of dioxin and furan analogues indicated that these compounds should be effective competitive binding ligands at the T_4-specific binding sites of prealbumin. The importance of lateral halogen substituents to binding to thyroxine-binding prealbumin also was indicated by computer modeling. From the accumulated evidence, McKinney et al. (1985c, d) proposed a cooperative receptor mechanism for

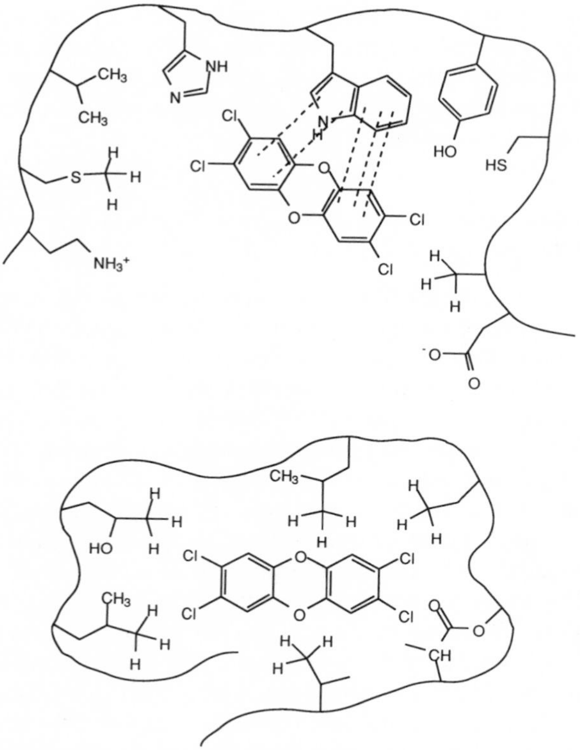

Figure 8-15. Models for two separate receptors. The Ah receptor (*top*) involves a stacking interaction and is dependent on polarization of the molecule through π complexation. The "toxic" TCDD receptor (*bottom*) involves a "pocket" type interaction and involves specific polarizabilities of the lateral substituents. Adapted, with permission, from McKinney *et al.* (1985a).

TCDD toxicity and suggested that proteins with a role in thyroid action may be involved. In this model, the Ah receptor is considered a storage and translocating protein in equilibrium with the nuclear binding events, with equilibrium movement in and out of the nucleus dependent on available binding ligands and capacity. Binding (possibly charge transfer in nature) of the ligand to the Ah receptor is followed by translocation to the nucleus, where the ligand–protein complex may associate with a second receptor protein associated with chromatin—an event assisted by the presence of lateral halogens on the ligand. Subsequent nuclear events leading to the biological response would be activated by the complex resulting from cooperative binding of the two proteins to the ligand. This model is illustrated diagrammatically in Fig. 8-16 (McKinney *et al.*, 1985d).

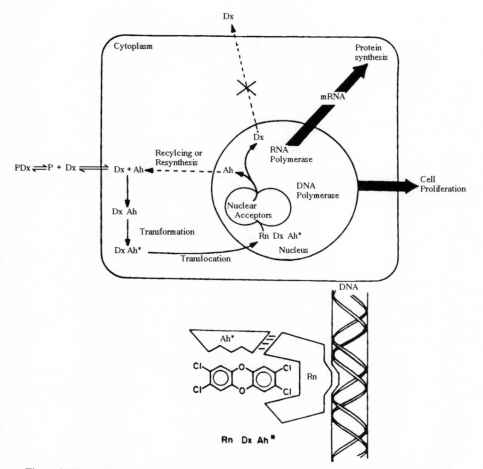

Figure 8-16. A cooperative protein receptor mechanism for TCDD toxicity. Binding of the ligand to the Ah receptor is followed by translocation to the nucleus, where the ligand–protein complex may associate with a second chromatin-associated receptor. Binding to the nuclear receptor is aided by the presence of the lateral halogens. Nuclear events leading to the biological responses would be triggered by the complex resulting from binding of the two proteins to the ligand (lower diagram) (McKinney *et al.*, 1985d; used with permission).

8.3.5.6 TCDD as a Potent and Persistent Thyroid Hormone

Structural requirements implicit in the cooperative protein binding model are the presence of an accessible planar face for binding to the Ah receptor and of lateral substituents to facilitate binding to the nuclear receptor. As shown in Fig. 8-17, the thyroid hormones meet such

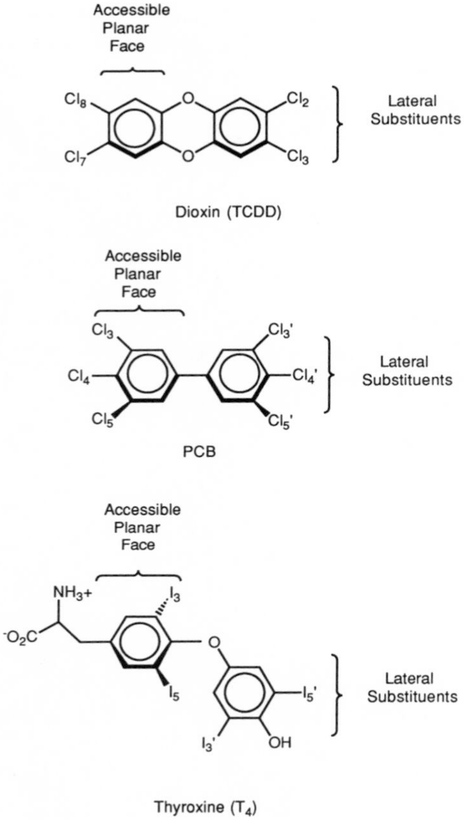

Figure 8-17. Schematic comparison of TCDD, PCB, and thyroid hormones, emphasizing the common features of an accessible planar face for binding to the Ah receptor and lateral substituents to facilitate binding to a nuclear receptor. Adapted, with permission, from J. D. McKinney, R. Fannin, S. Jordan, K. Chae, U. Rickenbacher, and L. Pedersen, Polychlorinated biphenyls and related compound interactions with specific binding sites for thyroxine in rat liver nuclear extracts, *J. Med. Chem.* 30:79–86. Copyright 1987 American Chemical Society.

requirements (McKinney *et al.*, 1987). Direct evidence for the involvement of thyroid function in TCDD toxicity came from the demonstration that T_3 binds to the Ah receptor, having about one-tenth the affinity of TCDD, while T_4 has somewhat less affinity (McKinney, 1985). Indirect evidence for a relationship between TCDD toxicity and thyroid function was discussed in Section 8.3.5.1. More recently, McKinney *et al.* (1987) have shown that PCBs and related compounds can bind specifically to

T_4-specific sites in rat liver nuclear extracts and that thyroid hormone tyrosyl rings form molecular complexes with aromatic donors, supporting the existence of a possible charge transfer interaction with the Ah receptor (Chae and McKinney, 1988).

If the model for TCDD action is valid, and if this activity is mediated by thyroid hormone receptors, several implications arise, and these have been discussed by McKinney *et al.* (1985d). Tissue T_4 depletion has been correlated with replacement of T_4 by TCDD at the level of tissue receptors. TCDD toxicity thus could be related in part to the expression of persistent thyroid hormone activity. It is noteworthy that a possible physiological role for reverse T_3 (rT_3) is suggested by this proposal. The model predicts that rT_3 would bind weakly to the Ah receptor, and nuclear rT_3 thus could arise primarily only from deiodination of nuclear T_4. rT_3 so generated, however, should bind to the nuclear receptor—but not simultaneously to the nuclear receptor and the Ah receptor, as required for a response—and therefore could function as an endogenous T_3 antagonist.

8.3.5.7 TCDD Toxicity—Summary

Accumulated evidence suggests strongly that thyroid hormone function is related to toxicity of TCDD and related halogenated aromatic compounds, but continued research will be required to define precisely the molecular events initiated by these toxins (for a recent discussion of possible relationships between thyroid hormones and TCDD toxicity, see Hong *et al.*, 1987). As McKinney (1985) has noted, "the investigator who proposes a molecular mechanism of action almost always extends himself beyond the data available to make the claim." In any event, the research directed toward addressing the critical issue of toxic mechanisms of halogenated aromatic hydrocarbons has produced valuable information on cellular functions in general.

8.4 Medical Applications of Halogenated Hydrocarbons

8.4.1 Halogenated Volatile Anesthetics

8.4.1.1 A Brief History of Inhalation Anesthetics

The use of diethyl ether as an anesthetic in 1842 marked the beginning of the use of general inhalation anesthesiology. In 1844, a painless tooth extraction was achieved under the influence of nitrous oxide, and, in 1847, the clinical use of chloroform was initiated. Despite problems associated with these agents—for example, the explosive flammability of diethyl ether and toxic side effects, including the frequently fatal "delayed chloroform poisoning"—early progress in the development of improved volatile anesthetics was slow. Several hydrocarbon anesthetics—cyclopropane being a notable example—were studied in the 1920s. Research during World War II produced many new fluorinated compounds, and in the early 1950s a concerted effort was made to find among them nonflammable volatile compounds suitable for general anesthesia (for a more detailed historical account, see Dobkin, 1979a). Of hundreds of such compounds evaluated, most, with notable exceptions, have been eliminated from consideration for a variety of reasons, including low potency or high toxicity (Wilson and Fabian, 1979). The development of halogenated anesthetics and their current status in clinical applications will be reviewed briefly in this section, along with a discussion of biochemical mechanisms of anesthetic action and toxicity, where relevant.

8.4.1.2 Development of Halogenated Anesthetics

The now well-documented hepatotoxicity of chloroform has brought a halt to its use in anesthesiology. The mechanism of this toxicity is related to biodehalogenation and production of reactive intermediates and will be discussed in Chapter 9. Other simple halogenated compounds that have been investigated and found to be unsatisfactory for use as anesthetics include ethyl chloride (flammable) and trichloroethylene (incompatible with soda lime, used as a carbon dioxide absorbent). Wilson and Fabian (1979) have reviewed the properties and problems associated with these and other compounds that have been evaluated and found to be unsatisfactory, including the ether analogue fluroxene, the halogenated ethane teflurane, halopropanes, and sevoflurane, a halogenated methyl isopropyl ether derivative (Fig. 8-18).

8.4.1.3 Halothane

The fluorinated hydrocarbon halothane ($CF_3CHClBr$), developed and first used clinically in the 1950s, is a nonexplosive, potent inhalation anesthetic. In particular, it is at least five times more potent than diethyl ether, and induction of anesthesia is rapid (for a discussion of the

$CF_3CH_2OCH=CH_2$	CF_3CHFBr	$CFH_2OCH(CF_3)_2$
FLUROXENE	TEFLURANE	SEVOFLURANE

Figure 8-18. Examples of halogenated anesthetics.

physiological responses to deepening planes of surgical anesthesia, see Denson *et al.*, 1976).

Halothane has enjoyed wide-scale clinical use, despite causing a mild form of hepatotoxicity, characterized by minor elevation of serum transaminase, in about 20% of patients. The results of recent experiments using rat hepatocyte monolayers suggest that several factors are involved in cell death—hypoxia, phenobarbital induction, and halothane exposure all contribute to hepatocyte damage. The results of these experiments are relevant to the transient mild hepatotoxicity observed in patients receiving halothane anesthesia (Schieble *et al.*, 1988).

Halothane also causes a much rarer and more severe form of toxicity, often fatal, producing a high level of serum transaminases and hepatic necrosis. Clinical features—including association with multiple halothane exposure, fever, rash, and other signs of an allergic response—are symptoms of a drug-induced hypersensitivity reaction. Extensive research has provided additional evidence that this acute toxicity has an immunological basis and is initiated by metabolic enzymes. Under aerobic conditions, liver microsomes oxidize halothane to trifluoroacetyl halide, which can acylate tissue molecules or be hydrolyzed to trifluoroacetic acid (Fig. 8-19). Several lines of evidence indicate that the bound trifluoroacetyl (TFA) moiety can function as a hapten in certain individuals to trigger an immune response. For example, two of six patients with halothane-associated liver cell necrosis had serum antibodies that reacted with trifluoracetylated rabbit serum albumin. Radical intermediates may also be involved. Under anaerobic conditions, halothane is reduced by cytochrome P-450 to the 1-chloro-2,2,2-trifluorethyl radical, which can form adducts with liver macromolecules or be reduced further to 1-chloro-2,2,2-trifluorethane (Fig. 8-19). It has not been determined whether the CF_3CHCl radical can induce an immune response by binding to cellular protein. However, halothane irreversibly inactivates cytochrome P-450 under conditions of low oxygen concentration that promote formation of the radical metabolite of halothane. The destruction of cytochrome P-450 by

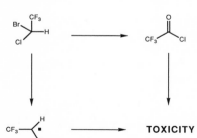

Figure 8-19. Metabolism of halothane. Trifluoroacetyl chloride is produced under aerobic conditions whereas the chlorotrifluoroethyl radical is produced under anaerobic conditions.

the CF_3CHCl radical, or lipid hydroperoxides formed therefrom, may be mediated by formation of heme-derived protein adducts, adducts that could also function as halothane-derived immunogens (Satoh *et al.*, 1987, and references therein).

The immunological basis of acute toxicity of halothane is quite complex. However, recent progress in identifying halothane-derived immunogens has been substantial and promises to have important practical benefits, including the detection of halothane-sensitized individuals (Satoh *et al.*, 1987). Gelman and Van Dyke (1988) have provided a recent editorial discussion of the complexities arising in the study of halothane toxicity and of possible factors that may be involved.

8.4.1.4 *Methoxyflurane*

Methoxyflurane ($CH_3OCF_2CHCl_2$) was introduced into clinical use in 1959. After initial enthusiastic acceptance, and several years of clinical use, a dose-related nephrotoxicity became increasingly apparent, and methoxyflurane now is little used as a surgical anesthetic (reviewed by Van Poznak, 1979). Other problems included potential flammability, high lipid and blood solubility that caused slow onset of and recovery from anesthesia, and an enhancement of the arrhythmogenic effects of epinephrine (Stoelting, 1987).

Hepatic microsomal metabolism of methoxyflurane degrades 75% of inhaled methoxyflurane, and studies have implicated metabolically released fluoride ion as the causative agent of renal toxicity. Two metabolic pathways have been implicated—a demethylation pathway produces dichloroacetate and fluoride, and a dechlorination pathway gives methoxydifluoroacetate, which can form oxalate by release of fluoride (Fig. 8-20). The relative importance of these pathways has been unclear. Results of a recent *in vivo* NMR investigation in the rat indicate that most methoxyflurane is metabolized via dechlorination, while a small percentage

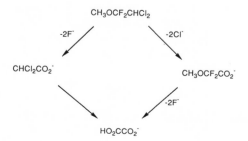

Figure 8-20. Two metabolic pathways implicated for the metabolism of methoxyflurane (Selinsky *et al.*, 1988).

is metabolized by an alternative pathway, presumed to be the demethylation pathway (Selinsky et al., 1988).

8.4.1.5 Enflurane and Isoflurane

Another fluorinated ether, enflurane (CHF_2OCF_2CHClF), has good anesthetic properties, is metabolized only to a small degree (ca. 2%), and elicits little hepatotoxicity. Intermediate solubility in blood and high potency permit rapid onset of and rapid recovery from anesthesia, and, unlike halothane, enflurane does not enhance arrhythmogenicity of epinephrine. Results of extensive evaluation by several groups in the 1960s and 1970s indicated that it should be a useful inhalation anesthetic for adults and children (reviewed by Dobkin, 1979b). Case reports of "enflurane hepatitis" have appeared (see below).

The isomeric ether isoflurane ($CHF_2OCHClCF_3$) is even more metabolically stable than enflurane, with very little metabolism detectable *in vivo*. Isoflurane has several other distinct advantages over enflurane, and initial tests in 1969 indicated excellent anesthetic properties. Following a suspension of human trials caused by indications that enflurane might be carcinogenic (reviewed by Dobkin, 1979c), isoflurane was introduced into clinical use in 1981 (Stoelting, 1987).

The decreased potential for toxicity of enflurane and isoflurane reflects their decreased degrees of metabolism (ca. 2% and 0.2%, respectively). However, cases of "enflurane hepatitis" and isoflurane-induced hepatic dysfunction have been reported, albeit much less frequently than cases of halothane hepatotoxicity. Recent studies suggest that these rare cases of hepatotoxicity may, in fact, be linked to metabolites. Thus, immunoreactive protein adducts that are recognized by specific anti-TFA antibodies are formed in liver following enflurane, isoflurane, and halothane administration. The relative amounts formed by the anesthetics decreased in the order halothane ≫ enflurane ≫ isoflurane, correlating directly with the relative extents of metabolism. These results suggest that sensitization to halothane anesthesia may produce cross-sensitization to flurane and isoflurane. Schemes postulated for the formation of acylated proteins by metabolic processing of halothane, enflurane, and isoflurane are shown in Fig. 8-21 (Christ et al., 1988). The metabolism of enflurane and isoflurane has been the subject of several investigations (Stoelting, 1987, and references therein).

8.4.1.6 Pharmacological Effects of Halogenated Anesthetics

Among the pharmacological actions of inhaled anesthetics are effects on the central nervous system, cerebral blood flow and general circulation,

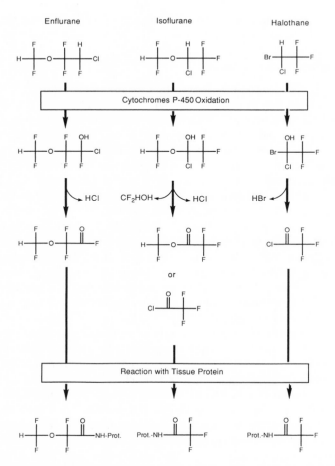

Figure 8-21. Metabolic schemes for the metabolism of enflurane, isoflurane, and halothane. Adapted, with permission, from D. D. Christ, H. Satoh, J. G. Kenna, and L. R. Pohl, Potential metabolic basis for enflurane hepatitis and the apparent cross-sensitization between enflurane and halothane. *Drug Metab. Dispos.* 16:135–140. Copyright 1988 The American Association of Pharmacology and Experimental Therapeutics.

including blood pressure and several cardiac parameters, and breathing, hepatic, renal, skeletal muscle, and several other functions (for a review, see Stoelting, 1987). The scope of these effects reflects the pervasive biodistribution of these lipophilic agents; however, a review of this important aspect of the science of anesthesiology is beyond the scope of this chapter.

8.4.1.7 Halogenated Anesthetics as Probes of
the Mechanism of Anesthesia

A plot of the logarithm of anesthetic potency versus the logarithm of oil/gas or membrane/gas partition coefficients of general anesthetics produces a straight line with unit slope (the Meyer–Overton rule). This empirical rule holds for wide concentrations of anesthetics having divergent structures, including noble gases, nitrous oxide, halogenated alkanes (e.g., chloroform, halothane), halogenated ethers (e.g., enflurane), alkanes, cyclopropanes, and alcohols. This correlation has been used to support the unitary hypothesis of anesthetic action, that is, that all inhalation anesthetics operate by one and the same physical or biochemical mechanism. Nonetheless, while this relationship between lipid solubility and anesthetic potency strongly implicates lipid membranes as the site of action, the identity of these membranes and the molecular bases of anesthetic action have remained a matter of dispute. Furthermore, there is evidence that anesthetics can affect several aspects of synaptic transmission and that these effects differ quantitatively for different drugs. Despite these differences, the effects appear to summate to produce the Meyer–Overton correlation (reviewed by Urban, 1985).

Of fundamental importance in theories of anesthetic action are the mechanisms by which dissolution of anesthetics in lipid bilayers can alter membrane excitability. Most theories assume that resultant changes in the physical properties of the bilayers, such as changes in membrane fluidity, membrane volume, and lipid phase transitions, in turn cause changes in lipid–protein interactions and functional properties of the proteins. Thus, such membrane protein-mediated processes as transmembrane transport, receptor–ligand binding, and electrophysiological channel dynamics may be altered (Forman et al., 1985, and references therein; reviewed by Miller, 1985).

The relationships of these processes to anesthetic action have been the subject of extensive research, and, reflecting their clinical importance, halogenated anesthetics have been included in most of these studies. In particular, advantage has been taken of the presence of fluorine through several ^{19}F-NMR studies. No attempt will be made to provide a comprehensive review of this work, but specific examples will be given to illustrate current areas of interest.

The observation of distinct ^{19}F-NMR signals for halothane, methoxyflurane, and isoflurane in the brains of intact rabbits demonstrated the potential of fluorinated anesthetics as in vivo biological probes. The relatively narrow linewidths observed in this study suggested that the anesthetics possessed a fair amount of mobility in the rabbit brain.

The time course of clearance of halothane from the rabbit brain, followed by ^{19}F-NMR spectroscopy, was significantly longer than anticipated (Wyrwicz *et al.*, 1983). In subsequent studies, enflurane and halothane were found to reside in two compartments in the rabbit brain, characterized by different time constants for clearance of the anesthetics. In the first, enflurane and halothane had comparable clearance half-lives, 25 min and 26 min, respectively, but in the second, isoflurane (174 min) was cleared significantly more rapidly than was halothane (320 min) (Wyrwicz *et al.*, 1987). ^{19}F-NMR spectroscopy *in vitro* also revealed that halothane exists in two distinct environments in brain tissue, characterized by different spin–spin relaxation times (T_2), chemical shifts, and kinetics of occupancy. In one of these compartments, inspired halothane is much more highly immobilized $(T_2 = 3.6 \text{ ms})$ than in the other $(T_2 = 43 \text{ ms})$. Furthermore, at concentrations leading to anesthesia, this compartment becomes saturated, suggesting that occupancy of this saturable compartment might be responsible for the anesthetic effect of halothane (Evers *et al.*, 1987). Further support for a relationship between this compartment and anesthetic activity comes from the demonstration that a direct correlation exits between ^{19}F T_2 values (between 0.5 and 4.5 ms) and the potency of a series of fluorinated ether anesthetics. In addition, T_2 values for these drugs in adipose tissue were much larger (200–400 ms) and did not correlate with anesthetic potency, suggesting that volatile anesthetic molecules have a specific affinity for, and become immobilized in, a membrane component present in neural tissue and absent in adipose tissue. Hexafluoroethane, a volatile nonanesthetic, achieved lower brain concentrations than did the anesthetics and had a significantly larger T_2 value (18.5 ms) (Evers *et al.*, 1988).

As noted above, the effects of anesthetics on membranes are thought to translate ultimately into altered functional properties of proteins involved in synaptic transmission. Proposed interactions of anesthetics with membrane proteins include binding at hydrophobic sites within membrane proteins, binding at boundary lipids adjacent to proteins, and inhibition of ligand binding to water-soluble protein (Forman *et al.*, 1985, and references therein). Recently, Forman *et al.* (1985) proposed that general anesthetics may act by reducing membrane protein conformational transition energies, facilitating the diversion of proteins to nonfunctional states.

Calcium is an important second messenger in signal transduction in various cells, including nerve cells, and many calcium-dependent processes are influenced by volatile anesthetics (Kress *et al.*, 1987, and references therein). In PC 12 cells, clinical concentrations of halothane, isoflurane, enflurane, and methoxyflurane were found to depress calcium fluxes through voltage-gated channels. Thus, voltage-sensitive calcium channels may be one membrane component with which volatile anesthetics interact (Kress *et al.*, 1987).

The above are examples of current research designed to elucidate the mechanism(s) of action of volatile anesthetics. While the actions on the CNS have been stressed in this summary, extensive research has also been done on the peripheral activities of these agents (see, e.g., Prokocimer *et al.*, 1988). The broad biodistribution and ready interaction of these lipid-soluble anesthetics with cell membranes produce a wide spectrum of effects, many of which are undesirable (cardiac effects, for example). The sorting out of the molecular events that produce these effects will require continued research in a number of disciplines.

8.4.2 Perfluorocarbons as Artificial Oxygen Transporters

Complete substitution of fluorine for hydrogen in aliphatic hydro-carbons imparts several unique physical and chemical properties to the resulting perfluorocarbon (PFC). Included is a dramatic solubility of gases in PFCs—50 vol % of oxygen and greater than 400 vol % of carbon dioxide (Riess and Le Blanc, 1982). The physiological availability of gases dissolved in PFCs was demonstrated impressively by Clark and Gollan (1966), who showed that mice completely submerged in perfluorobutyl-tetrahydrofuran (Fig. 8-22) survived unharmed for extended periods. The extensive research that has stemmed from the recognition of the medical potential of oxygenated PFCs has focused on several biomedical applications, including use of oxygenated PFCs in liquid ventilation (directly related to Clark and Gollan's experiment), treatment of decompression sickness, delivery of oxygen to ischemic tissues, oxygen perfusion and preservation of organs prior to transplantation, oxygen enhancement of tumors to increase sensitivity to cytotoxic agents and radiation, and use as contrast media and for NMR imaging. However, the use of emulsified PFCs as a replacement for red blood cells to transport oxygen and carbon dioxide has attracted the most attention (Faithfull, 1987; Lowe, 1987; and

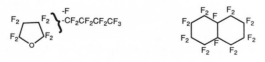

PERFLUOROBUTYLTETRAHYDROFURAN   PERFLUORODECALIN

Figure 8-22. Examples of perfluorocarbons used as artificial oxygen carriers.

references therein). Biochemistry and biomedical applications of PFCs will be reviewed briefly in this section.

8.4.2.1 PFC Emulsions

Intravenous infusion into animals of plasma-insoluble PFC liquids results in embolisms and other circulatory problems leading to death. Extensive research, in particular by the Green Cross Co. of Osaka, Japan, has produced PFC emulsions dispersed in isotonic solutions that are effective transporters of oxygen (Clark *et al.*, 1976; Clark and Moore, 1982; Yokoyama *et al.*, 1982; and references therein). Many PFCs initially screened for this purpose were subject to unacceptable retention in the body, largely in the liver. However, *cis–trans* perfluorodecalin (Fig. 8-22) was shown to be cleared from the liver in a reasonable time, had no apparent acute toxicity to mice, and had a vapor pressure compatible with intravenous administration. The first commercial PFC emulsion, Fluosol-DA 20% (F-DA), became available in Japan in 1978. This is a formulation of perfluorodecalin (14%) and perfluorotripropylamine (6%), along with emulsifiers and other agents to improve stability (for details, see Lowe, 1987). A second commercial product, Fluosol-43, based on perfluorotributylamine (20%), suffers from the long retention of perfluorotributylamine in the body.

The solubility of a gas in PFCs and in PFC emulsions depends on the molecular volume of the gas ($CO_2 \gg O_2 \gg N_2$) and increases linearly with partial pressure according to Henry's law. While the solubility of O_2 in pure PFCs is high, the solubility of O_2 in PFC emulsions, estimated to be between 5.7 and 7.6 ml/atm at 37 °C per 100 ml of commercial F-DA, is low compared to oxygenated arterial blood, and animal and human subjects undergoing transfusion or hemodilution with PFCs breathe supplementary O_2. On the other hand, the absence of a chemical bond between O_2 and PFC results in a more rapid O_2 exchange between the PFC emulsion and tissues compared to O_2 exchange between blood and tissues (Lowe, 1987).

8.4.2.2 Animal Studies with PFC Emulsions

Several animal species were used to examine the physiological effects of varying degrees of blood replacement with PFC emulsions (Lowe, 1987, and references therein). Initial experiments indicated that animals could survive with no ill effects when allowed to recover from almost complete blood replacement with PFCs (Geyer, 1975). However, more thorough studies using exchange transfusions to produce hematocrits of 2% have revealed possible immunological disturbances (Hardy *et al.*, 1983) and lethal side effects related to altered plasma enzyme concentrations (Lowe and McNaughton, 1986, 1985). The potential harmful effects of PFCs on immune response are the subject of much current research. There is speculation that some adverse effects of F-DA, including effects on immune response, may be caused by a surfactant (Pluronic F-68) used in preparation of the emulsion (for a further discussion of the adverse effects of PFCs and related issues, see Lowe, 1987).

8.4.2.3 Clinical Applications of PFC Emulsions

Clinical interest in the use of PFC emulsions has grown rapidly, although adverse side effects, similar to those seen in animals, have delayed widespread acceptance. A major benefit of hemodilution with PFC emulsions appears to result from the low viscosity and extremely small particle size of the emulsion, which leads to a higher tissue blood flow. For example, under conditions brought about by a myocardial infarction, hemodilution may be expected to improve perfusion through available collaterals. In addition, the suggestion has been made that PFC particles may be able to penetrate deeply into hypoxic tissue beds, reoxygenating sludged cells. Faithfull (1987) has reviewed clinical applications of PFC emulsions under these and other situations in which improvement of microcirculation is critical—examples of therapeutic uses of PFCs include protection against cerebral damage following cerebral arterial occlusion, treatment of sickle cell crisis, treatment of incipient gangrene, and oxygenation of solid tumors to improve sensitivity to radiation and cytotoxic agents.

Oxygenated PFC emulsions have been used effectively for perfusion and preservation of vital organs prior to transplantation, both in animals and humans. In addition, traumatically amputated human extremities have been perfused successfully with PFC emulsions prior to reimplantation. Other clinical applications include the use of PFC-containing liquids in liquid membrane oxygenators during partial coronary bypass and the application of the radiopaque perfluorocarbons as contrast agents in radiography (Faithfull, 1987; Lowe, 1987).

The biomedical applications of the "first-generation" PFC emulsions have been impressive. While research into the biochemistry and clinical applications of these continues at a rapid pace, there is also a considerable effort being made to produce later-generation PFC emulsions, with improved biological properties (see, e.g., Clark and Moore, 1982). Continued exploitation of the unique gas-transporting properties of PFCs seems likely to produce even more striking results.

8.5 Summary

The biochemical behavior of polyhalogenated compounds to a large degree reflects their extreme lipophilicity. Thus, halogenated insecticides and halogenated volatile anesthetics interact with and perturb the functioning of lipid membrane components. This is translated into alterations in neural functions, producing in some instances increased (e.g., the action of DDT) or decreased (volatile anesthetics) neural activity. Pervasive distribution in the subject produces a spectrum of effects. The ability of PCBs, dioxins, and related compounds to induce a variety of nuclear-regulated events, together with their stability and lipid solubility, makes these agents particularly serious environmental hazards. PFCs, on the other hand, are virtually insoluble in biological material, including lipids, and are finding increasing medical applications based on their gas-dissolving abilities. Certain aspects of biodegradation of polyhalogenated compounds have been discussed in this chapter. This will be considered more comprehensively in Chapter 9.

References

Ali, S. F., Hong, J.-S., Wilson, J. E., Lamb, J. C., Moore, J. A., Mason, G. A., and Bondy, S. C., 1982. Subchronic dietary exposure of rats to chlordecone (Kepone[R]) modifies levels of hypothalamic β-endorphine, *Neurotoxicology* 3:119–124.

Bastomsky, C. H., 1977. Enhanced thyroxine metabolism and high uptake goiters in rats after a single dose of 2,3,7,8-tetrachloro-*p*-dioxin, *Endocrinology* 101:292–296.

Bickel, M. H., and Muehlebach, S., 1980. Pharmacokinetics and ecodisposition of polyhalogenated hydrocarbons: Aspects and concepts, *Drug Metab. Rev.* 11:149–190.

Birnbaum, L. S., Weber, H., Harris, M. W., Lamb, J. C., IV, and McKinney, J. D., 1985. Toxic interaction of specific polychlorinated biphenyls and 2,3,7,8-tetrachlorodibenzo-*p*-dioxin: Increased incidence of cleft palate in mice, *Toxicol. Appl. Pharmacol.* 77:292–302.

Bloomquist, J. R., and Soderlund, D. M., 1985. Neurotoxic insecticides inhibit GABA-dependent chloride uptake by mouse brain vesicles, *Biochem. Biophys. Res. Commun.* 133:37–43.

Bloomquist, J. R., Adams, P. M., and Soderlund, D. M., 1986. Inhibition of γ-aminobutyric acid-stimulated chloride flux in mouse brain vesicles by polychlorocycloalkane and pyrethroid insecticides, *Neurotoxicology* 7:11–20.

Bondy, S. C., and Hong, J. S., 1987. Modulation of adrenal ornithine decarboxylase by chlordecone, *p*, *p'*-DDT and permethrin, *Neurotoxicology* 8:15–22.

Brinkman, U. A. Th., ad De Kok, A., 1980. Production, properties and usage, in *Halogenated Biphenyls, Terphenyls, Naphthalenes, Dibenzodioxins and Related Products* (R. D. Kimbrough, ed.), Elsevier/North-Holland Biomedical Press, Amsterdam, pp. 1–40.

Brooks, G. T., 1986. Insecticide metabolism and selective toxicity, *Xenobiotica* 16:989–1002.

Chae, K., and McKinney, J. D., 1988. Molecular complexes of thyroid hormone tyrosyl rings with aromatic donors. Possible relationship to receptor protein interactions, *J. Med. Chem.* 31:357–362.

Chen, P. H., Tilson, H. A., Marbury, G. D., Karboum, F., and Hong, J. S., 1985. Effect of chlordecone (Kepone) on the rat brain concentration of 3-methoxy-4-hydroxyphenyl glycol: Evidence for a possible involvement of the norepinephrine system in chlordecone-induced tremor, *Toxicol. Appl. Pharmacol.* 77:158–164.

Christ, D. D., Satoh, H., Kenna, J. G., and Pohl, L. R., 1988. Potential metabolic basis for enflurane hepatitis and the apparent cross-sensitization between enflurane and halothane, *Drug Metab. Dispos.* 16:135–140.

Clark, L. C., Jr., and Gollan, F., 1966. Survival of mammals breathing organic liquids equilibrated with oxygen at atmospheric pressure, *Science* 152:1755–1756.

Clark, L. C., Jr., and Moore, R. E., 1982. Basic and experimental aspects of oxygen transport by highly fluorinated organic compounds, in *Biomedicinal Aspects of Fluorine Chemistry* (R. Filler and Y. Kobayashi, eds.), Kodansha Ltd., Tokyo, and Elsevier Biomedical Press, Amsterdam, pp. 213–226.

Clark, L. C., Jr., Wesseler, E. P., Kaplan, S., Emory, C., Moore, R., and Denson, D., 1976. Intravenous infusion of cis–trans perfluorodecalin emulsions in the Rhesus monkey, in *Biochemistry Involving Carbon–Fluorine Bonds* (R. Filler, ed.), ACS Symposium Series 28, American Chemical Society, Washington, D.C., pp. 135–170.

Coats, J. R., Metcalf, R. L., and Kapoor, I. P., 1977. Effective DDT analogues with altered aliphatic moieties. Isobutanes and chloropropanes, *J. Agric. Food Chem.* 25:859–868.

Cole, L. M., and Casida, J. E., 1986. Polychlorocycloalkane insecticide-induced convulsions in mice in relation to disruption of the GABA-regulated chloride ionophore, *Life Sci.* 39:1855–1862.

Costa, L. G., 1987a. Toxicology of pesticides: a brief history, in *Toxicology of Pesticides: Experimental, Clinical and Regulatory Perspectives* (L. G. Costa, C. L. Galli, and S. D. Murphy, eds.), Spinger-Verlag, Berlin, pp. 1–10.

Costa, L. G., 1987b. Interaction of insecticides with the nervous system, in *Toxicology of Pesticides: Experimental, Clinical and Regulatory Perspectives* (L. G. Costa, C. L. Galli, and S. D. Murphy, eds.), Springer-Verlag, Berlin, pp. 77–91.

Denson, D. D., Uyeno, E. T., Simon, R. L., Jr., and Peters, H. M., 1976. Preparation and physiological evaluation of some new fluorinated volatile anesthetics, in *Biochemistry Involving Carbon–Fluorine Bonds* (R. Filler, ed.), ACS Symposium Series 28, American Chemical Society, Washington, D.C., pp. 190–208.

Desaiah, D., 1982. Biochemical mechanisms of chlordecone neurotoxicity: A review, *Neurotoxicology* 3:103–110.

Dobkin, A. B., 1979a. Anesthetic history, in *Development of New Volatile Inhalation Anaesthetics* (A. B. Dobkin, ed.), Elsevier/North-Holland Press, pp. 1–4.

Dobkin, A. B., 1979b. Enflurane (Ethrane), in *Development of New Volatile Inhalation Anaesthetics* (A. B. Dobkin, ed.), Elsevier/North-Holland Press, Amsterdam, pp. 155–228.

Dobkin, A. B., 1979c. Forane (isoflurane, compound 469), in *Development of New Volatile Inhalation Anaesthetics* (A. B. Dobkin, ed.), Elsevier/North-Holland Press, pp. 229–264.

Dubois, J. M., and Bergman, C., 1977. Asymmetrical currents and sodium currents in Ranvier nodes exposed to DDT, *Nature* 266:741–742.

Evers, A. S., Berkowitz, B. A., and d'Avignon, D. A., 1987. Correlation between the anaesthetic effect of halothane and saturable binding in brain, *Nature* 328:157–160.

Evers, A. S., Haycock, J. C., and d'Avignon, D. A., 1988. The potency of fluorinated ether anesthetics correlates with their ^{19}F spin–spin relaxation times in brain tissue, *Biochem. Biophys. Res. Commun.* 151:1039–1045.

Faithfull, N. S., 1987. Fluorocarbons, current status and future applications, *Anaesthesia* 42:234–242.

Firestone, D., 1984. Chlorinated aromatic compounds and related dioxins and furans: Production, uses, and environmental exposure, in Banbury Report 18. Biological Mechanisms of Dioxin Action (A. Poland and R. D. Kimbrough, eds.), Cold Spring Harbor Laboratory, pp. 3–16.

Forman, S. A., Verkman, A. S., Dix, J. A., and Solomon, A. K., 1985. n-Alkanols and halothane inhibit red cell anion transport and increase band 3 conformational change rate, *Biochemistry* 24:4859–4866.

Fukuta, T. R., 1976. Physicochemical aspects of insecticidal action, in *Insect Biochemistry and Physiology* (C. F. Wilkenson, ed.), Plenum, New York, pp. 397–428.

Gandolfi, O., Cheney, D. L., Hong, J. S., and Costa, E., 1984. On the neurotoxicity of chlordecone: A role for γ-aminobutyric acid and serotonin, *Brain Res.* 303:117–123.

Gelman, S., and Van Dyke, R., 1988. Mechanism of halothane-induced hepatotoxicity: Another step on a long road, *Anesthesiology* 68:479–482.

Geyer, R. P., 1975. "Bloodless" rats through the use of artificial blood substitutes, *Fed. Proc.* 34:1499–1505.

Goldstein, J. A., 1980. Structure–activity relationships for the biochemical effects and the relationships to toxicity, in *Halogenated Biphenyls, Terphenyls, Naphthalenes, Dibenzodioxins and Related Products* (R. D. Kimbrough, ed.), Elsevier/North-Holland, Biomedical Press, Amsterdam, pp. 151–190.

Hammond, B., Katzenellenbogen, B. S., Krauthammer, N., and McConnell, J., 1979. Estrogenic activity of the insecticide chlordecone (Kepone) and interaction with uterine estrogen receptors, *Proc. Natl. Acad. Sci. USA* 76:6641–6645.

Hardy, R. N., Lowe, K. C., and McNaughton, D. C., 1983. Acute responses during blood substitution in the conscious rat, *J. Physiol.* 338:451–461.

Hart, M. M., Reagan, R. L., and Adamson, R. H., 1973. The effects of isomers of DDD on the ACTH-induced steroid output, histology and ultrastructure of the dog adrenal cortex, *Toxicol. Appl. Pharmacol.* 24:101–113.

Hayes, W. J., Jr., 1982. Chlorinated hydrocarbon insecticides, in *Pesticides Studied in Man* (W. J. Hayes, ed.), Williams and Wilkins, Baltimore, pp. 172–283.

Herr, D. W., Gallus, J. A., and Tilson, H. A., 1987. Pharmacological modification of tremor and enhanced acoustic startle by chlordecone and p, p'-DDT, *Psychopharmacology (Berlin)* 91:320–325.

Hille, B., 1977. Ionic channels of nerve: Questions for theoretical chemists, *BioSystems* 8:195–199.

Holan, G., 1969. New halocyclopropane insecticides and the mode of action of DDT, *Nature* 221:1025–1029.

Holan, G., 1971. Rational design of degradable insecticides, *Nature* 232:644–647.

Hong, J. S., Tilson, H. A., Uphouse, L. L., Gerhart, J., and Wilson, W. E., 1984. Effects of chlordecone exposure on brain neurotransmitters: Possible involvement of the serotonin system in chlordecone-elicited tremor, *Toxicol. Appl. Pharmacol.* 73:336–344.

Hong, J. S., Herr, D. W., Hudson, P. M., and Tilson, H. A., 1986. Neurochemical effects of DDT in rat brain in vivo, *Arch. Toxicol., Suppl.* 9:14–26.

Hong, L. H., McKinney, J. D., and Luster, M. I., 1987. Modulation of 2,3,7,8-tetrachlorodibenzo-p-dioxin (TCDD)-mediated myelotoxicity by thyroid hormones, *Biochem. Pharmacol.* 36:1361–1365.

Hrdina, P. D., Singhal, R. L., Peters, D. A. V., and Ling, G. M., 1973. Some neurochemical alterations during acute DDT poisoning, *Toxicol. Appl. Pharmacol.* 25:276–288.

Hudson, P. M., Chen, P. H., Tilson, H. A., and Hong, J. S., 1985. Effects of p, p'-DDT on the rat brain concentrations of biogenic amine and amino acid neurotransmitters and their association with p, p'-DDT induced tremor and hyperthermia, *J. Neurochem.* 45:1349–1355.

Jensen, B. L., Caldwell, M. W., French, L. G., and Briggs, D. G., 1987. Toxicity, ultrastructural effects, and metabolic studies with 1-(o-chlorophenyl)-1-(p-chlorophenyl)-2,2-dichloroethane (o, p'-DDD) and its methyl analog in the guinea pig and rat, Toxicol. Appl. Pharmacol. 87:1–9.

Jordon, J. E., Grice, T., Mishra, S. K., and Desaiah, D., 1981. Acute chlordecone toxicity in rats: A relationship between tremor and ATPase activities, Neurotoxicology 2:355–364.

Joy, R. M., 1982. Chlorinated hydrocarbon insecticides, in Pesticides and Neurological Diseases (D. J. Ecobichon and R. M. Joy, eds.), CRC Press, Boca Raton, Florida, pp. 91–150.

Kaddus, A. A., Ghiasuddin, S. M., Matsumura, F., Scott, J. G., and Tanaka, K., 1983. Difference in the picrotoxinin receptor between the cyclodiene-resistant and susceptible strains of the German cockroach, Pestic. Biochem. Physiol. 19:157–166.

Kimbrough, R. D., 1980. Environmental pollution of air, water and soil, in Halogenated Biphenyls, Terphenyls, Naphthalenes, Dibenzodioxins and Related Products (R. D. Kimbrough, ed.), Elsevier/North-Holland Biomedical Press, Amsterdam, pp. 77–80.

Kimbrough, R. D., 1984. Skin lesions in animals and humans: A brief overview, in Banbury Report 18. Biological Mechanisms of Dioxin Action (A. Poland and R. D. Kimbrough, eds.), Cold Spring Harbor Laboratory, pp. 357–363.

Knutson, J. C., and Poland, A., 1982. Response of murine epidermis to 2,3,7,8-tetrachlorodibenzo-p-dioxin: Interaction of the Ah and hr loci, Cell 30:225–234.

Komulainen, H., and Bondy, S. C., 1987. Modulation of levels of free calcium within synaptosomes by organochlorine insecticides, J. Pharmacol. Exp. Ther. 241:575–581.

Korach, K. S., Sarver, P., Chae, K., McLachlan, J. A., and McKinney, J. D., 1988. Estrogen receptor-binding activity of polychlorinated hydroxybiphenyls: Conformationally restricted structural probes, Mol. Pharmacol. 33:120–126.

Kress, H. G., Eckhardt-Wallasch, H., Tas, P. W. L., and Koschel, K., 1987. Volatile anesthetics depress the depolarization-induced cytoplasmic calcium rise in PC 12 cells, FEBS Lett. 221:28–32.

Kupfer, D., 1975. Effects of pesticides and related compounds on steroid metabolism and functions, CRC Crit. Rev. Toxicol. 4:83–124.

Lawrence, L. J., and Casida, J. E., 1984. Interactions of lindane, toxaphene and cyclodienes with brain-specific t-butylbicyclophosphorothionate receptor, Life Sci. 35:171–178.

Lowe, K. C., 1987. Perfluorocarbons as oxygen-transport fluids, Comp. Biochem. Physiol. 87A:825–838.

Lowe, K. C., and McNaughton, D. C., 1985. Intravascular fluid composition following blood replacement with fluosol-DA in the rat, Br. J. Pharmacol. 84:116P.

Lowe, K. C., and McNaughton, D. C., 1986. Changes in plasma enzyme concentrations in response to blood substitution with perfluorocarbon emulsion in the conscious rat, Experientia 42:1228–1231.

Marshall, J. S., and Tompkins, L. S., 1968. Effects of o, p'-DDD and similar compounds on thyroxine binding globulin, J. Clin. Endocrin. Metab. 28:386–392.

Matsumura, F., 1975. Toxicology of Insecticides, Plenum Press, New York, pp. 165–251.

Matsumura, F., and Ghiasuddin, S. M., 1983. Evidence for similarities between cyclodiene type insecticides and picrotoxinin in their action mechanisms, J. Environ. Sci. B 18:1–14.

McBlain, W. A., 1987. The levo enantiomer of o, p'-DDT inhibits the binding of 17β-estradiol to the estrogen receptor, Life Sci. 40:215–221.

McKinney, J. D., 1985. The molecular basis of chemical toxicity, Environ. Health Perspect. 61:5–10.

McKinney, J. D., Long, G. A., and Pedersen, L., 1984. PCB and dioxin binding to cytosol receptors: A theoretical model based on molecular parameters, Quant. Struct.-Act. Relat. 3:99–105.

McKinney, J. D., Darden, T., Lyerly, M. A., and Pedersen, L. G., 1985a. PCB and related compound binding to the Ah receptor(s). Theoretical model based on molecular parameters and molecular mechanics, *Quant. Struct.-Act. Relat.* 4:166–172.

McKinney, J. D., Chae, K., McConnell, E. E., and Birnbaum, L. S., 1985b. Structure–induction versus structure–toxicity relationships for polychlorinated biphenyls and related aromatic hydrocarbons, *Environ. Health Perspect.* 60:57–68.

McKinney, J. D., Chae, K., Oatley, S. J., and Blake, C. C. F., 1985c. Molecular interactions of toxic chlorinated dibenzo-*p*-dioxins and dibenzofurans with thyroxine binding prealbumin, *J. Med. Chem.* 28:375–381.

McKinney, J. D., Fawkes, J., Jordan, S., Chae, K., Oatley, S., Coleman, R. E., and Briner, W., 1985d. 2,3,7,8-Tetrachlorodibenzo-*p*-dioxin (TCDD) as a potent and persistent thyroxine agonist: A mechanistic model for toxicity based on molecular reactivity, *Environ. Health Perspect.* 61:41–53.

McKinney, J. D., Fannin, R., Jordan, S., Chae, K., Rickenbacher, U., and Pedersen, L., 1987. Polychlorinated biphenyls and related compound interactions with specific binding sites for thyroxine in rat liver nuclear extracts, *J. Med. Chem.* 30:79–86.

Miller, K. W., 1985. Specific and nonspecific actions of general anesthetic agents, in *Effects of Anesthesia* (B. G. Covino, H. A. Fozzard, K. Rehder, and G. Strichartz, eds.), American Physiological Society, Bethesda, Maryland, pp. 29–37.

Narahashi, T., and Haas, G. H., 1967. DDT: Interaction with nerve membrane conductance changes, *Science* 157:1438–1440.

Narahashi, T., and Haas, G. H., 1968. Interaction of DDT with the components of lobster nerve membrane conductance, *J. Gen. Physiol.* 51:177–198.

Narahashi, T., and Yamasaki, T., 1960. Mechanisms of increase in negative afterpotential by dicophanum (DDT) in the giant axons of the cockroach, *J. Physiol.* 152:122–140.

Neal, R. A., Beatty, P. W., and Gasiewicz, T. A., 1979. Studies on the mechanisms of toxicity of 2,3,7,8-tetrachlorodibenzo-*p*-dioxin (TCDD), *Ann. N.Y. Acad. Sci.* 320:204–213.

Nebert, D. W., Eisen, H. J., Negishi, M., Lang, M. A., and Hjelmeland, L. M., 1981. Genetic mechanisms controlling the induction of polysubstrate monooxygenase (P-450) activities, *Annu. Rev. Pharmacol. Toxicol.* 21:431–462.

Nelson, J. A., 1974. Effects of dichlorodiphenyltrichloroethane (DDT) analogs and polychlorinated biphenyl (PCB) mixtures on 17β-[^{3}H]estradiol binding to rat uterine receptors, *Biochem. Pharmacol.* 23:447–451.

Poland, A., and Glover, E., 1973. Chlorinated dibenzo-*p*-dioxins: Potent inducers of δ-aminolevulinic acid synthetase and aryl hydrocarbon hydroxylase, *Mol. Pharmacol.* 9:736–747.

Poland, A., and Glover, E., 1977. Chlorinated biphenyl induction of aryl hydrocarbon hydroxylase activity: A study of the structure–activity relationship, *Mol. Pharmacol.* 13:924–938.

Poland, A., and Knutson, J. C., 1982. 2,3,7,8-Tetrachlorodibenzo-*p*-dioxin and related halogenated aromatic hydrocarbons: Examination of the mechanism of toxicity, *Annu. Rev. Pharmacol. Toxicol.* 22:517–554.

Poland, A., Glover, E., and Kende, A. S., 1976. Stereospecific, high affinity binding of 2,3,7,8-tetrachlorodibenzo-*p*-dioxin by hepatic cytosol. Evidence that the binding species is receptor for aryl hydrocarbon hydroxylase, *J. Biol. Chem.* 251:4936–4946.

Prokocimer, P. G., Maze, M., Vickery, R. G., Kraemer, F. B., Gandjei, R., and Hoffman, B. B., 1988. Mechanism of halothane-induced inhibition of isoproterenol-stimulated lipolysis in rat adipocytes, *Mol. Pharmacol.* 33:338–343.

Quraishi, M. S., 1977. *Biochemical Insect Control, Its Impact on Economy, Environment, and Natural Selection*, John Wiley & Sons, New York, pp. 98–123.

Reggiani, G., 1980. Localized contamination with TCDD—Seveso, Missouri and other areas, in *Halogenated Biphenyls, Terphenyls, Naphthalenes, Dibenzodioxins and Related Products* (R. D. Kimbrough, ed.), Elsevier/North-Holland Biomedical Press, Amsterdam, pp. 303–371.

Riess, J. G., and Le Blanc, M., 1982. Solubility and transport phenomena in perfluorochemicals relevant to blood substitution and other biomedical applications, *Pure Appl. Chem.* 54:2382–2406.

Rozman, K., Rozman, T., and Greim, H., 1984. Effects of thyroidectomy and thyroxine on 2,3,7,8-tetrachloro-*p*-dioxin (TCDD) induced toxicity, *Toxicol. Appl. Pharmacol.* 72:372–376.

Safe, S., Bandiera, S., Sawyer, T., Robertson, L., Safe, L., Parkinson, A., Thomas, P. E., Ryan, D. E., Reik, L. M., Levin, W., Denomme, M. A., and Fujita, T., 1985. PCBs: Structure–function relationships and mechanism of action, *Environ. Health Perspect.* 60:47–56.

Satoh, H., Davies, H. W., Takemura, T., Gillette, J. R., Maeda, K., and Pohl, L. R., 1987. An immunochemical approach to investigating the mechanism of halothane-induced hepatotoxicity, *Prog. Drug Metab.* 10:187–206.

Schieble, T. M., Costa, A. K., Heffel, D. F., and Trudell, J. R., 1988. Comparative toxicity of halothane, isoflurane, hypoxia, and phenobarbital induction in monolayer cultures of rat hepatocytes, *Anesthesiology* 68:485–494.

Selinsky, B. S., Perlman, M. E., and London, R. E., 1988. *In vivo* nuclear magnetic resonance studies of hepatic methoxyflurane metabolism. II. A reevaluation of hepatic metabolic pathways, *Mol. Pharmacol.* 33:567–573.

Stoelting, R. K., 1987. *Pharmacology and Physiology in Anesthetic Practice*, J. B. Lippincott Co., Philadelphia, pp. 35–68.

Stryer, L., 1988. *Biochemistry*, 3rd ed., W. H. Freeman and Co., New York, p. 1011.

Tilson, H. A., Emerich, D., and Bondy, S. C., 1986a. Inhibition of ornithine decarboxylase alters neurological responsiveness to a tremorigen, *Brain Res.* 379:147–150.

Tilson, H. A., Hudson, P. M., and Hong, J. S., 1986b. 5,5-Diphenylhydantoin antagonizes the neurochemical and behavioral effects of *p, p'*-DDT but not of chlordecone, *J. Neurochem.* 47:1870–1878.

Urban, B. W., 1985. Modifications of excitable membranes by volatile and gaseous anesthetics, in *Effects of Anesthesia* (B. G. Covino, H. A. Fozzard, K. Rehder, and G. Strichartz, eds.), American Physiological Society, Bethesda, Maryland, pp. 13–28.

van den Bercken, J., 1970. The effect of DDT and dieldrin on myelinated nerve fibres, *Eur. J. Pharmacol.* 20:205–214.

Van Poznak, A., 1979. Methoxyflurane (Penthrane[R]): Seeking its niche, in *Development of New Volatile Inhalation Anaesthetics* (A. B. Dobkin, ed.), Elsevier/North-Holland Press, Amsterdam, pp. 113–153.

Van Woert, M. H., Plaitakis, A., and Hwang, E. C., 1982. Neurotoxic effects of DDT [1,1,1-trichloro-2,2-*bis*(*p*-chlorophenyl)ethane]: Role of serotonin, in *Mechanisms of Action of Neurotoxic Substances* (K. N. Prasad and A. Vernadakis, eds.), Raven Press, New York, pp. 143–154.

Vos, J. G., 1984. Dioxin-induced thymic atrophy and the suppression of thymus-dependent immunity, in Banbury Report 18. Biological Mechanisms of Dioxin Action (A. Poland and R. D. Kimbrough, eds.), Cold Spring Harbor Laboratory, pp. 401–410.

Whitlock, J. P., Jr., and Galeazzi, D. R., 1984. 2,3,7,8-Tetrachlorodibenzo-*p*-dioxin receptors in wild type and variant mouse hepatoma cells. Nuclear location and strength of nuclear binding, *J. Biol. Chem.* 259:980–985.

Wilson, S. L., and Fabian, L. W., 1979. Inhalation anaesthetics: Going, going, gone, in *Development of New Volatile Inhalation Anaesthetics* (A. B. Dobkin, ed.), Elsevier/North-Holland Press, Amsterdam, pp. 5–57.

Woolley, D. E., 1982. Neurotoxicity of DDT and possible mechanisms of action, in

Mechanisms of Action of Neurotoxic Substances (K. N. Prasad and A. Vernadakis, eds.), Raven Press, New York, pp. 95–141.

Wu, C. H., Oxford, G. S., Narahashi, T., and Holan, G., 1980. Interaction of a DDT analogue with the sodium channel of lobster axon, *J. Pharmacol. Exp. Ther.* 212:287–293.

Wyrwicz, A. M., Schofield, J. C., Tillman, P. C., Gordon, R. E., and Martin, P. A., 1983. Noninvasive observations of fluorinated anesthetics in rabbit brain by fluorine-19 nuclear magnetic resonance, *Science* 222:428–430.

Wyrwicz, A. M., Conboy, C. B., Ryback, K. R., Nichols, B. G., and Eisele, P., 1987. In vivo ^{19}F-NMR study of isoflurane elimination from brain, *Biochim. Biophys. Acta* 927:86–91.

Yeager, J., and Munson, S., 1945. Physiological evidence of a site of action of DDT in an insect, *Science* 102:305–307.

Yokoyama, K., Suyama, T., and Naito, R., 1982. Development of perfluorochemical (PFC) emulsions as an artificial blood substitute, in *Biomedicinal Aspects of Fluorine Chemistry* (R. Filler and Y. Kobayashi, eds.), Kodansha Ltd., Tokyo, and Elsevier Biomedical Press, Amsterdam, pp. 191–212.

Metabolism of Halogenated Compounds—Biodehalogenation

9.1 Biodehalogenation—Introduction

The metabolism of halogenated hydrocarbons *in vivo* is initiated predominantly by two classes of enzymes—the cytochrome P-450-dependent monooxygenases and the glutathione *S*-transferases. These enzymes, in general, convert nonexcretable lipophilic compounds ultimately to hydrophilic metabolites that can be eliminated in urine and/or bile. For example, oxidative replacement of halogen with hydroxyl followed by glucuronide formation is a major pathway for detoxification and excretion. On the other hand, processes initiated by these enzymes are often responsible for conversion of a relatively harmless substrate into a more toxic or carcinogenic intermediate. Bioactivation of xenobiotics to toxic and/or carcinogenic metabolites has been the subject of extensive research and will be discussed in this chapter. The role of deiodinases in thyroid hormone function was discussed in some detail in Chapter 6. This will be reconsidered briefly in this chapter in the context of biodehalogenation mechanisms. Metabolism and detoxification of special classes of halogenated compounds (pesticides, TCDD, etc.) will also be considered.

9.1.1 Cytochrome P-450-Linked Monooxygenases

The heme-containing cytochrome P-450 monooxygenases are responsible for the metabolism of a host of endogenous and exogenous substances. Several forms of cytochrome P-450 exist in the liver, having different but sometimes overlapping substrate specificities. These isozymes have been grouped into three broad classes, based on their selective inducibility by certain classes of compounds, exemplified by phenobarbital (PB), 3-methylcholanthrene (3-MC), and pregnenolone-16α-carbonitrile (PCN). Based on studies with rat liver microsomes and purified enzymes, the PBinducible enzymes [P-450 (PB)] appear to be responsible for most halocarbon metabolism. Halocarbons are effective inducers of cytochrome P-450 enzymes, an induction that can facilitate not only the metabolism of the inducer, but also that of endogenous substrates. Nebert *et al.* (1981) have reviewed the genetic mechanisms controlling the induction of various forms of P-450 monooxygenase activities. In sections below, mechanisms of monooxygenase-catalyzed reactions will be discussed, along with how these reactions determine the metabolic fate of various classes of halogenated compounds (reviewed by Anders and Pohl, 1985; Macdonald, 1984; Henschler, 1985).

9.1.2 Glutathione S-Transferases

Among the functions of glutathione (GSH) (Fig. 9-1)—usually the most abundant sulfhydryl compound present in animal tissues—is the protection of cellular macromolecules from the actions of electrophilic carcinogenic and mutagenic agents. This protection is mediated by GSH S-transferase-catalyzed conjugation of the electrophilic agent with GSH. Multiple GSH S-transferases have been identified, all having almost absolute specificity for GSH as the thiol substrate, but showing broad and overlapping specificities for the second substrate. GSH S-transferase activity, like GSH, is found primarily in the cytosol and generally is greater in hepatic than in other tissues (Chasseaud, 1979, and references therein). A protein-bound microsomal GSH S-transferase, which differs from cytosolic enzyme in molecular weight and immunological reactions, has been characterized (Morgenstern *et al.*, 1982). The role of the GSH S-transferases appears to be to promote ionization of GSH to GS$^-$ and to bind the electrophilic substrate to promote preferential reaction of the electrophile with GS$^-$ rather than with other cellular nucleophiles. In the initial step of metabolic processing, GSH conjugates formed in the liver are transported to bile and plasma (Anders *et al.*, 1988, and references therein; for a review of reactions and properties of GSH S-transferases, see Chasseaud, 1979).

GSH conjugates are generally less toxic than the electrophilic precursor and are eliminated in the bile or, after biotransformation to mercapturic acids, are excreted in the urine (Fig. 9-1). However, enzymatic processing of certain GSH conjugates produces metabolites that are responsible for the toxic nature of the parent compound (Anders *et al.*, 1988; Elfarra and Anders, 1987). Examples of this phenomenon will be given below.

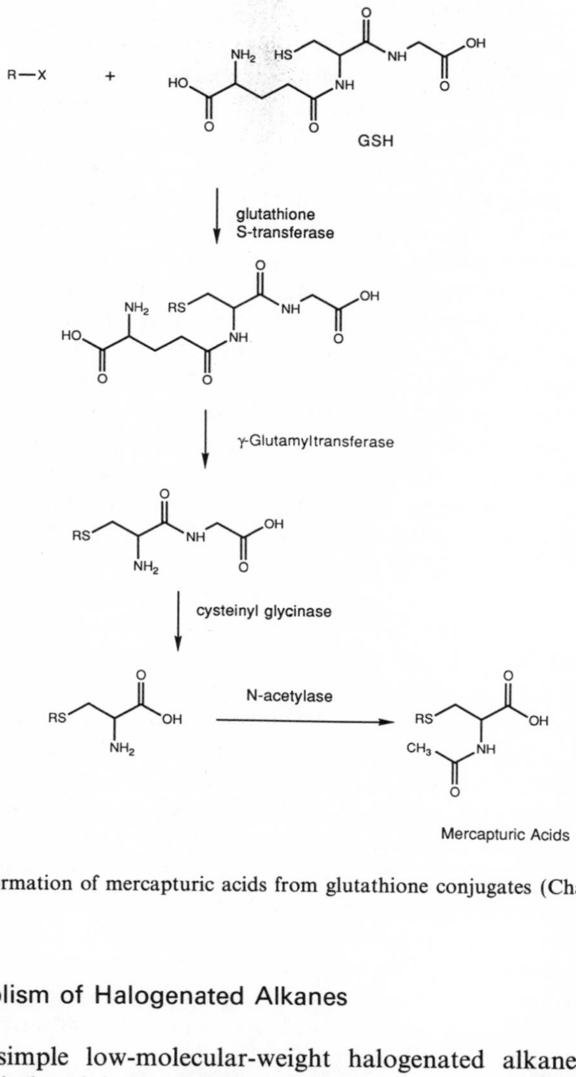

Figure 9-1. Formation of mercapturic acids from glutathione conjugates (Chasseaud, 1979).

9.2 Metabolism of Halogenated Alkanes

Many simple low-molecular-weight halogenated alkanes have had important industrial applications. Thus, carbon tetrachloride and chloroform are used as industrial and laboratory solvents and are formed as by-products of sewage and drinking water treatment (Chapter 1). 1,2-Dibromoethane was used in the past as a pesticide and lead scavenger in gasoline, and 1,2-dichloroethane has applications as a lead scavenger, an industrial solvent, and a grain fumigant. The recognition that many of these simple compounds exhibit hepatotoxicity, nephrotoxicity, carcinogenicity, and/or mutagenicity has led to more rigid control over their use, as well as to extensive research into the mechanisms of toxicities.

Biodehalogenation of aliphatic halocarbons can occur by several enzyme-catalyzed processes, including substitution, elimination, reduction, oxidation, and reductive oxygenation. The route(s) to which a given substrate is vulnerable is dependent on substrate structure, including the identity of the halogen substituent. Anders and Pohl (1985) have reviewed important physical and chemical properties of the carbon–halogen bond that influence the course of these reactions. For example, the electronegativity of the bound halogen induces electrophilic character at the α carbon, making this carbon susceptible to nucleophilic attack. In addition, the ability of halogens to stabilize α-carbon carbonium ions, free radicals, carbanions, and carbenes plays an important role in the targeting of the carbon–halogen bond as a site for metabolic processing. Subsequent events are often driven by the ability of the halide ion to serve as a good leaving group in substitution or elimination reactions. Behavior of the different halogens is, in general, determined by physicochemical properties associated with the electronegativity trend within the series ($F > Cl > Br > I$).

9.2.1 Monooxygenase-Catalyzed Oxidations of Halogenated Alkanes

Several reviews on the metabolism of halogenated alkanes are available that include excellent discussions of the fundamental physicochemical principles which influence the course of monooxygenase-catalyzed reactions of these substrates (e.g., Anders and Pohl, 1985; Macdonald, 1984). These principles will be summarized briefly here.

9.2.1.1 Mechanism of Cytochrome P-450 Hydroxylation

A general scheme for the mechanism of cytochrome P-450 hydroxylation is shown in Fig. 9-2, in which RH represents the substrate and ROH the product (White and Coon, 1980). Key to the oxidation process is the formation of a transient $RH(Fe=O)^{3+}$ species that contains the highly electrophilic perferryl oxygen, which is capable of insertion into a carbon–hydrogen bond of the substrate.

9.2.1.2 Oxidative Dehydrohalogenation of Halogenated Alkanes

Oxidative metabolism of simple alkyl chlorides, bromides, and iodides (e.g., RCH_2X) is associated primarily with insertion of an oxygen atom

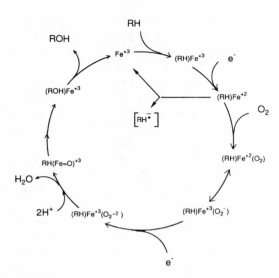

Figure 9-2. Proposed scheme for the mechanism of cytochrome P-450-catalyzed hydroxylation reactions (RH is substrate and ROH is product). Adapted from White and Coon, 1980; reproduced, with permission, from *Annual Review of Biochemistry*, Vol. 49. Copyright 1980 Annual Reviews Inc.

into one of the α C$-$H bonds. The product, a *gem*-halohydrin, will suffer rapid loss of hydrogen halide to produce a carbonyl compound (e.g., an aldehyde, ketone, acid chloride, or carbonyl chloride, depending on substrate structure) (Fig. 9-3). Subsequent reactions of the carbonyl compound determine the metabolites observed (see below). The rate of oxidative metabolism of haloalkanes possessing at least one α C$-$H bond increases with increasing halogen substitution ($CH_3X < CH_2X_2 < CHX_3$) and with decreasing electronegativity of the halide (e.g., $RCH_2F \ll RCH_2Cl < RCH_2Br < RCH_2I$).

By this process of oxidative dehydrohalogenation, simple alkyl halides (RCH_2X) are converted to aldehydes, which may be subject to subsequent oxidative or reductive transformations. Acid halides (RCOX) formed from dihaloalkanes can react with available nucleophiles (including GSH) or undergo hydrolysis. Cytochrome P-450-dependent oxidation of dihalomethanes produces carbon monoxide and hydrogen halide (formed by the rapid decomposition of the intermediate formyl halide) with relative rates consistent with substituent electronegativities ($CH_2I_2 > CH_2Br_2 > CH_2BrCl \gg CH_2Cl_2$), with no concomitant formation of formaldehyde, formic acid, or carbon dioxide (Fig. 9-3) (Macdonald, 1984; Stevens *et al.*, 1980, and references therein).

Because of the toxic and possible carcinogenic properties of tri-halomethanes, their metabolism has received special attention. Phosgene, a major initial product of the metabolism of chloroform (Fig. 9-3) (still widely used as an industrial and laboratory solvent), reacts with cellular proteins and has been implicated as the causative agent of *in vivo* liver toxicity. While the rate of metabolism of trihalomethanes follows the expected order $(CHI_3 > CHBr_3 > CHCl_3)$, the hepatotoxicity appears to follow the order $CHBr_3 > CHCl_3 \gg CHI_3$. The lack of hepatotoxicity of iodoform may reflect alternative modes of metabolism, or, because of extreme reactivity or thermodynamic instability, carbonyl diiodide (COI_2) may suffer hydrolysis to carbon dioxide or decomposition to carbon monoxide at such a rate as to preclude reaction with tissue macromolecules (Macdonald, 1984, and references therein).

The α-hydroxylation metabolic pathway has been observed in more complex systems. The toxicity of halothane, discussed in Chapter 8, is also related to the formation of reactive intermediates by α-hydroxylation.

The mechanism by which oxygen is transferred from the perferryl oxygen species to the substrate has been a matter of much recent research.

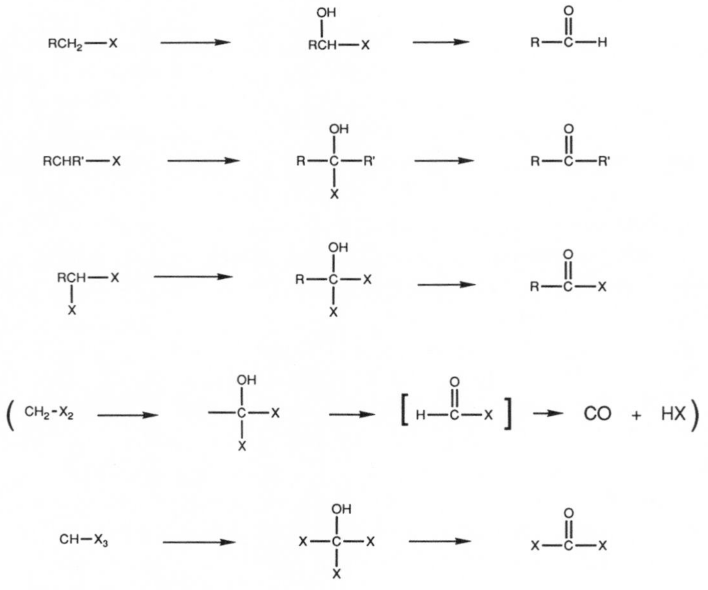

Figure 9-3. Formation and subsequent reactions of geminal halohydrins (Anders and Pohl, 1985).

Debate has focused on whether this involves a concerted transfer of an "oxenoid" species or whether a sequential process involving radical intermediates intervenes. Data supporting the latter process (Fig. 9-4) include a large deuterium isotope effect ($k_{RH}/k_{RD} = 11.5$) and a loss of stereochemistry, attributed to radical epimerization, during the cytochrome P-450-dependent hydroxylation of tetradeuteronorbornane (Groves *et al.*, 1978).

There also is evidence that the initial step in cytochrome P-450 oxidation of heteroalkanes, possibly including haloalkanes, may be a one-electron oxidation of the heteroatom. For example, cyclopropyl and 1-methylcyclopropyl iodides and bromides were among a series of inactivators of cytochrome P-450 heme. The logarithm of the inactivation constant was related inversely to the single-electron oxidation potential, supporting a single-electron transfer–cyclopropane ring opening sequence, shown in Fig. 9-5 (Guengerich *et al.*, 1984; Guengerich and Macdonald, 1984; and references therein). In the 1-methyl series, direct α-hydroxylation is, of course, blocked. From these and other results, Guengerich and Macdonald (1984) suggested that an alternative mechanism for α-carbon hydroxylation of heteroatom-substituted alkanes, including haloalkanes, may involve single-electron transfer from the heteroatom to the perferryl species ($Fe^V = O$) to give an intermediate halogen radical cation. Transfer of a labile proton (or hydrogen atom) to the reduced oxygenated heme species ($Fe^{IV} = O$) generates a carbon-centered radical (or cation). Oxygen rebound from iron hydroxide to the radical or cation would produce the geminal halohydrin (or, more generally, a carbon substituted geminally with hydroxide and a heteroatom) (Fig. 9-6). This mechanism provides a ready explanation for the regioselectivity of halocarbon hydroxylations. Indeed, Guengerich and Macdonald (1984) have proposed that this initial single-electron transfer step and the heteroatom radical cation produced therefrom are key to a unified mechanism for all oxidative processes catalyzed by cytochrome P-450.

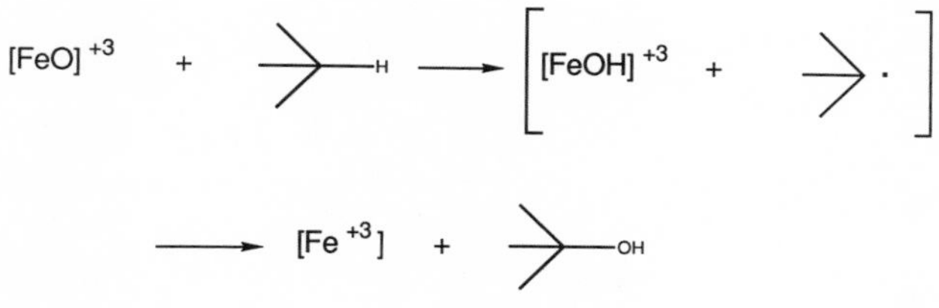

Figure 9-4. Proposed free-radical mechanism for the transfer of oxygen from a perferryl oxygen species to the substrate (Guengerich and Macdonald, 1984).

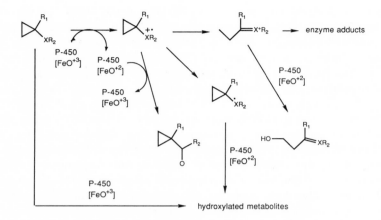

Figure 9-5. Proposed mechanism for the cytochrome P-450-catalyzed oxidation and ring opening of heteroatomic cyclopropyl substrates. Adapted, with permission, from F. P. Guengerich, R. J. Willard, J. P. Shea, L. E. Richard, and T. L. Macdonald. Mechanism-based inactivation of cytochrome P-450 by heteroatom-substituted cyclopropanes and formation of ring-opened compounds, *J. Am. Chem. Soc.* 106:6446–6447. Copyright 1984 American Chemical Society.

9.2.1.3 Halogen Oxidation

In the mechanistic proposal of Guengerich and Macdonald, if oxygen transfer from heme-bound iron ($Fe^{IV} = O$; Fig. 9-6) to the heteroatom radical cation occurs prior to transfer of the α hydrogen, oxidation of the heteroatom results. Cytochrome P-450-catalyzed formation of amine oxides and sulfoxides provides examples of such product formation. Product

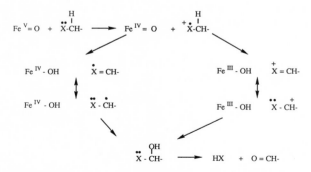

Figure 9-6. Proposed free-radical mechanism for α-hydroxylation of heteroatom-substituted alkanes (see text for details) (Guengerich and Macdonald, 1984).

distribution profiles from the oxidation of a series of 1,2-dihaloethanes and 1-halopropanes were initially interpreted to result from the partitioning of an initial metabolic intermediate (the halogen radical cation) between hydrogen abstraction leading to halogen release and halogen oxygenation (Guengerich et al., 1980; Tachizawa et al., 1982). According to this mechanism, C_1–C_2 products were postulated to be formed from a haloso intermediate which would undergo β-elimination or nucleophilic substitution, while C_1 functionalized products would be produced by oxidative dehydrohalogenation (Fig. 9-7). Direct evidence for the ability of cytochrome P-450 monooxygenases to carry out halogen oxygenation was found in the cytochrome P-450-catalyzed equilibration of unlabeled iodosobenzene and [^{125}I]iodobenzene (Burka et al., 1980), while Macdonald et al. (1980) described evidence for the intermediacy of alkyliodosyl compounds in the peracid oxidation of alkyl iodides. However, subsequent studies on the oxidative metabolism of 1,2-haloethanes using ^{2}H and ^{18}O isotopes gave results inconsistent with a major role for haloso intermediates (Koga et al., 1986).

9.2.1.4 Reductive Oxygenation of Halogenated Alkanes

Carbon tetrachloride and bromotrichloromethane are metabolized by rat microsomal cytochrome P-450 to phosgene and "electrophilic chlorine."

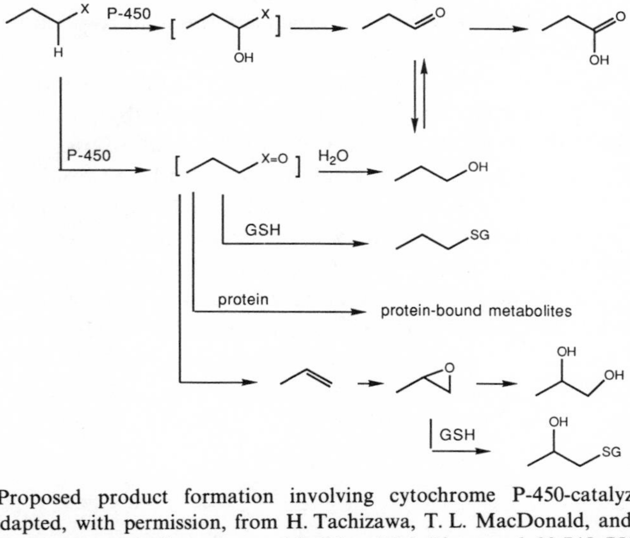

Figure 9-7. Proposed product formation involving cytochrome P-450-catalyzed halogen oxidation. Adapted, with permission, from H. Tachizawa, T. L. MacDonald, and R. A. Neal, Rat liver microsomal metabolism of propyl halides, *Mol. Pharmacol.* 22:745–751. Copyright 1982 The American Association of Pharmacology and Experimental Therapeutics.

The rates of formation of these products increase as dioxygen decreases from 100% to 5% and become negligible in the absence of dioxygen. These results are consistent with an initial cytochrome P-450-catalyzed conversion of the substrates to the trichloromethyl radical followed by reaction of this radical with dioxygen. Thus, at high dioxygen concentrations, the substrate is unable to compete with dioxygen for the electrons supplied by cytochrome P-450. At low concentrations, this competition is successful, reduction occurs, and the derived radical reacts with dioxygen, if present in sufficient concentration, to form the trichloromethylperoxyl radical from which are derived phosgene and electrophilic chlorine (Fig. 9-8). In the absence of dioxygen, the radical abstracts a hydrogen atom from the medium to form chloroform, is reduced by cytochrome P-450 to dichlorocarbene, or reacts with microsomal lipids (Anders and Pohl, 1985, and references therein).

The nature of the "electrophilic chlorine" and the mechanism of its formation from the trichloromethylperoxyl radical have been the subjects of much research. Direct oxidation of chloride to hypochlorite is not supported by experimental evidence, nor is insertion of oxygen into carbon tetrachloride to form trichloromethyl hypochlorite. Pathways under consideration include dimerization of the peroxyl radical to form a tetroxide, followed by decomposition to phosgene and two chlorine radicals (Fig. 9-9). Alternatively, a σ complex formed between the peroxyl radical and iron porphyrins could decompose either homolytically or by rearrangement to give the chlorine radical or the iron hypochlorite species (Fig. 9-9) (Anders and Pohl, 1985).

Cytotoxicity of carbon tetrachloride has been associated with lipid peroxidation chain reactions, and the trichloromethylperoxyl radical is more electrophilic than the trichloromethyl radical (Slater et al., 1985). Together with the fact that carbon tetrachloride-induced lipid peroxidation is maximal at a dioxygen concentration of 5%, this suggests that the trichloromethylperoxyl radical may be responsible for this manifestation of carbon tetrachloride toxicity (Kieczka and Kappus, 1980). Slater et al. (1985) have reviewed the hepatotoxic effects of carbon tetrachloride and its use as a model compound in the study of free-radical-mediated liver injury.

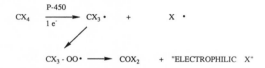

Figure 9-8. Formation of "electrophilic chlorine" during cytochrome P-450-mediated metabolism of carbon tetrachloride and bromotrichloromethane (Anders and Pohl, 1985).

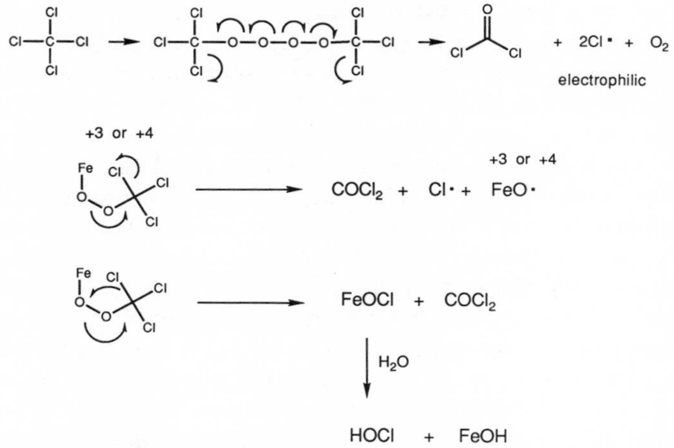

Figure 9-9. Two proposed mechanisms for the formation of electrophilic chlorine from the trichloromethylperoxyl radical. Adapted, with permission, from Anders and Pohl, 1985.

There is debate concerning the relationship between lipid peroxidation and irreversible toxicity of hepatotoxic compounds, including carbon tetrachloride. For example, in the presence of carbon tetrachloride, cellular damage in isolated hepatocytes is evident before onset of lipid peroxidation (Stacey and Priestly, 1978). Other studies have implicated inhibition of Ca^{2+} transport, leading to a lethal increase in intracellular Ca^{2+}, as an important event in cell death caused by xenobiotics (Younes and Siegers, 1984, and references therein). In this regard, carbon tetrachloride has been shown to inhibit microsomal Ca^{2+} pump activity *in vivo* and *in vitro* (Moore *et al.*, 1976; Lowrey *et al.*, 1981).

9.2.1.5 Monooxygenase-Catalyzed Reductive Dehalogenations

Highly halogenated aliphatic compounds, for example, carbon tetrachloride and halothane, undergo cytochrome P-450-catalyzed reductive metabolism that involves transfer of either one or two electrons from cytochrome P-450 to the substrate. The electrons are supplied to this process by an associated NADPH-linked enzyme. Since the halocarbon substrate must compete with dioxygen for the electrons supplied by cytochrome P-450, this process occurs most effectively under conditions of minimal oxygen concentration. Reduction is initiated by a rate-determining transfer of a single electron to the antibonding orbital (σ^* orbital) or d orbital of the carbon–halogen bond to produce a transient radical anion.

The ease of this electron transfer decreases with increasing electronegativity of the bound halogen; that is, $C-I > C-Br > C-Cl \gg C-F$. Elimination of halide from this intermediate produces a carbon-centered radical. Abstraction of a hydrogen atom from the medium at this stage produces the reduced alkane. While still associated with cytochrome P-450, the radical anion may accept a second electron to form a carbanionic intermediate. This carbanion can undergo α-elimination to produce a carbene or, if the route is available, β-elimination to form an alkene. These processes are shown in Fig. 9-10 (reviewed by Anders and Pohl, 1985; Brault, 1985; Macdonald, 1984). Cheeseman *et al.* (1985) have reviewed experimental procedures and results that have confirmed the identity of free radicals as intermediates in the metabolism of halogenated alkanes in general and carbon tetrachloride in particular. Brault (1985) studied reactions of pulse-generated carbon-centered radicals and peroxyl radicals derived from them with iron porphyrin and unsaturated fatty acids to develop a model for the subsequent toxic actions of radicals derived enzymatically from polyhalogenated alkanes.

While carbon tetrachloride has been studied most extensively as a model compound for these reductive metabolic processes, other compounds have also received attention. In particular, because of toxicity associated with its use as an inhalation anesthetic (Chapter 8), halothane has been studied extensively. Reduction of halothane to 2-chloro-1,1,1-tri-fluoroethane and 2-chloro-1,1-difluoroethylene is thought to involve a radical process, and the 1-chloro-2,2,2-trifluoroethyl radical may be responsible for the halothane-induced loss of cytochrome P-450 seen *in vitro* and *in vivo* (reviewed by Anders and Pohl, 1985). As discussed in Chapter 8, the

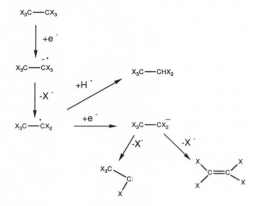

Figure 9-10. Reductive dehalogenation of highly halogenated aliphatic compounds. Adapted, with permission, from Anders and Pohl, 1985.

acute hepatotoxicity seen in rare instances apparently results from oxidative metabolism with formation of trifluoroacetylated macromolecules.

Reduction of DDT to DDD is another example of reductive metabolism of a polyhalogenated substrate (Chapter 8). This is discussed further in Section 9.2.3.1.

9.2.2 GSH-Dependent Metabolism of Haloalkanes

Conjugation of electrophilic agents is an important detoxification mechanism. Many halogenated hydrocarbons are substrates for GSH transferase-catalyzed nucleophilic substitution reactions that produce S-substituted glutathione derivatives. As discussed above, and shown in Fig. 9-1, these can be processed to nontoxic mercapturic acids and eliminated from the body. The formation of S-methylglutathione from chloromethane or iodomethane and of S-propylglutathione from 1-chloropropane are examples of this detoxification mechanism. These reactions in general are S_N2 displacements of halide with thiolate anion and proceed with inversion of configuration in chiral substrates (Anders and Pohl, 1985; Chasseaud, 1979; and references therein). Detoxification of fluoroacetate in mammals has been shown to be associated with a specific GSH S-transferase activity, and free fluoride ion and S-carboxymethylcysteine have been identified as products of fluoroacetate metabolism (Mead et al., 1985, and references therein). A fluoroacetate-specific defluorinase purified from mouse liver cytosol is immunodistinct from GSH transferase enzymes found in the same preparation (Soiefer and Kostyniak, 1984).

Although GSH conjugation normally is associated with detoxification, certain halogenated alkanes are among substrates that are activated to toxic metabolites by this process. For example, 1,2-dichloroethane is hepatotoxic and nephrotoxic and has been classified as a carcinogen by the National Cancer Institute. 1,2-Dichloroethane is metabolized by both microsomal cytochrome P-450 monooxygenases and cytosolic GSH-S-transferases, but activation by the latter is apparently responsible for carcinogeniticity and toxicity (Anders et al., 1988; Webb et al., 1987; and references therein). Recent research has identified S-(2-chloroethyl)-L-cysteine as the likely nephrotoxic metabolite of 1,2-dichloroethane. This product, through formation of an episulfonium intermediate, is capable of reacting with nucleophiles at physiological pH (Fig. 9-11) (Anders and Pohl, 1985; Elfarra et al., 1985; and references therein). S-(2-Chloroethyl)-L-cysteine was also implicated as the toxic metabolite in similar studies with isolated rat hepatocytes. In this work, depletion of intracellular GSH

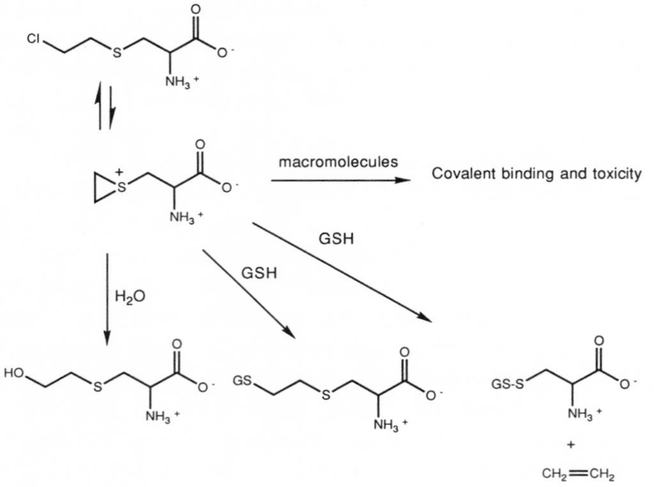

Figure 9-11. Formation of a reactive episulfonium intermediate during GSH-dependent metabolism of 1,2-dichloroethane. Reprinted, with permission, from M. W. Anders, L. Lash, W. Dekant, A. A. Elfarra, and D. R. Dohn, Biosynthesis and biotransformation of glutathione S-conjugates to toxic metabolites, *Crit. Rev. Toxicol.* 18:311:341. Copyright 1988 CRC Press, Inc., Boca Raton, Florida.

and inhibition of Ca^{2+} transport and Ca^{2+}-ATPase preceded cell death. While the biochemical lesion(s) leading to toxicity has not been established unequivocally, these and other data suggest that protein alkylation or depletion of GSH may lead to a lethal alteration of Ca^{2+} homeostasis (Webb *et al.*, 1987).

The commercial use of 1,2-dibromoethane, formerly employed as a pesticide and lead scavenger, has been restricted due to its tumorigenic potential. Although the metabolism of 1,2-dibromoethane is primarily oxidative (van Bladeren *et al.*, 1981), the carcinogenic effects of this compound also appear to be mediated through GSH conjugation. Thus, although 2-bromoacetaldehyde, the major product of oxidative dehydrohalogenation, can react rapidly with GSH and with cellular protein thiols, it alkylates DNA only slowly. With the use of radiolabeled GSH, efficient irreversible binding of equimolar amounts of 1,2-dibromoethane and GSH to intracellular DNA and RNA and extracellular DNA was demonstrated *in vitro* (Ozawa and Guengerich, 1983). Although not demonstrable directly, the intermediate episulfonium ion was suggested as the likely active electrophile in this process. The major adduct formed *in vivo* and *in vitro* has been identified as S-[2-(N^7-guanyl)ethyl]glutathione (Fig. 9-12) (Koga *et al.*, 1986). These data support the theory that GSH conjugation is more important in DNA alkylation and mutagenesis than is the oxidative metabolic pathway (Guengerich *et al.*, 1987; Ozawa and Guenferich, 1983; and references therein). The identification of this adduct also indicates that γ-glutamyltransferase-catalyzed cleavage of glutamic acid to produce S-(2-chloroethyl)cysteine is not required for bioactivation in this instance, as had been suggested by other work (Vadi *et al.*, 1985). The relevance of the guanyl N^7-adduct is not clear, particularly since this position is not in a base-pairing region and since N^7 purine lesions are thought to be rarely associated with mutations. Thus, other DNA adducts formed in small amounts may be responsible for the mutagenic activity of 1,2-dibromoethane (Guengerich *et al.*, 1987). A similar situation pertains with the mutagenic activity of vinyl chloride (Section 9.3.1.1).

1,2-Dichlorothane and 1-bromo-2-chloroethane appear to be activated by the same mechanism. Higher homologues, for example, 1,3-dichloropropane, although excellent substrates for GSH S-transferase, give no DNA adducts (Guengerich *et al.*, 1987, and references therein). 1,2-Dibromo-3-chloropropane was at one time an important soil fumigant but is now banned in the United States because of its toxicity. This compound was the subject of a recent detailed investigation using ^{1}H and ^{13}C two-dimensional NMR analysis of metabolites in urine of rats injected with ^{13}C-enriched material. Results of these studies indicate that direct reaction of 1,2-dibromo-3-chloropropane with GSH to yield reactive sulfonium intermediates is not a significant factor in acute toxicity or genotoxity (Dohn *et al.*, 1988).

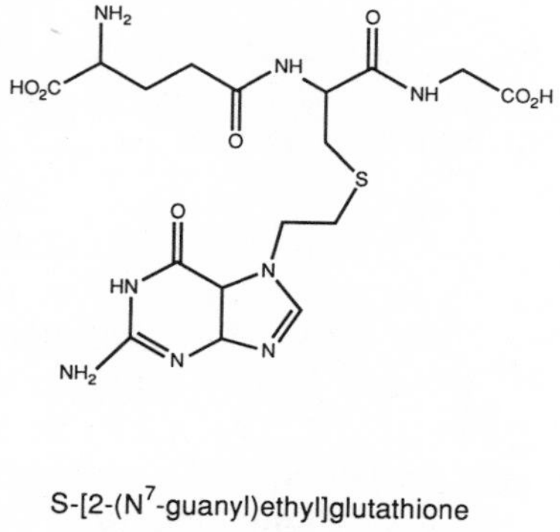

S-[2-(N^7-guanyl)ethyl]glutathione

Figure 9-12. S-[2-(N^7-Guanyl)ethyl]glutathione, the major DNA-derived adduct formed during metabolism of 1,2-dibromoethane (Koga *et al.*, 1986).

9.2.3 Metabolism of Environmentally Persistent Compounds Having Aliphatic Halogen–Carbon Bonds

The principal biochemical mechanisms available for biodehalogenation of aliphatic halocarbons have been reviewed in the above sections. The bedhavior of selected potential environmental pollutants, in particular, industrially produced toxic compounds, will be summarized below.

9.2.3.1 Metabolism of DDT

The metabolism of DDT, the subject of a vast amount of research, will be discussed only briefly. Several reviews are available for further information (e.g., Fukami, 1980; Brooks, 1986; Quraishi, 1977).

The DDT dehydrochlorinase-catalyzed dehydrochlorination of DDT to DDE, discussed in Section 8.2.1.5, is a major cause of DDT resistance in insects. In vertebrates, metabolism is initiated either by a similar dehydrochlorination to DDE or by a cytochrome P-450-dependent enzymatic reductive dechlorination to 2,2-(p, p'-dichlorodiphenyl)-1,1-dichloroethane (DDD). A one-electron reduction, producing a dichloroethyl radical, is consistent with an observed metabolism-dependent covalent binding of DDT and DDD to liver microsomal protein, especially under anaerobic conditions (Kelner et al., 1986; Baker and Van Dyke, 1984). The major metabolites found in vertebrate feces and bile are 2,2-bis(p-chlorophenyl)-acetic acid (DDA) and conjugates thereof, a consequence of further enzymatic processing of DDD and DDE. Recent evidence indicates that DDA is formed from DDD through C-1 hydroxylation, formation of an unstable acid chloride by spontaneous loss of hydrogen chloride, and hydrolysis of the acid chloride to DDA (Fig. 9-13) (Gold and Brunk, 1984). This mechanism supplants a more elaborate scheme, formulated much earlier, involving formation of 1-chloro-2,2-bis(p-chlorophenyl)-ethene (DDMU). Formation of DDA from DDE through a putative epoxide intermediate has been suggested (Fawcett et al., 1987).

9.2.3.2 Metabolism of Lindane

The metabolism of lindane has been studied extensively in several animal species and in humans. Under aerobic conditions, lindane is subject to cis-dehydrogenation to 3,6/4,5-1,2,3,4,5,6-hexachlorocyclohexane (3,6/4,5-HCCH), dehydrochlorination to isomers of 1,3,4,5,6-pentachlorocyclohexene (PCCH), and trans-elimination to 3,6/4,5-PCCH in human liver microsomes, rat liver, and housefly abdomen. Under anaerobic conditions, rat liver monooxygenases trans-dechlorinate lindane to a fourth

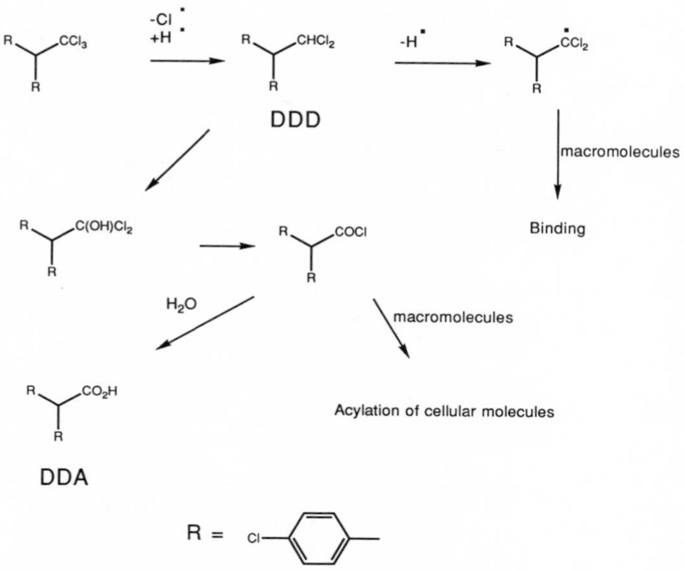

Figure 9-13. Proposed mechanism for the metabolism of DDT (Gold and Brunk, 1984).

primary metabolite, 3,4,6/5-tetrachlorocyclohexene (3,4,6/5-TCCH). Several secondary metabolites include tetrachlorophenol, pentachlorobenzene, and various conjugates, including GSH-derived conjugates (Fig. 9-14) (Fitzloff et al., 1982; Fukami, 1980; and references therein). Metabolism of lindane, and other isomers of hexachlorocyclohexane present in commercial preparations of the insecticide, was the subject of a recent study using negative-ion mass spectrometry as a very sensitive method for detection of metabolites (as little as 60 femtograms was detectable). Different isomers were cleared from the brain by different metabolic pathways (Artigas et al., 1988).

9.2.3.3 Metabolism of Toxaphene Components

Metabolic studies of toxaphene have been made difficult by the complexity of this mixture of chlorinated terpenes. Reductive dechlorination at C-2 of 2,2,5-endo,6,exo,8,9,10-heptachlorobornane, a particularly toxic component of toxaphene, to isomeric heptachlorobornanes and dehydrochlorination of these to hexachorobornanes has been demonstrated in several systems, including rat liver microsomes under anaerobic conditions (Fig. 9-15) (Saleh and Casida, 1978). Oxidative metabolism as well as pro-

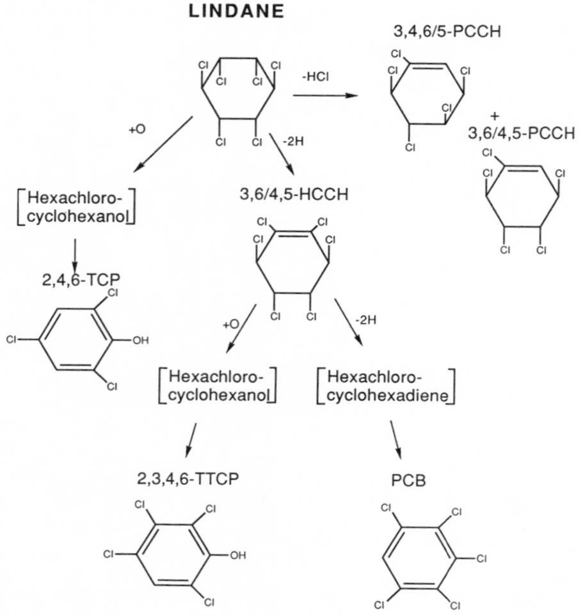

LINDANE

Figure 9-14. Metabolism of lindane. After Fitzloff *et al.*, 1982, with permission.

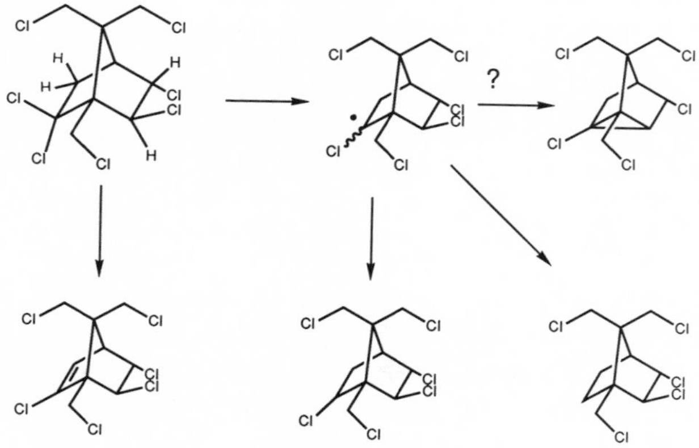

Figure 9-15. Metabolism of toxaphene. Adapted, with, permission from M. A. Saleh, and J. E. Casida, Reductive dechlorination of the toxaphene component 2,2,5-*endo*,6-*exo*,8,9,10,- heptachlorobornane in various chemical, photochemical, and metabolic systems, *J. Agric. Food Chem.* 26:583–590. Copyright 1978 American Chemical Society.

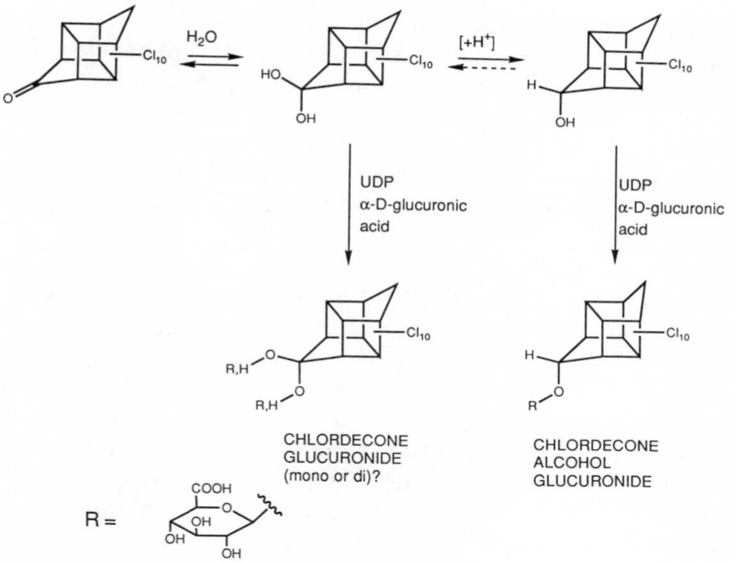

Figure 9-16. Metabolism of chlordecone. Adapted, with permission, from M. W. Fariss, R. V. Blanke, J. J. Saady, and P. S. Guzelian, Demonstration of major metabolic pathways for chlordecone (Kepone) in humans, *Drug Metab. Dispos.* 8:434–438. Copyright 1980 The American Association of Pharmacology and Experimental Therapeutics.

cesses involving GSH *S*-transferases are important in the total metabolism of toxaphene components to more polar metabolites (Chandurkar and Matsumura, 1979).

9.2.3.4 Metabolism of Chlordecone

Bioreduction of chlordecone to chlordecone alcohol represents the major metabolic route of chlordecone in man. Chlordecone (as the hydrate) and chlordecone alcohol are excreted predominantly as glucuronide conjugates (Fig. 9-16). Reduction of chlordecone in humans has been shown to occur in the liver (Farris *et al.*, 1980).

9.3 Metabolism of Halogenated Alkenes

The oxidative metabolism of halogenated alkenes to a large extent involves cytochrome P-450-catalyzed formation and subsequent reactions of oxiranes. Halogen substitution on a double bond increases the rate of oxirane formation as seen in the order of reactivity ethylene < vinyl chloride < trichloroethylene < tetrachloroethylene. Subsequent reactions of halogenated oxiranes include alkylation of cellular macromolecules (e.g., DNA and RNA bases and proteins), conjugation with such low-molecular-weight nucleophiles as GSH, hydrolysis to vicinal diols (spontaneously or catalyzed by epoxide hydrolase), or rearrangement (Fig. 9-17) (reviewed by Henschler, 1985).

While the mechanism of cytochrome P-450-catalyzed oxidation of alkanes is well established, details of the mechanism of alkene epoxidation are still unclear. Concerted mechanisms involving direct insertion of oxene into the C=C bond are given support by the transfer of stereochemistry from olefin to epoxide often observed. Although the metabolism of unsaturated compounds primarily involves epoxide formation, other transformations, including aldehyde formation and alkylation of heme nitrogen, also can occur and must be explained by proposed mechanisms. For example, certain unsaturated compounds are capable of suicidal inactivation of cytochrome P-450. Ortiz de Montellano *et al.* (1983) have identified heme–olefin adducts that correspond formally to alkylation of the heme nitrogen with olefin-derived epoxide. However, alkylation stereochemistry requires addition of the pyrrolic nitrogen and oxygen to

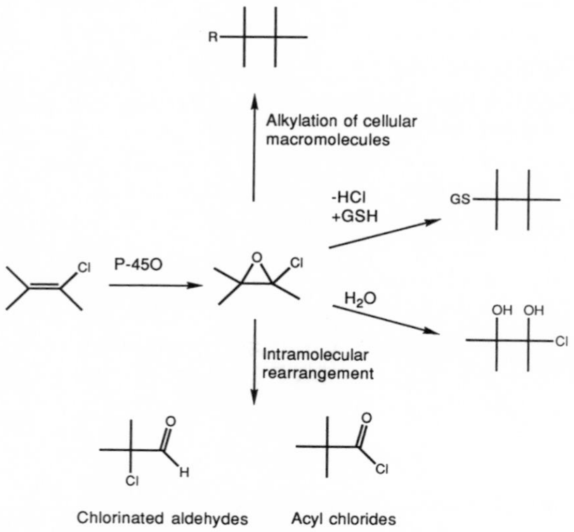

Figure 9-17. Epoxide formation as a key step in the cytochrome P-450-dependent metabolism of halogenated alkenes. Adapted, with permission, from Henschler, 1985.

the same side of the double bond, a result inconsistent with *trans* ring opening of the epoxide, but consistent with *cis* addition of activated oxygen and a pyrrolic nitrogen to the double bond (Fig. 9-18). Guengerich and Macdonald (1984) have interpreted these and other results in terms of their unified mechanism for cytochrome P-450 catalysis. According to this proposal, an intermediate cation is formed by addition of the $Fe^V = O$ form of cytochrome P-450 to the double bond. This addition may be preceded by a single-electron transfer to form a transient $[Fe^{IV} - O]^{2+}$-olefin complex, followed by collapse to the $[Fe^{III} - O]$-olefin addition product. This intermediate can (1) collapse to the epoxide, (2) undergo 1,2-group migration to give an aldehyde, or (3) alkylate the pyrrolic nitrogen (Fig. 9-19).

An alternative proposal consistent with the stereochemistry observed involves a $2a + 2s$ cycloaddition of olefin with hypervalent $Fe^V = 0$ to give an intermediate metallaoxetane (Collman *et al.*, 1985) (Fig. 2-20). Subsequent rearrangements of this intermediate would account for epoxide formation, rearrangement to aldehydic products, heme nitrogen alkylation, and stereospecific hydrogen exchange (Collman *et al.*, 1986; Groves *et al.*, 1986). On the other hand, Castellino and Bruice (1988) found no evidence for a metallocycle intermediate and favored formation of the acyclic cationic intermediate described above, either directly or from an initially formed cation radical (Fig. 9-19).

Dolphin *et al.* (1989) have prepared and studied the reactions of model β-hydroxyalkyl σ-metalloporphyrins (Fig. 9-21). The metallocycle is an alternative representation of the σ-alkyliron porphyrin, and Dolphin *et al.* have suggested that the latter complex could readily account for the results that had been rationalized by postulating the existence of the cyclic intermediate (Groves *et al.*, 1986).

The above represents a brief sampling of the ongoing research into the nature of cytochrome P-450-catalyzed olefin epoxidation. Space does not permit adequate discussion of the many skillfully designed experiments that

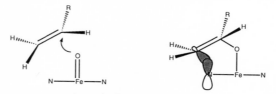

Figure 9-18. *cis*-Addition of activated oxygen and a pyrrolic nitrogen to double bond during alkylation of heme nitrogen during metabolism of unsaturated compounds. Adapted, with permission, from P. R. Ortiz de Montellano, B. L. K. Mangold, C. Wheeler, K. L. Kunze, and N. O. Reich, Stereochemistry of cytochrome P-450-catalyzed epoxidation and prosthetic heme alkylation, *J. Biol. Chem.* 258:4208–4213. Copyright 1983 American Chemical Society.

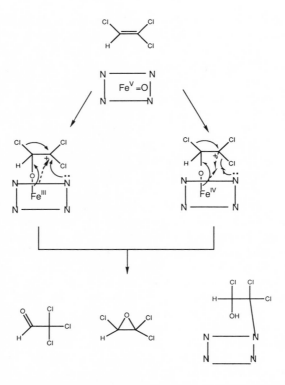

Figure 9-19. Proposed mechanism of halogenated alkene metabolism involving formation of an [Fe]–olefin addition product. Adapted, with permission, from F. P. Guengerich and T. L. Macdonald, Chemical mechanism of catalysis by cytochrome P-450: A unified view: *Acc. Chem. Res.* 17:9–16. Copyright 1984 American Chemical Society.

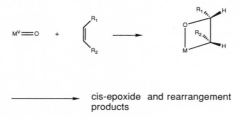

cis-epoxide and rearrangement products

Figure 9-20. Proposed formation of a metallocycle intermediate by cycloaddition of an olefin with Fe′=O (Collman *et al.*, 1985).

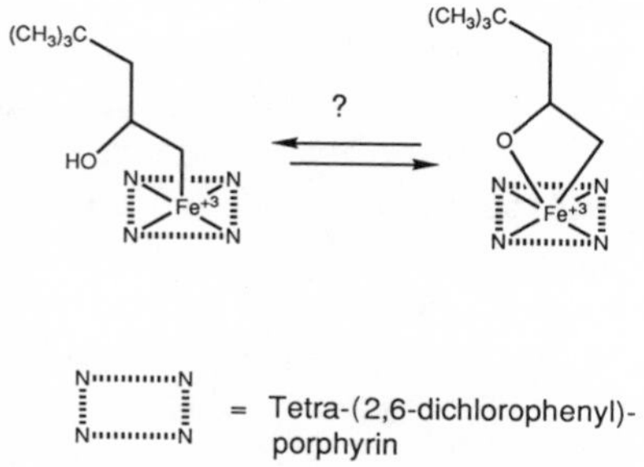

Figure 9-21. A model β-hydroxyalkyl σ-metalloporphyrin (Dolphin *et al.*, 1989).

have added much information on the nature of this important metabolic process. The interested reader is urged to explore the recent literature cited in this section. The above discussion is pertinent to the following sections, in which specific examples of oxidation of halogenated olefins will be considered, and certain mechanistic issues will be revisited in these examples.

9.3.1 Vinyl Halides

Similar to the situation with halogenated alkanes, wherein the simplest member, carbon tetrachloride, has been studied most extensively, vinyl chloride has been the most thoroughly examined of unsaturated chlorocarbons. As with carbon tetrachloride, adverse biological effects and the potential health and environmental problems assoicated with these effects have prompted the intense scrutiny of the biochemical behavior of vinyl chloride. Toxicity to experimental animals was noted as early as 1930, shortly after the commercial introduction of vinyl chloride—used extensively in the production of poly(vinyl chloride). An association between vinyl chloride and cancer in workers employed in vinyl chloride polymerization factories was discovered in 1974. Subsequent epidemiological studies confirmed this association and established the liver, brain, and lungs as target organs for vinyl chloride carcinogenic action (reviewed by Vainio and Saracci, 1984).

9.3.1.1 Bioactivation of Vinyl Chloride

A large volume of research has established that vinyl chloride is activated by cytochrome P-450 monooxygenase to chlorooxirane. This intermediate rearranges spontaneously to chloroacetaldehyde, which can be reduced enzymatically to chloroethanol or oxidized to chloroacetic acid, and is converted by epoxide hydrolase to a geminal halohydrin, which loses hydrogen chloride to form glycol aldehyde. The main pathway for deactivation of the chlorooxirane is through conjugation with GSH and subsequent formation of metabolites derived through mercapturic acid formation. Toxic, mutagenic, and carcinogenic responses elicited by vinyl chloride derive from reactions of the electrophilic chlorooxirane and chloroacetaldehyde. The more reactive chlorooxirane reacts mainly with DNA bases whereas chloroacetaldehyde primarily alkylates protein nucleophiles. The primary role of chlorooxirane in toxicity is shown by the fact that synthetic chlorooxirane is much more mutagenic *in vitro* and carcinogenic in animals than is chloroacetaldehyde (Henschler, 1985, and references therein). Furthermore, while reaction of chloroacetaldehyde (and bromoacetaldehyde) with DNA has been demonstrated in cell-free systems, no such reaction has been detected *in vivo*, and the major metabolite isolated from biological chloroacetaldehyde precursors (chloroethanol and 2,2'-dichlorodiethyl ether), other than vinyl chloride, is thiodiglycolic acid (Bolt, 1988, and references therein). Guengerich *et al.* (1981) have constructed a semiquantitative scheme to describe the sequential steps in vinyl chloride metabolism (Fig. 9-22).

9.3.1.2 The Role of Etheno DNA Adducts in Vinyl Halide Carcinogenicity

Formation of cyclic nucleic acid adducts has been implicated in the carcinogenesis of several chemical classes, including vinyl halides. Certain of these cyclic structures, when present in DNA, have been shown to lead to misincorporation of bases during DNA replication and thus may represent promutagenic lesions (reviewed by Bolt, 1988; Bartsch, 1986). DNA and RNA etheno adducts of vinyl chloride and vinyl bromide are

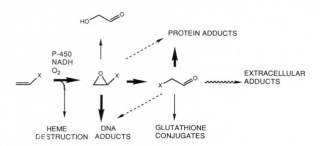

Figure 9-22. A semiquantitative scheme for the cytochrome P-450-mediated metabolism of vinyl chloride. After Guengerich *et al.*, 1981, with permission.

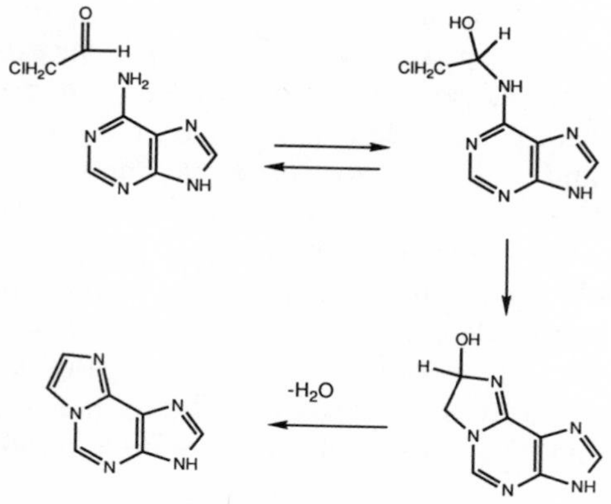

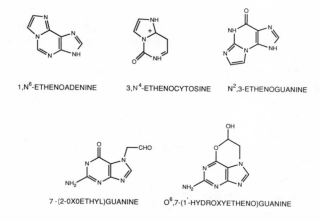

Figure 9-23. Mechanism of formation of etheno DNA adducts (Bolt, 1988).

formed by cyclization reactions of the halooxirane, or the corresponding haloaldehyde, with exocyclic and endocyclic nitrogens of the nucleotide bases (Fig. 9-23). Thus $1,N^6$-ethenoadenosine and $3,N^4$-ethenocytidine are formed in rat liver RNA *in vivo* and *in vitro*. In contrast, demonstration of DNA-adduct formation *in vivo* has been more difficult, and 7-(2-oxoethyl)guanosine appears to be the major DNA adduct formed *in vivo* (Fig. 9-24). Etheno-adduct formation is discussed further in Chapter 2 of Vol. 9B of this series.

1,N⁶-ETHENOADENINE   3,N⁴-ETHENOCYTOSINE   N²,3-ETHENOGUANINE

7-(2-OXOETHYL)GUANINE   O⁶,7-(1´-HYDROXYETHENO)GUANINE

Figure 9-24. Major DNA adducts formed during vinyl chloride metabolism (reviewed by Bolt, 1988; Bartsch, 1986).

The contrasting behavior of RNA and DNA toward electrophilic attack by chlorooxirane apparently reflects the greater ease of alkylation of single-stranded RNA as compared to alkylation of positions on DNA participating in base-pair processes. Alkylation of single-stranded DNA segments, loops, and replication forks may explain low levels of formation of etheno adducts of deoxyadenosine and deoxycytidine after long-term exposure of rats to vinyl chloride. Incorporation during short-term exposure thus could occur, but would be below detection levels. Supporting this view is the recent demonstration of the formation of N^2,3-ethenon-guanine (Fig. 9-24), at 1 % of the level of 7-(2-oxoethyl)guanine formation, after two succesive exposures of 12-day-old rats (an age at which DNA replication is rapid) to ^{14}C–vinylchloride (Bolt, 1988, and references therein).

Identification of 7-(2-oxoethyl)guanine as the major DNA adduct produced *in vivo* by vinylchloride exposure suggested that formation of this adduct was the likely promutagenic event in vinyl chloride carcinogenicity. Thus, the hypothesis was advanced that, as the hemiacetal [O^6,7-(1'-hydroxyetheno)guanine; Fig. 9-24], this adduct could simulate adenine and miscode for thymine (Schere *et al.*, 1981). However, a direct determination of the replicative behavior of a chlorooxirane-modified poly-DNA [poly(dG-dA)] revealed that 7-(2-oxoethyl)guanine did not miscode for either thymine or adenine. These data indicate that this adduct may play only a minor role in the carcinogenicity of vinyl chloride (Barbin *et al.*, 1985). Accordingly, recent research has concentrated on determination of the efficiency with which other etheno-DNA adducts—potentially present as minor vinyl chloride-produced lesions—induce miscoding. Of particular significance is the demonstration by Singer *et al.* (1987) that N^2,3-ethenoguanosine is read as adenosine and causes mutagenic dT incorporation. Thus, identification of this adduct in low concentrations *in vivo* (see above) also has increased significance. Other research has shown that ethenocytosine can cause misincorporation of thymine, whereas ethenoadenosine is a poor mutagen (Bolt, 1988).

There is strong evidence that formation of chlorooxirane is the initial activating step in carcinogenesis of vinyl chloride. While N-7 of guanine is the preferential site of alkylation in DNA by this electrophile, this appears not to be the event that induces mutagenesis. Instead, etheno adducts, the formation of which may be as much as two orders of magnitude lower than that of the major adduct, appear to be responsible for DNA miscoding. Of these, 3,N^4-ethenocytosine and especially N^2,3-ethenoguanine are the subjects of much current attention as likely promutagenic lesions. For further discussions of these issues, the reviews by Bolt (1988) and Bartsch (1986) are recommended.

Vinyl bromide and vinyl fluoride have not been investigated to the same extent as has vinyl chloride. There are indications that both are metabolized through electrophilic oxiranes, and there is direct evidence for DNA- and RNA-adduct formation. The order of carcinogenicity vinyl chloride > vinyl fluoride > vinyl bromide has been suggested (Henschler, 1985, and references therein).

9.3.2 1,1-Dichloroethylene (Vinylidene Chloride)

Because of the unsymmetrical halogen substitution pattern, vinylidene chloride and its epoxide are both exceptionally reactive. In fact, the evidence for the metabolic formation of the oxirane is indirect, since, unlike chorooxirane, 1,1-dichlorooxirane cannot be synthesized by conventional methods. Thus, the sole product from peracid oxidation of vinylidene chloride is the rearrangement product chloroacetyl chloride (Fig. 9-25). The metabolic formation of chloroacetyl chloride and derived secondary products has also been confirmed *in vivo*. Several products corresponding to the reaction of chloroacetic acid with GSH and subsequent enzymatic processing have been detected. Methylthioacetylaminoethanol may originate from the reaction of chloroacetyl chloride or 1,1-dichlorooxirane with phosphatidylethanolamine. This process may have significance with respect to hepatotoxicity of vinylidene chloride if it occurs on the phospholipid in lipid membranes. While the alternate product of rearrangement, dichloroacetaldehyde (Fig. 9-25), has been demonstrated *in vitro*, it has not been detected in animal studies (reviewed by Henschler, 1985).

9.3.3 *cis*- and *trans*-1,2-Dichloroethylene

Both isomers of 1,2-dichloroethylene are metabolized by cytochrome P-450 monooxygenases to unstable dichlorooxiranes. The principal

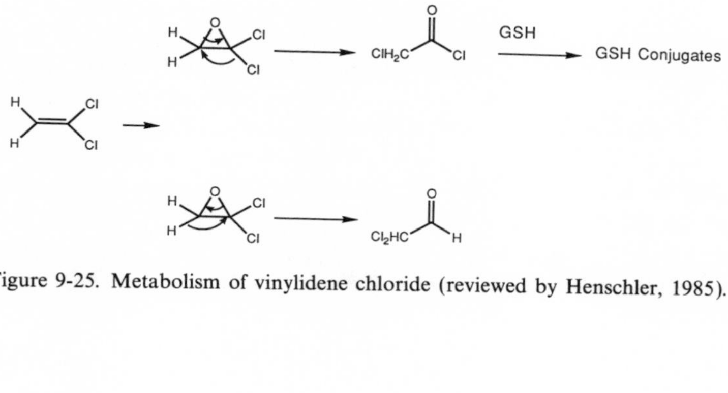

Figure 9-25. Metabolism of vinylidene chloride (reviewed by Henschler, 1985).

metabolites of each isomer are dichloroacetic acid and dichloroethanol, formed enzymatically from dichloroacetaldehyde, the rearrangement product of 1,2-dichlorooxiranes. No monochloroacetic acid-derived products have been detected, reflecting the unlikelihood of a 1,2-hydride shift during rearrangement of the oxiranes. Neither *cis-* nor *trans-*1,2-dichloroethylene is mutagenic (reviewed by Henschler, 1985).

9.3.4 Trichloroethylene

Trichloroacetic acid, trichloroethanol, and the glucuronide of trichloroethanol are important metabolites of trichloroethylene. These are derived from oxirane formation, followed by chloride shift from the least substituted carbon to the most substituted to give chloral as a metabolic intermediate. The course of oxirane rearrangement of trichlorooxirane produced by bioactivation of trichloroethylene differs, however, from the thermal rearrangement of synthetic trichlorooxirane, which produces dichloroacetyl chloride as the predominant product (Fig. 9-26). The thermal formation of dichloroacetylchloride has been rationalized in terms of the fact that rearrangement in this mode proceeds through the more stable monochlorosubstituted carbonium-ion intermediate. To explain the alternate mode of rearrangement of cytochrome P-450-derived trichlorooxirane, the proposal has been made that heme Fe^{3+} functions as a Lewis acid catalyst to facilitate rearrangement. The steric course of the rearrangement is controlled by approach of activated oxygen to the "bay region" (see below). In this configuration, complexation of heme Fe^{3+} with newly formed oxirane and the nearest chlorine substituent (the 2-chloro substituent) leads to rearrangement to chloral (Fig. 9-26) (Henschler, 1985, and references therein).

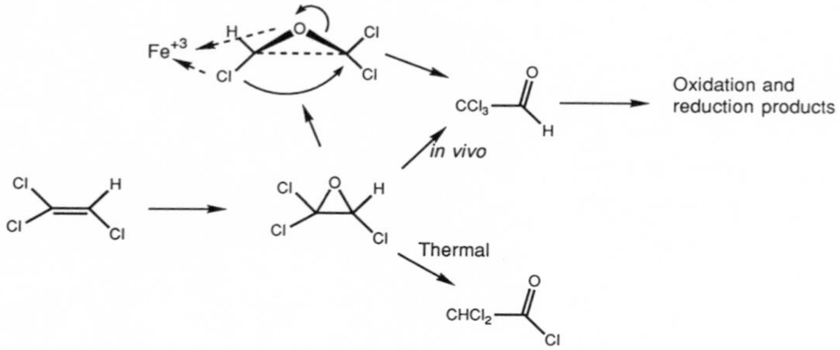

Figure 9-26. Proposed mechanism to explain regioselectivity of rearrangement in the cytochrome P-450-catalyzed oxidation of trichloroethylene (Henschler, 1985).

The formation of chloral rather than dichloroacetyl chloride is reflected in the minimal toxicity of trichloroethylene, since the former is much less reactive toward nucleophiles than is the latter. Absence of trichlorooxirane-derived products suggests that rearrangement to chloral occurs before the oxirane can escape from the immediate environment of the enzyme (reviewed by Henschler, 1985).

9.3.5 Tetrachloroethylene

Tetrachloroethylene is metabolized primarily to products derived from trichloroacetyl chloride. As with the case of trichloroethylene, rapid rearrangement of the monooxygenase-produced tetrachlorooxirane in the enzyme environment appears to represent a defensive mechanism, since tetrachlorooxirane is mutagenic while tetrachloroethylene in the presence of activating enzymes is not. Available evidence indicates that tetrachloroethylene is not carcinogenic (reviewed by Henschler, 1985).

9.3.6 Halogenated Allylic Compounds

Henschler (1985) has also reviewed the metabolic behavior of halogenated allylic compounds. In this case, metabolic activation through cytochrome P-450-catalyzed oxirane formation can no longer be assumed to be the primary mode of metabolism. Instead, the ability of many compounds of this class to act as direct alkylating agents is the major determining factor in their potential toxicity. A good correlation has been found between alkylating ability and mutagenic potency of a series of allylic halides (Eder *et al.*, 1980).

9.4 GSH-Dependent Metabolism and Toxicity of Halogenated Alkenes

The role of GSH *S*-transferase-catalyzed GSH conjugation in the toxicity of a series of 1,2-dihaloethanes has been discussed in Section 9.2.2. In the case of halogenated alkenes, the normal protective function of GSH conjugation is subverted by a second major mechanism of bioactivation. By this mechanism, hepatic cytosolic and microsomal GSH *S*-transferase catalyze the formation of *S*-alkyl or *S*-alkenyl GSH conjugates. These are metabolized to the corresponding *S*-cysteine adducts, which may be detoxified through formation of mercapturic acids or processed further by cysteine conjugate β-lyase (β-lyase) to give unstable thiols that form reactive and toxic electrophilic intermediates or stable toxic products (Fig. 9-27). The active transport of cysteine S-conjugates to the kidney by the renal anion transport system and localization of β-lyase in the kidney account for the selective nephrotoxicity. Examples of bioactivation of nephrotoxic halogenated alkenes will be given in the following sections, along with a discussion of molecular mechanisms of toxicity (reviewed by Anders *et al.*, 1988).

9.4.1 Renal β-Lyase and Nephrotoxicity

Certain polyhalogenated alkenes are potent nephrotoxins—2-chloro-1,1,2-trifluoroethylene (CTFE), trichloroethylene (TCE), and hexachloro-1,3-butadiene (HCB) are notable examples (Fig. 9-28). GSH S-transferase-catalyzed addition of GSH to CTFE forms S-(2-chloro-1,1,2-trifluoro-ethyl)glutathione (CTFG), with no evidence of vinyl product. Enzyme-catalyzed conjugation of TCE and HCB occurs by the addition–elimination mechanism to form S-(1,2-dichlorovinyl)glutathione (DCVG) and S-(penta-chloro-1,3-butadienyl)glutathione (PCBG), respectively (Fig. 9-28). These conjugates, formed in the liver, are transported to the kidneys and further metabolized by γ-glutamyltransferase and cysteinyl glycine dipeptidase to the corresponding S-cystienyl conjugates. As noted above, these can be acetylated to nontoxic mercapturic acids and excreted in the urine or can be processed further by β-lyase (Fig. 9-27). Much data have been accumulated that clearly implicate β-lyase-catalyzed extrusion of reactive thiol intermediates from the cysteine conjugates as the ultimate activating step in nephrotoxicity of these olefins. This work, the subject of a recent review (Anders *et al.*, 1988), will be summarized below.

The postulated mechanism of action of β-lyase, a pyridoxal phosphate (PLP)-dependent enzyme (an enzyme that has been shown to be identical to glutamine transaminase), is shown in Fig. 9-29 (Elfarra *et al.*, 1987).

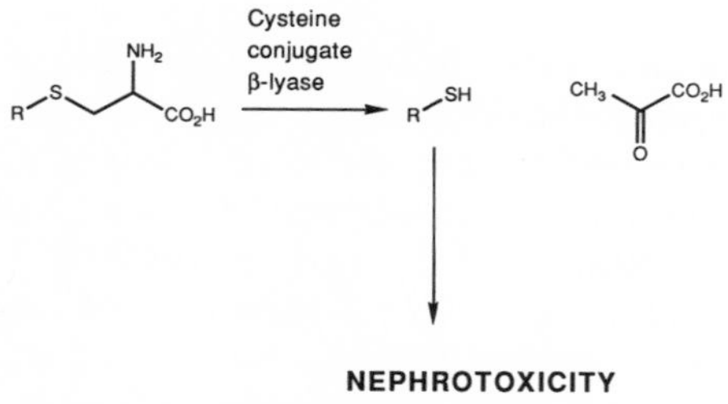

Figure 9-27. Cysteine conjugate β-lyase and nephrotoxicity.

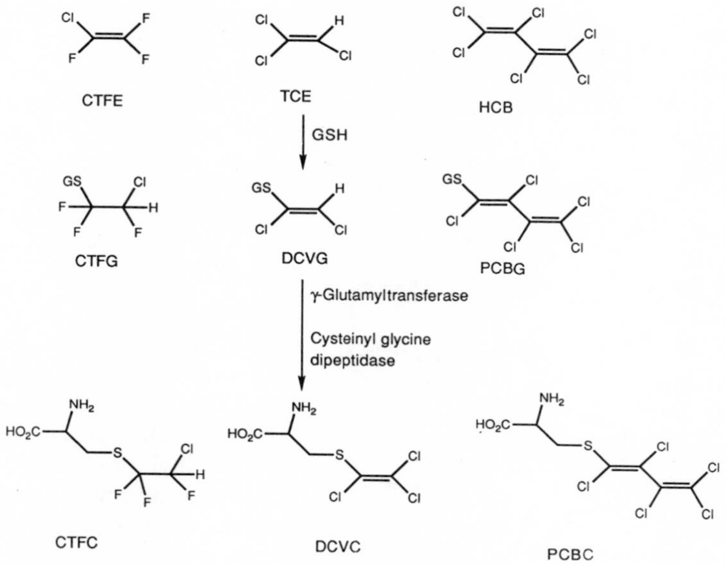

Figure 9-28. Examples of nephrotoxic haloalkenes and their derived GSH conjugates (reviewed by Anders *et al.*, 1988).

Biochemical evidence for the role of β-lyase in nephrotoxicity includes the absence of such toxicity in the case of synthetic S-(2-chloro-1,1,2-tri-fluoroethyl)-D,L-α-methylcysteine and S-(1,2-dichlorovinyl)-D,L-α-methyl-cysteine, analogs not capable of undergoing β-elimination. In contrast, the homocysteine conjugate S-(1,2-dichlorovinyl)-L-homocysteine (DCVHC) is an extremely potent nephrotoxin, as a result of renal γ-lyase-catalyzed elemination of the toxic thiol component.

Compounds that modulate the activity of β-lyase have the expected result on nephrotoxicity. Thus, inhibitors, such as aminooxyacetic acid, an inhibitor of PLP-dependent enzymes, block toxicity, whereas α-ketoacids, which enhance β-lyase activity by converting the pyridoxamine form of the cofactor to the pyridoxal form, increase nephrotoxicity. The requirement for γ-glutamyltransferase for processing of the initial GSH conjugates to cysteine conjugates is shown by the ability of γ-glutamyltransferase inhibitors to protect against CTFG-induced toxicity.

Biochemical data supporting the role of β-lyase in the nephrotoxicity of these halogenated alkenes are validated further by other evidence, including a close correlation of subcellular localization of β-lyase activity and patterns of toxicity. For example, localization of β-lyase in renal

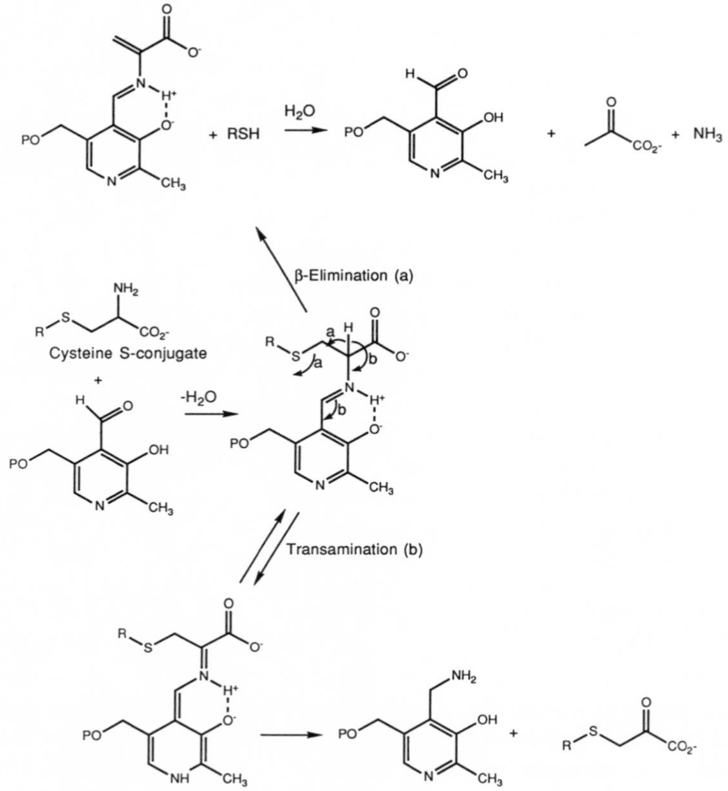

Figure 9-29. Proposed mechanism of cysteine conjugate β-lyase (Elfarra et al., 1987).

tubular mitochondria is consistent with selective mitochondrial toxicity of nephrotoxic cysteine S-conjugates. A more detailed discussion of the extensive research that has defined the processes involved in bioactivation of these nephrotoxic olefins is given in the review by Anders et al. (1988).

9.4.2. Molecular Mechanism of β-Lyase-Dependent Toxicity

9.4.2.1 Ultimate Toxic Metabolites of Nephrotoxic Cysteine S-Conjugates

A major urinary metabolite of $[^{35}S]S$-(1,2-dichlorovinyl)-L-cysteine $[^{35}S]$-DCVC) is inorganic sulfate. This suggested that decomposition of thiol products of β-lyase-catalyzed cleavage might lead to the formation of hydrogen sulfide, a likely metabolic precursor of sulfate. That hydrogen sulfide is a mitochondrial poison further implicated this species as a potential toxic product of thiol decomposition. However, a study of S-(2-chloro-1,1,2-trifluoroethyl)-L-cysteine (CTFC) metabolism has revealed that CTFC and hydrogen sulfide inhibit mitochondrial respiration at different sites. Although a reactive thiol formed by β-lyase cleavage of CTFC does partition to hydrogen sulfide and thiosulfate, rapid oxidation of the former to sulfate apparently protects the mitochondria from hydrogen sulfide toxicity (Banki et al., 1986). Cleavage of CTFC would be expected initially to produce 2-chloro-1,1,2-trifluoroethanethiol, a nucleophilic species which could eliminate fluoride to give chlorofluoro-thionoacetyl fluoride, a potent acylating agent. The intermediate β-lyase-produced thiol was trapped as the stable benzyl thioether by incubating CTFC with β-lyase in a model PLP system. Formation of chlorofluoro-thionoacetyl fluoride was confirmed in the same system by formation of the corresponding diethyl thioamide in the presence of diethylamine. Chloro-fluoroacetic acid, hydrogen sulfide, and fluoride are formed by hydrolysis of chlorofluorothionoacetyl fluoride (Fig. 9-30) (Dekant et al., 1987). Thiols from other halogenated S-conjugates can likewise produce reactive acylating agents. 1,1,2,2-Tetrafluoroethanethiol is dehydrofluorinated to 1,1,2-trifluorothionoacetyl fluoride, while enethiols formed from halovinyl S-conjugates may form thioketenes (Anders et al., 1987).

From this research, the bioactivation mechanism for nephrotoxicity of CTFC has been defined. However, the relative roles of the reactive inter-mediates and terminal metabolites have not been determined. Whereas hydrogen sulfide appears not to contribute to toxicity (see above), fluoride is nephrotoxic and may play a role. Acylation of cellular nucleophiles with chlorofluorothionoacetyl fluoride and potential toxicity of chlorofluoro-acetic acid are factors which must be considered (Dekant et al., 1987). The greater toxicity of homocysteine S-conjugates relative to cysteine S-conjugates may result from formation of 2-oxo-3-butenoic acid, a potential Michael acceptor of cellular nucleophiles, during γ-lyase-catalyzed elimination (Anders et al., 1988).

9.4.2.2 Biochemical Lesions in β-Lyase-Dependent S-Cysteine Conjugate Nephrotoxicity and Cytotoxicity

Mitochondria have been identified as important cellular targets of toxic cysteine S-conjugates. The mechanism of mitochondrial toxicity of dichlorovinyl S-cysteinyl (DCVC) and dichlorovinyl S-homocysteinyl (DCVHC) conjugates has been investigated with respect to the time

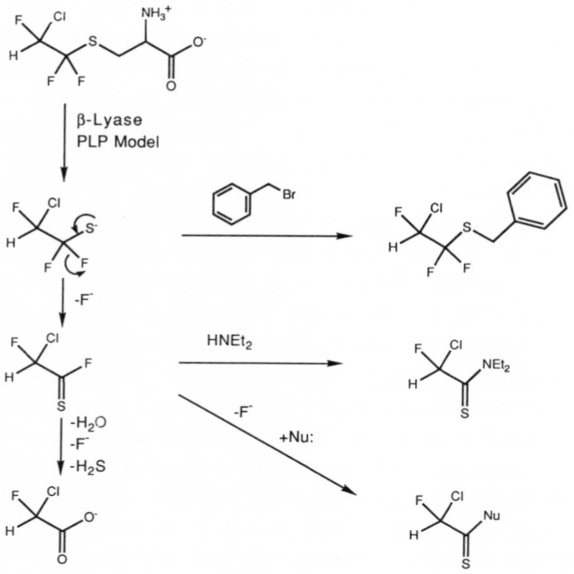

Figure 9-30. Formation of reactive intermediates from 2-chloro-1,1,2-trifluoroethyl S-cysteinyl conjugate (CTFC). After Dekant et al., 1987, with permission.

course of alterations of mitochondrial function, structural integrity, and metabolism. Effects on mitochondrial function included inhibition of succinate-dependent state 3 mitochondrial respiration, inhibition of Ca^{2+} sequestration, and a collapse of the mitochondrial membrane potential. Structural effects included an increased membrane permeability, matrix swelling, and protein leakage from the matrix. DCVC and DCVHC inhibited the sulfhydryl-dependent citric acid cycle enzymes succinate: cytochrome c oxidoreductase and isocitrate dehydrogenase, causing an alteration in energy metabolism, as reflected in a decreased ATP concentration. Increased oxidation of GSH to GSSG and lipid peroxidation indicated oxidative stress, but these apparently are secondary effects. The inhibition of succinate:cytochrome c oxidoreductase and the selective inhibition of succinate-dependent state 3 mitochondrial respiration suggest that complex II (the succinate–Q reductase complex) may be the primary site of action of DCVC and DCVHC, possibly by acylation of critical sulfhydryl groups by S-conjugate metabolites such as a thioketene. The loss of ability of mitochondria to retain Ca^{2+}, an energy-dependent process, may play a critical role in S-conjugate-induced cytotoxicity (reviewed by Anders et al., 1988).

9.4.3 Metabolism of Cyclodiene Insecticides

Carbon–carbon double bonds are targets for metabolism of cyclodiene insecticides containing sites of unsaturation. Aldrin (and the derived epoxide dieldrin) and chlordane will be considered. Additional details and further examples are reviewed by Matsumura (1975).

9.4.3.1 Aldrin

Aldrin is converted to its toxic epoxide, dieldrin, by mammals, soil microorganisms, plants, and insects (Matsumura, 1975, and references therein). This transformation is mediated predominantly by cytochrome P-450-dependent monooxygenases. However, in certain tissues having low cytochrome P-450 content, other available oxidative pathways can carry out this epoxidation. Thus, aldrin epoxidation occurs by a prostaglandin endoperoxide synthase (PES)-dependent process in seminal vesicle microsomes, a tissue having high PES activity and low monooxygenase activity. PES catalyzes the synthesis of an intermediate hydroperoxide [prostaglandin G_2 (PGG$_2$)] which is reduced by a hydroperoxidase to PGH$_2$. Aldrin (and other xenobiotics) can function as a cofactor in this

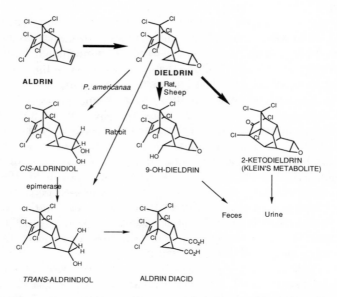

Figure 9-31. Metabolilsm of aldrin and dieldrin. Adapted, with permission, from Matsumura, 1975.

reduction, serving as an oxygen acceptor (Lang *et al.*, 1986). Similarly, aldrin epoxidation can be triggered by iron-initiated lipid peroxidation in liver microsomes (Lang and Maier, 1986). While dieldrin is quite stable metabolically, the epoxide functionality provides a center for metabolic inactivation. Examples of derived polar products are shown in Fig. 9-31 (Matsumura, 1975).

9.4.3.2 Chlordane

Both α- and β-chlordane can be oxidized to the epoxide, oxychlordane. While oxychlordane occupies a key position in chlordane metabolism, other pathways are available, and a spectrum of products are produced (reviewed by Brimfield and Street, 1979).

9.5 Metabolism of Halogenated Aromatic Compounds

9.5.1 Haloaromatic Compounds and the NIH Shift

Ring-substituted aromatic compounds are subject to cytochrome P-450-catalyzed hydroxylation by a process that is analogous to olefin epoxidation. As discussed in Chapter 7 of Vol. 9B of this series, rearrangements observed in certain of the hydroxylation reactions, for example, formation of 3-chlorotyrosine from 4-chlorophenylalanine, were important observations in the identification of arene-oxide intermediates and discovery of the "NIH shift" that can accompany these oxidations. Therefore, metabolic hydroxylation reactions of aromatic compounds do not invariably lead to dehalogenation, even when the halogen is at the locus of enzymatic attack. Arene oxides and their oxepin tautomers are highly reactive and have not been isolated as products of metabolism of substituted benzenoid compounds. However, their existence has been inferred from the isolation of phenols, dihydrodiols, and dihydrophenolic GSH conjugates derived from them (reviewed by Macdonald, 1984).

The metabolism of chloro- and bromobenzene has been investigated extensively because of the toxicity of these compounds and also because they serve as models for more complex compounds, such as PCBs, PBBs, and dioxins. Three pathways for the oxidative metabolism of chloro-benzene have been revealed. One pathway produces chlorobenzene 3,4-epoxide, which rearranges exclusively to 4-chlorophenol or reacts with nucleophiles (water, GSH, or other cellular nucleophiles); a second route produces chlorobenzene 2,3-epoxide, which isomerizes to 2-chlorophenol or reacts with nucleophiles, while a third pathway involves direct insertion of oxygen at the 3-position to give 3-chlorophenol (Fig. 9-32). This latter path was postulated to account for the fact that neither the 2,3- nor the 3,4-epoxide of chlorobenzene rearranges to 3-chlorophenol (Selander et al., 1975). Guengerich and Macdonald (1984) have proposed that all products are formed from a halobenzene substituted in the *meta* position by an iron oxygen species. This may either collapse to either of the two epoxides or, by a hydride shift, give the 3-halophenol (Fig. 9-33).

Reactive metabolites that are formed from bromobenzene cause liver GSH depletion, liver necrosis, and nephrotoxicity. The multiple products of metabolism include the 2,3-epoxide, the 3,4-epoxide, phenol-oxide(s), and the quinone/semiquinone forms of 4-bromocatechol and 2-bromohydroquinone. Extensive research has established bromobenzene 3,4-epoxide as the likely hepatotoxic agent, while GSH conjugates of 2-bromohydroquinone appear responsible for nephrotoxicity. Activation of these GSH conjugates by renal γ-glutamyltransferases and β-lyase are required for nephrotoxicity (reviewed by Lau and Monks, 1988).

9.5.2 Reductive Dehalogenation of Iodothyronines

As discussed in Chapter 6, the selective reduction of T_3 and T_4 plays a critical physiological role in thyroid hormone function. These monodeiodination reactions are reductive enzymatic processes that appear

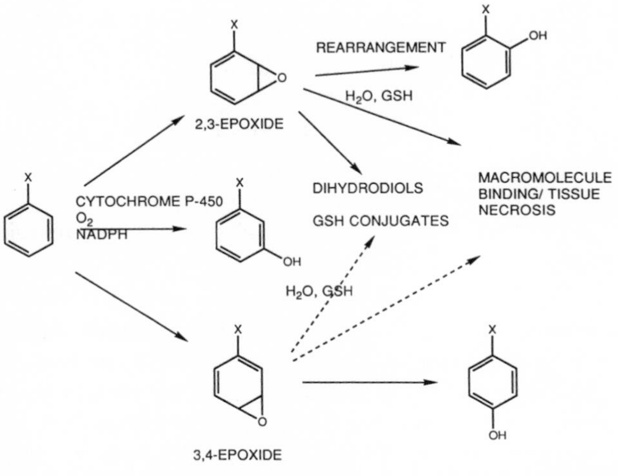

Figure 9-32. Metabolic transformations of chloro- and bromobenzene. Adapted, with permission, from T. L. Macdonald, Chemical mechanisms of halocarbon metabolism, *Crit. Rev. Toxicol.* 11:85–120. Copyright 1984 CRC Press, Inc., Boca Raton, Florida.

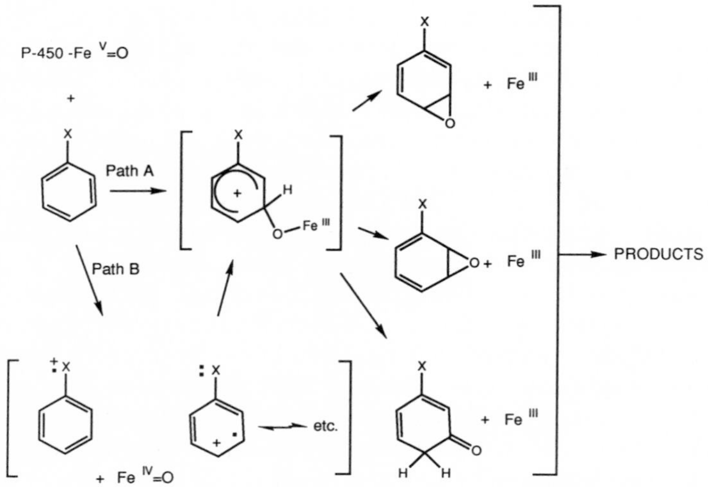

Figure 9-33. Proposed formation of halobenzene metabolites through a common inter-mediate. Adapted, with permission, from F. P. Guengerich and T. L. Macdonald, Chemical mechanism of catalysis by cytochrome P-450: A unified view, *Acc. Chem. Res.* 17:9–16. Copyright 1984 American Chemical Society.

to involve two half-reactions which require the presence of a sulfhydryl-containing cofactor. In the first reaction, illustrated by reduction of T_4, an iodinium ion is transferred from the substrate to a sulfhydryl group of the enzyme, resulting in an intermediate E–SI moiety. In the second step, this intermediate is reduced by a sulfhydryl-containing cofactor to regenerate active enzyme (reviewed by Engler and Burger, 1984).

9.5.3 Metabolism of Fluorinated Polycyclic Hydrocarbons—Modulation of Carcinogenicity

Members of the polycyclic aromatic hydrocarbon (PAH) class of com-pounds exhibit a wide range of carcinogenicities. Extensive research into the structural requirements for carcinogenicity has culminated recently in the formulation of the "bay region" theory of PAH carcinogenesis. Thus, Jerina and Daly (1976) suggested that a "bay region" is a structural requirement for carcinogenicity and that the ultimate carcinogen (which covalently binds to DNA) is a vicinal dihydrodiol epoxide in which the epoxide group is adjacent to a bay region. The two-stage activation

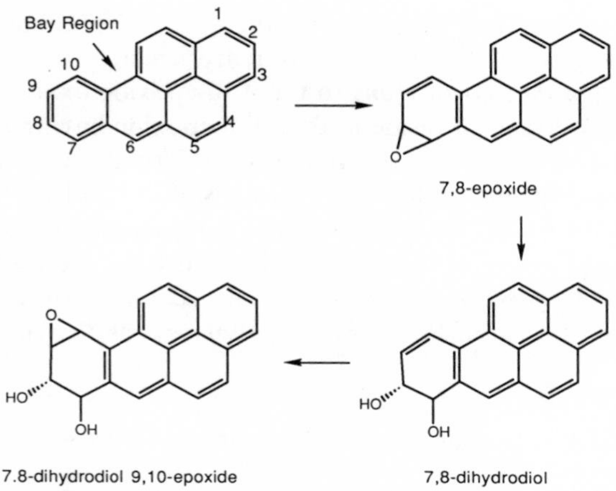

Figure 9-34. The relationship between the "bay region" and polycyclic aromatic hydrocarbon carcinogenicity (Jerina and Daly, 1976; reviewed by Lehr et al., 1985).

sequence is illustrated for benzo[a]pyrene in Fig. 9-34. This theory provides a mechanistic basis for carcinogenesis, as well as an explanation for the widely divergent activities within this class (reviewed by Lehr et al., 1985).

The tumorigenic activity of several fluorinated analogues of PAH has been examined. In general, tumorigenic activity descreased if fluorine was substituted on an angular ring adjacent to a bay region. These results are consistent with the assumption that such substitution blocks formation of the angular ring bay region dihydrodiol epoxide. Indeed, metabolic studies with fluorinated analogues of a number of carcinogenic PAHs have confirmed that fluorine effectively blocks oxidation at the double bond to which it is attached. Hecht et al. (1985) have reviewed these data and have tabulated numerous examples. The decreased tumorigenicity of 7-, 8-, 9-, and 10-fluorobenzo[a]pyrene (Fig. 9-35) is illustrative of these results.

A more subtle effect of fluorine substitution on carcinogenicity of

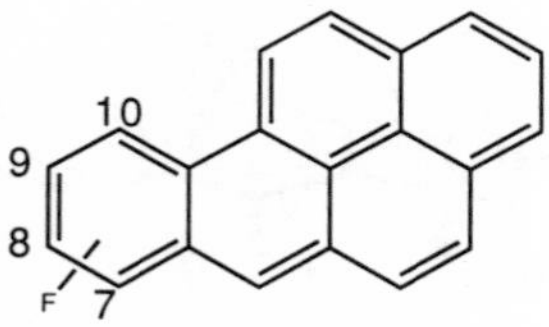

Figure 9-35. Example of noncarcinogenic fluorinated PAHs.

PAHs has been described recently (Chang *et al.*, 1987). The diol epoxides derived from *trans*-7,8-dihydroxy-7,8-dihydrobenzo[*a*]pyrene have the 7-hydroxyl group either *cis* or *trans* to the 9,10-epoxide oxygen. Tumorigenic activity has been found to reside in that diol epoxide having the diols in the *trans*-diequatorial conformation, and this conformation is favored for the 7-hydroxyl *trans* isomer. However, electrostatic repulsion of a fluorine substituted at the 6-position (6-fluorobenzo[*a*]-pyrene diol epoxide) of this same 7-hydroxyl *trans* isomer causes the hydroxyl groups to prefer the pseudoaxial configuration (Fig. 9-36). This change in conformation is reflected in an absence of tumorigenic activity in the fluorinated analogue.

9.5.4 Metabolism of Environmentally Persistent Polyhalogenated Aromatic Compounds

Chemical and thermal stability are characteristic of many commercially important halogenated aromatic compounds, and it is the resulting environmental persistence that has created the major ecological and environmental problems. These compounds are also biologically very stable, but, in most cases, they are subject to slow biodegradation. Of critical importance are mechanisms by which vertebrates can detoxify and excrete ingested material.

9.5.4.1 TCDD

In rats and hamsters, radioactivity found in various tissues following administration of [^{3}H]-TCDD is present as TCDD. On the other hand, eliminated radioactivity, primarily in feces, was present as metabolites. The

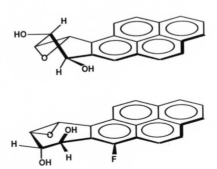

Figure 9-36. Effect of fluorine substitution on the conformation of diol epoxides. Adapted, with permission, from Chang *et al.*, 1987.

absence of metabolites in liver and fat indicates that, once formed, metabolites are quickly cleared (Neal *et al.*, 1984). There are quantitative differences in metabolism of TCDD in various mammalian species. Thus, the main metabolite in the dog, 1,3,7,5-tetrachloro-2-hydroxydibenzo-*p*-dioxin, is a minor metabolite in the rat. Tetrachlorodihydroxydiphenyl ether and trichlorodihydroxydibenzo-*p*-dioxin were identified as the major metabolites in the rat, and a trichloro-trihydroxydiphenyl ether was found only in the rat. Isomer distributions in minor metabolites also differed in rat and dog. In addition, the metabolites were excreted as glucuronide conjugates in the rat, whereas the dog excreted free phenolic compounds. A proposed scheme for TCDD metabolism is shown in Fig. 9-37 (Poiger and Buser, 1984, and references therein). There in no apparent consistent correlation between differences in metabolism in various species and sensitivity to TCDD (Neal *et al.*, 1984).

9.5.4.2 PCBs

The degree of metabolic stability of PCBs is related directly to the degree and position of chlorination. For this reason, PCBs found in adipose tissue are considerably more highly chlorinated than the original commercial mixtures. Therefore, as environmental PCBs progress through the food chaim, the low-chlorine PCBs are metabolized and disappear, while unmetabolizable components accumulate. Metabolism is facilitated

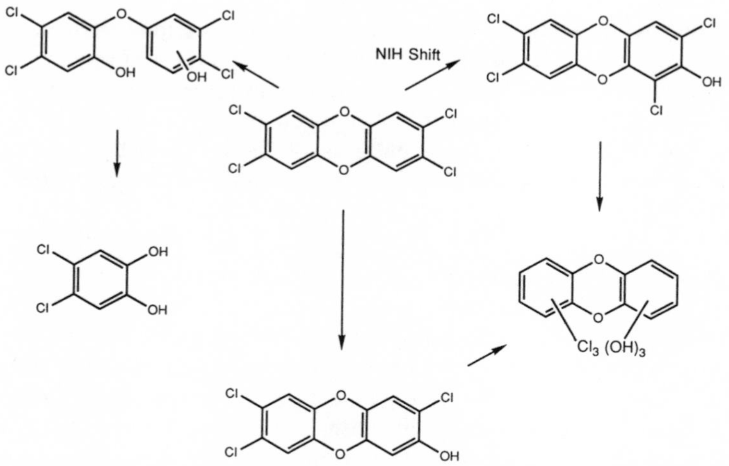

Figure 9-37. Metabolism of TCDD. After Poiger and Buser, 1984, with permission; copyright 1984 Cold Spring Harbor Laboratory.

greatly if there is at least one pair of adjacent, unsubstituted carbon atoms. The most abundant isomer found in human adipose tissue is 2,4,5,2′,4′5′-hexachlorobiphenyl (21.5% of the total) (Bickel and Muehlebach, 1980).

9.6 Biodehalogenation Mediated by Microorganisms

The emphasis to this point has been primarily on mammalian metabolism of halogenated compounds. The adaptability of micro-organisms to environmental pressures has made their metabolic machinery available for biodegradation of many halogenated compounds, including environmentally persistent molecules (reviewed by Neidleman and Geigert, 1986). Of particular importance to removal of environmental pollutants is the recent demonstration that anaerobic microbial organisms can reductively dehalogenate halogenated aromatic compounds. Suflita et al. (1982) demonstrated the reductive dehalogenation of halobenzoates by methanogenic bacteria and noted in this report the potential of anaerobic microbes as tools for the removal of xenobiotics from the environment. In a more stringent test, anaerobic microorganisms from Hudson River sediments have been found to dechlorinate most PCBs found in a commercial mixture (aroclor 1242) (Quensen et al., 1988). Neidleman and Geigert (1986) have tabulated several naturally occurring microorganisms that carry out dehalogenation along with the potential environmental pollutants that these can dehalogenate.

Bacteria also carry out facile hydrolytic dehalogenation of halogenated carboxylic acids. Haloacetate dehalogenase acts on haloacetates, and 2-haloacid dehalogenase degrades haloacetates as well as other short-chain halogenated carboxylic acids. The reaction has been shown to involve a base-catalyzed S_N2 displacement of halogen by hydroxide. Noteworthy is the facility with which fluoroacetate is degraded by halohydrases from certain microorganisms (reviewed by Neidleman and Geigert, 1986).

9.7 Metabolism and Biodehalogenation—Summary

An emphasis in this chapter has been placed on metabolic activation of halogen-containing xenobiotics to toxic metabolites. Such activation obviously is not limited to halogenated compounds, since members of other classes of compounds likewise can be activated to toxic or carcinogenic compounds, to the detriment of the host. Moreover, the protective function of such metabolic enzymes as the cytochrome P-450 monooxygenases and GSH S-transferases normally is carried out, and metabolism usually leads to the production of polar metabolites that can be excreted. Nor is biodehalogenation invariably related only to metabolism—the essential role of thyroxine deiodination in thyroid function can be cited again. Nonetheless, a vast amount of research has been done on elucidating the mechanisms of toxicities of halogenated organic compounds, particularly those that pose an environmental threat. An attempt has been made in this chapter to give an overview of recent progress in this research.

References

Anders, M. W., and Pohl, L. R., 1985. Halogenated alkanes, in *Bioactivation of Foreign Compounds* (M. W. Anders, ed.), Academic Press, Orlando, Florida, pp. 283–315.

Anders, M. W., Elfarra, A. A., and Lash, L. H., 1987. Cellular effects of reactive intermediates: Nephrotoxicity of S-conjugates of amino acids, *Arch. Toxicol.* 60:103–108.

Anders, M. W., Lash, L., Dekant, W., Elfarra, A. A., and Dohn, D. R., 1988. Biosynthesis and biotransformation of glutathione S-conjugates to toxic metabolites, *Crit. Rev. Toxicol.* 18:311–341.

Artigas, F., Martinez, E. Camón, L. Gelpí, E., and Rodríguez-Farré, E., 1988. Brain metabolites of lindane and related isomers: Identification by negative ion mass spectrometry, *Toxicology* 49:57–63.

Baker, M. T., and Van Dyke, R. A., 1984. Metabolism-dependent binding of the chlorinated insecticide DDT and its metabolite, DDD, to microsomal protein and lipids, *Biochem. Pharmacol.* 33:255–260.

Banki, K., Elfarra, A. A., Lash, L. H., and Anders, M. W., 1986. Metabolism of S-(2-chloro-1,1,2-trifluoroethyl)-L-cysteine to hydrogen sulfide and the role of hydrogen sulfide in S-(2-chloro-1,1,2-trifluoroethyl)-L-cysteine-induced mitochondrial toxicity, *Biochem. Biophys. Res. Commun.* 138:707–713.

Barbin, A., Laib, R. J., and Bartsch, H., 1985. Lack of miscoding properties of 7-(2-oxoethyl)guanine, the major vinyl chloride–DNA adduct, *Cancer Res.* 45:2440–2444.

Bartsch, H., 1986. The role of cyclic nucleic acid base adducts in carcinogenesis and mutagenesis, in *The Role of Cyclic Nucleic Acid Adducts in Carcinogenesis and Mutagenesis* (B. Singer and H. Bartsch, eds.), IARC Scientific Publications, No. 70, Oxford University Press, New York, pp. 3–14.

Bickel, M. H., and Muehlebach, S., 1980. Pharmacokinetics and ecodisposition of polyhalogenated hydrocarbons: Aspects and concepts, *Drug Metab. Rev.* 11:149–190.

Bolt, H. M., 1988. Roles of etheno-DNA adducts in tumorigenicity of olefins, *CRC Crit. Rev. Toxicol.* 18:299–309.

Brault, D., 1985. Model studies in cytochrome P-450-mediated toxicity of halogenated compounds: Radical processes involving iron porphyrins, *Environ. Health Perspect* 64:53–60.

Brimfield, A. A., and Street, J. C., 1979. Mammalian biotransformation of chlordane: *In vivo* and primary hepatic comparisons, *Ann. N.Y. Acad. Sci.* 320:247–256.

Brooks, G. T., 1986. Insecticide metabolism and selective toxicity, *Xenobiotica* 16:989–1002.

Burka, L. T., Thorsen, A., and Guengerich, F. P., 1980. Enzymatic monooxygenation of halogen atoms: Cytochrome P-450 catalyzed oxidation of iodobenzene by iodosobenzene, *J. Am. Chem. Soc.* 102:7615–7616.

Castellino, A. J., and Bruice, T. C., 1988. Intermediates in the epoxidation of alkenes by cytochrome P-450 models. 1. cis-Stilbene as a mechanistic probe, J. Am. Chem. Soc. 110:158–162.

Chandurkar, P. S., and Matsumura, F., 1979. Metabolism of toxaphene components in rats, Arch. Environ. Contam. Toxicol. 8:1–24.

Chang, R. L., Wood, A. W., Conney, A. H., Yagi, H. Sayer, J. M., Thakker, D. R., Jerina, D. M., and Levin, W., 1987. Role of diaxial versus diequatorial hydroxyl groups in the tumorigenic activity of a benzo[a]pyrene bay-region diol epoxide, Proc. Natl. Acad. Sci. USA 84:8633–8636.

Chasseaud, L. F., 1979. The role of glutathione and glutathione S-transferases in the metabolism of chemical carcinogens and other electrophilic agents, Adv. Cancer Res. 29:175–274.

Cheeseman, K. H., Albano, E. F., Tomasi, A., and Slater, T. F., 1985. Biochemical studies on the metabolic activation of halogenated alkanes, Environ. Health Perspect 64:85–101.

Collman, J. P., Brauman, J. I., Meunier, B., Hayashi, T., Kodaked, T., and Raybuck, S. A., 1985. Epoxidation of olefins by cytochrome P-450 model compounds. Kinetics and stereochemistry of oxygen atom transfer and origin of shape selectivity, J. Am. Chem. Soc. 107:2000–2005.

Collman, J. P., Kodadek, T., and Brauman, J. I., 1986. Oxygenation of styrene by cytochrome P-450 model systems. A mechanistic study, J. Am. Chem. Soc. 108:2588–2594.

Dekant, W., Lash, L. H., and Anders, M. W., 1987. Bioactivation mechanism of the cytotoxic and nephrotoxic S-conjugate S-(2-chloro-1,1,2-trifluoroethyl)-L-cysteine, Proc. Natl. Acad. Sci. USA 84:7443–7447.

Dohn, D. R., Garziano, M. J., and Casida, J. E., 1988. Metabolites of [3-^{13}C]1,2-dibromo-3-chloropropane in male rats studied by ^{13}C and ^{1}H-^{13}C correlated two-dimensional NMR spectroscopy, Biochem. Pharmacol. 37:3485–3495.

Dolphin, D., Matsumoto, A., and Shortman, C., 1989. β-Hydroxyalkyl σ-metalloporphyrins: Models for epoxide and alkene generation from cytochrome P-450, J. Am. Chem. Soc. 111:411–413.

Eder, E., Neudecker, T., Lutz, D., and Henschler, D., 1980. Mutagenic potential of allyl and allylic compounds. Structure–activity relationship as determined by alkylating and direct in vitro mutagenic properties, Biochem. Pharmacol. 29:993–998.

Elfarra, A. A., and Anders, M. W., 1987. Renal processing of glutathione conjugates. Role in nephrotoxicity, Biochem. Pharmacol. 33:3729–3732.

Elfarra, A. A., Baggs, R. B., and Anders, M. W., 1985. Structure–nephrotoxicity relationships of S-(2-chloroethyl)-DL-cysteine and analogs: Role for an episulfonium ion, J. Pharmacol. Exp. Ther. 233:512–516.

Elfarra, A. A., Lash, L. H., and Anders, M. W., 1987. α-Ketoacids stimulate rat renal cystein conjugate β-lyase activity and potentiate the cytotoxicity of S-(1,2-dichlorovinyl)-L-cystein, Mol. Pharmacol. 31:208–212.

Engler, D., and Burger, A. G., 1984. The deiodination of the iodothyronines and of their derivatives in man, Endocrine Rev. 5:151–184.

Fariss, M. W., Blanke, R. V., Saady, J. J., and Guzelian, P. S., 1980. Demonstration of major metabolic pathways for chlordecone (Kepone) in humans, Drug. Metab. Dispos. 8:434–438.

Fawcett, S. C., King, L. J., Bunyan, P. J., and Stanley, P. I., 1987. The metabolism of ^{14}C-DDT, ^{14}C-DDD, ^{14}C-DDE and ^{14}C-DDMU in rats and Japanese quail, Xenobiotica 17:525–538.

Fitzloff, J. F., Portig, J., and Stein, K., 1982. Lindane metabolism by human and rat liver microsomes, Xenobiotica 12:197–202.

Fukami, J.-I., 1980. Metabolism of several insecticides by glutathione S-transferase, *Pharmacol. Ther.* 10:473–514.

Gold, B., and Brunk, G., 1984. A mechanistic study of the metabolism of 1,1-dichloro-2,2-bis(p-chlorophenyl)ethane (DDD) to 2,2-bis(p-chlorophenyl)acetic acid (DDA), *Biochem. Pharmacol.* 33:979–982.

Groves, J. T., McClusky, G. A., White, R. E., and Coon, M. J., 1978. Aliphatic hydroxylation of highly purified liver microsomal cytochrome P-450. Evidence for a carbon radical intermediate, *Biochem. Biophys. Res. Commun.* 81:154–160.

Groves, J. T., Avaria-Neisser, G. E., Fish, K. M., Imachi, M., and Kuczkowski, R. L., 1986. Hydrogen–deuterium exchange during propylene epoxidation by cytochrome P-450, *J. Am. Chem. Soc.* 108:3837–3838.

Guengerich, F. P., 1982. Metabolism of vinyl halides: *In vitro* studies on roles of potential activated metabolites, *Adv. Exp. Med. Biol.* 136A–136B:685–692.

Guengerich, F. P., and Macdonald, T. L., 1984. Chemical mechanism of catalysis by cytochrome P-450: A unified view, *Acc. Chem. Res.* 17:9–16.

Guengerich, F. P., Crawford, W. M., Jr., Domoradzki, J. Y., Macdonald, T. L., and Watanabe, P. G., 1980. *In vitro* activation of 1,2-dichloroethane by microsomal and cytosolic enzymes, *Toxicol. Appl. Pharmacol.* 55:303–317.

Guengerich, F. P., Mason, P. S., Scott, W. T., Fox, T. R., and Watanabe, P. G., 1981. Roles of 2-haloethylene oxides and 2-haloacetaldehydes derived from vinyl bromide and vinyl chloride in irreversible binding to protein and DNA, *Cancer Res.* 41:4391–4398.

Guengerich, F. P., Willard, R. J., Shea, J. P., Richards, L. E., and Macdonald, T. L., 1984. Mechanism-based inactivation of cytochrome P-450 by heteroatom-substituted cyclopropanes and formation of ring-opened compounds, *J. Am. Chem. Soc.* 106:6446–6447.

Guengerich, F. P., Peterson, L. A., Cmarik, J. L., Koga, N., and Inskeep, P. G., 1987. Activation of dihaloalkanes by glutathione conjugation and formation of DNA adducts, *Environ. Health Perspect* 76:15–18.

Hecht, S. S., Amin, S., Melikian, A. A., LaVoie, E. J., and Hoffmann, D., 1985. Effects of methyl and fluorine substitution on the metabolic activation and tumorigenicity of polycyclic aromatic hydrocarbons, in *Polycyclic Hydrocarbons and Carcinogenesis* (R. G. Harvey, ed.), American Chemical Society, Washington, D.C., pp. 85–105.

Henschler, D., 1985. Halogenated alkenes and alkynes, in *Bioactivation of Foreign Compounds* (M. W. Anders, ed.), Academic Press, Orlando, Florida, pp. 317–437.

Jerina, D. M., and Daly, J. W., 1976. Oxidation at carbon, in *Drug Metabolism—From Microbes to Man* (D. V. Parke and R. L. Smith, eds.), Taylor and Francis, London, pp. 13–32.

Kelner, M. J., McLenithan, J . C., and Anders, M. W., 1986. Thiol stimulation of the cytochrome P-450-dependent reduction of 1,1,1-trichloro-2,2-bis(p-chlorophenyl)ethane (DDT) to 1,1-dichloro-2,2-bis(p-chlorophenyl)ethane (DDD), *Biochem. Pharmacol.* 35:1805–1807.

Kieczka, H., and Kappus, H., 1980. Oxygen dependence of CCl_4-induced lipid peroxidation *in vitro* and *in vivo*, *Toxicol. Lett.* 5:191–196.

Koga, N., Inskeep, P. B., Harris, T. M., and Guengerich, F. P., 1986. S-[2-(N^7-Guanyl)ethyl]glutathione, the major DNA adduct formed from 1,2-dibromoethane, *Biochemistry* 25:2192–2198.

Lang, B., and Maier, P., 1986. Lipid peroxidation dependent aldrin epoxidation in liver microsomes, hepatocytes and granulation tissue cells, *Biochem. Biophys. Res. Commun.* 138:24–32.

Lang, B., Fred, K., and Maier, P., 1986. Prostaglanding synthase dependent aldrin epoxidation in hepatic and extrahepatic tissues of rats, *Biochem. Pharmacol.* 35:3643–3645.

Lau, S. S., and Monks, T. J., 1988. The contribution of bromobenzene to our current understanding of chemically-induced toxicities, *Life Sci.* 42:1259–1269.

Lehr, R. E., Kumar, S., Levin, W., Wood, A. W., Chang, R. L., Conney, A. H., Yagi, H.,
Sayer, J. M., and Jerina, D. M., 1985. The bay region theory of polycyclic aromatic
hydrocarbon carcinogenesis, in *Polycyclic Hydrocarbons and Carcinogenesis* (R. G.
Harvey, ed.), American Chemical Society, Washington, D.C., pp. 63–84.

Lowrey, K., Glende, E. A., Jr., and Recknagel, R. O., 1981. Destruction of liver microsomal
calcium pump activity by carbon tetrachloride and bromotrichloromethane, *Biochem.
Pharmacol.* 30:135–140.

Macdonald, T. L., 1984. Chemical mechanisms of halocarbon metabolism, *Crit. Rev. Toxicol.*
11:85–120.

Macdonald, T. L., Narasimhan, N., and Burka, L. T., 1980. Chemical and biological oxidation
of organohalides. Peracid oxidation of alkyl iodides, *J. Am. Chem. Soc.* 102:7760–7765.

Matsumura, F., 1975. *Toxicology of Insecticides*, Plenum Press, New York, pp. 165–251.

Mead, R. J., Moulden, D. L., and Twigg, L. E., 1985. Significance of sulfhydryl compounds
in the manifestation of fluoroacetate toxicity to rat, brush-tailed possum, woylie and
western grey kangaroo, *Aust. J. Biol. Sci.* 38:139–149.

Moore, L., Davenport, G. R., and Landon, E. J., 1976. Calcium uptake of a rat liver
microsomal subcellular fraction in response to *in vitro* administration of carbon
tetrachloride, *J. Biol. Chem.* 251:1197–1201.

Morgenstern, R., Guthenberg, C., and DePierre, J. W., 1982. Microsomal glutathione *S*-trans-
ferase. Purification, initial characterization and demonstration that it is not identical to
the cytosolic glutathione *S*-transferases A, B, and C, *Eur. J. Biochem.* 128:243–248.

Neal, R., Gasiewicz, T., Geiger, L., Olson, J., and Sawahata, T., 1984. Metabolism of 2,3,7,8-
tetrachlorodibenzo-*p*-dioxin in mammalian systems, in Banbury Report 18. Biological
Mechanisms of Dioxin Action (A. Poland and R. D. Kimbrough, eds.), Cold Spring
Harbor Laboratory, pp. 49–60.

Nebert, D. W., Eisen, H. J., Negishi, M., Lang, M. A., Hjelmeland, L. M., and Okey, A. B.,
1981. Genetic mechanisms controlling the induction of polysubstrate monooxygenase
(P-450) activities, *Annu. Rev. Pharmacol. Toxicol.* 21:431–462.

Neidleman, S. L., and Geigert, J., 1986. *Biohalogenation: Principles, Basic Roles and Application*, Ellis Horwood, Chichester, pp. 156–175.

Ortiz de Montellano, P. R., Mangold, B. L. K., Wheeler, C., Kunze, K. L., and Reich, N. O.,
1983. Stereochemistry of cytochrome P-450-catalyzed epoxidation and prosthetic heme
alkylation, *J. Biol. Chem.* 258:4208–4213.

Ozawa, N., and Guengerich, F. P., 1983. Evidence for the formation of an *S*-[2-(*N*[7]-
guanyl)ethyl]glutathione adduct in glutathione-mediated binding of the carcinogen 1,2-
dibromoethane to DNA, *Proc. Natl. Acad. Sci. USA* 80:5266–5270.

Poiger, H., and Buser, H.-R., 1984. The metabolism of TCDD in the dog and rat, in Banbury
Report 18. Biological Mechanisms of Dioxin Action (A. Poland and R. D. Kimbrough,
eds.), (A. Poland and R. D. Kimbrough, eds.), Cold Spring Harbor Laboratory,
pp. 39–47.

Quensen, J. F., III, Tiedje, J. M., and Boyd, S. A., 1988. Reductive dechlorination of
polychlorinated biphenyls by anaerobic microorganisms from sediments, *Science*
242:752–754.

Quraishi, M. S., 1977. *Biochemical Insect Control, Its Impact on Economy, Environment, and
Natural Selection*, John Wiley and Sons, New York, pp. 98–123.

Saleh, M. A., and Casida, J. E., 1978. Reductive dechlorination of the toxaphene component
2,2,5-*endo*,6-*exo*,8,9,10-heptachlorobornane in various chemical, photochemical, and
metabolic systems, *J. Agric. Food Chem.* 26:583–590.

Scherer, E., Van Der Laken, C. J. Gwinner, L. M. Laib, R. J., and Emmelot, P., 1981. Modification of deoxyguanosine by chloroethylene oxide, *Carcinogenesis* 2:671–677.

Selander, H. G., Jerina, D. M., and Daly, J. W., 1975. Metabolism of chlorobenzene with hepatic microsomes and solubilized cytochrome P-450 systems, *Arch. Biochem. Biophys.* 168:309–321.

Singer, B., Spengler, S. J., Chavez, F., and Kusmierek, J. T., 1987. The vinyl chloride-derived nucleoside, N^2,3-ethenoguanosine, is a highly efficient mutagen in transcription, *Carcinogenesis* 8:745–747.

Slater, T. F., Cheeseman, K. H., and Ingold, K. U., 1985. Carbon tetrachloride toxicity as a model for studying free-radical mediated liver injury, *Philos. Trans. Roy. Soc. London B* 311:633–645.

Soiefer, A. I., and Kostyniak, P. J., 1984. Purification of a fluoroacetate-specific defluorinase from mouse liver cytosol, *J. Biol. Chem.* 259:10787–10792.

Stacey, N., and Priestly, B. G., 1978. Lipid peroxidation in isolated rat hepatocytes: Relationship to toxicity of CCl_4, ADP/Fe^{3+}, and diethyl maleate, *Toxicol. Appl. Pharmacol.* 45:41–48.

Stevens, J. L., Ratnayaka, J. H., and Anders, M. W., 1980. Metabolism of dihalomethanes to carbon monoxide. IV. Studies in isolated rat hepatocytes, *Toxicol. Appl. Pharmacol.* 55:484–489.

Suflita, J. M., Horowitz, A., Shelton, D. R., and Tiedje, J. M., 1982. Dehalogenation: A novel pathway for the anaerobic biodegradation of haloaromatic compounds, *Science* 218:1115–1117.

Tachikzawa, H., MacDonald, T. L., and Neal, R. A., 1982. Rat liver microsomal metabolism of propyl halides, *Mol. Pharmacol.* 22:745–751.

Vadi, H. V., Schasteen, C. S., and Reed, D. J., 1985. Interactions of S-(2-haloethyl)mercapturic acid analogs with plasmid DNA, *Toxicol. Appl. Pharmacol.* 80:386–396.

Vainio, H., and Saracci, R., 1984. Carcinogenicity of selected vinyl compounds, some aldehydes, haloethyl nitrosoureas and furocoumarins: An overview, in *The Role of Cyclic Nucleic Acid Adducts in Carcinogenesis and Mutagenesis* (B. Singer and H. Bartsch, eds.), IARC Scientific Publications, No. 70, Oxford University Press, New York, pp. 15–29.

van Bladeren, P. J., Breimer, D. D., van Huijgevoort, J. A. T. C. M., Vermeulen, N. P. E., and van der Gen, A., 1981. The metabolic formation of N-acetyl-S-2-hydroxy-L-cysteine from tetradeutero-1,2-dibromoethane. Relative importance of oxidation and glutathione conjugation *in vivo, Biochem. Pharmacol.* 30:2499–2502.

Webb, W. W., Elfarra, A. A., Webster, K. D., Thom, R. E., and Anders, M. W., 1987. Role for an episulfonium ion in S-(2-chloroethyl)-DL-cysteine-induced cytotoxicity and its reactions with glutathione, *Biochemistry* 26:3017–3023.

White, R. E., and Coon, M. J., 1980. Oxygen activation by cytochrome P-450, *Annu. Rev. Biochem.* 49:315–356.

Younes, M., and Siegers, C.-P., 1984. Interrelation between lipid peroxidation and other hepatotoxic events, *Biochem. Pharmacol.* 33:3001–2003.

Index